Third Edition

General Biology I and II

Laboratory Manual

Mohawk Valley Community College

Edited by:

Robert B. Jubenville
Richard G. Thomas

KENDALL/HUNT PUBLISHING COMPANY
4050 Westmark Drive Dubuque, Iowa 52002

Copyright © 1994, 1998, 2008 by Robert B. Jubenville and Richard G. Thomas

ISBN 978-0-7575-5681-4

Kendall/Hunt Publishing Company has the exclusive rights to reproduce this work, to prepare derivative works from this work, to publicly distribute this work, to publicly perform this work and to publicly display this work.

All rights reserved. No part of this publication may be reproduced, stored in a retrieval system, or transmitted, in any form or by any means, electronic, mechanical, photocopying, recording, or otherwise, without the prior written permission of the copyright owner.

All rights reserved. No part of this publication may be reproduced, stored in a retrieval system, or transmitted, in any form or by any means, electronic, mechanical, photocopying, recording, or otherwise, without the prior written permission of the copyright owner.
Printed in the United States of America

10 9 8 7 6 5 4

Contents

TAXONOMIC KEYS

1

Laboratory Objectives

1. To emphasize the role of observation in scientific activity.
2. To introduce the concept of classification
3. To be able to use a dichotomous key for identification of trees
4. To understand the structure and construction of a dichotomous key.

Introduction

In order to study the great variety of organisms found in the living world, it is necessary to classify life forms along some systematic basis. This is the function of the field of biology known as systematics or taxonomy. Organisms are named and arranged into various categories, or taxa (singular, taxon), according to internationally recognized rules. These rules have been developed from the work of the Swedish naturalist Carolus Linnaeus, who in the mid-1700's produced his remarkable publication *Systema Naturae*. His system established the use of binomial nomenclature by which each type of organism is uniquely identified with a two-part name. The two parts, *genus* and *species,* represent the most restrictive of a series of taxa to which the organism has been assigned based upon physical, embryological, biochemical, and other traits. The categories to which life forms are assigned are listed below from the most inclusive to the most restrictive:

Kingdom
Phylum (or Division for plants)
Class
Order
Family
Genus
Species

All life forms are capable of performing the same basic set of activities which define the state of being "alive." These traits include metabolism, reproduction, growth, elimination, and others. By observing organisms around us, initially we see that life forms can be clustered into natural groupings which share common traits. Some life forms are able to produce their own food through the process of photosynthesis (producers) while other life forms depend upon the producers for their energy supply. With this cursory examination, life forms seem to be separated into two basic categories, the plants (autotrophic or self-feeding) and the animals (heterotrophic or feeding upon others).

Upon closer observation, it is apparent that more than two **kingdoms** of life exist. In this course we are going to follow Whittaker's 5-kingdom approach to classification (Monera, Protista, Fungi, Plantae, Animalia). The unique characteristics of each kingdom will become more clear as we progress through our study of general biology.

Within each major kingdom, life forms can be further sorted into smaller and more unique clusters (more restrictive taxa). Within the animal kingdom, there are animals that possess a dorsal nerve cord protected by vertebrae

(Phylum Vertebrata) and those other animals which lack vertebrae. The vertebrates can be further classified or assigned to still more restrictive taxa based upon differences in physical structures, embryology, biochemical pathways, etc.

In the best of situations, this approach produces a natural classification of order which reflects the most probable evolutionary history or phylogeny of the life forms involved. As you proceed from kingdom, phylum, class, order, family, genus, and species the taxa become more exclusive with the members of each succeeding taxa sharing more and more common traits.

Individual organisms can be identified by their unique two-part genus and species name. All humans belong to the genus *Homo* and the species *sapiens,* thus the scientific name of *Homo sapiens*. Note that the genus name is capitalized but the species name is not. Also notice that both words are *italicized* or underlined.

Activity 1

Using a Taxonomic Key to Identify Trees as to Their Genus and Species

One of the benefits of a classification scheme for organisms is that it can serve as a basis for the production of a dichotomous taxonomic key. A taxonomic key allows the identification of a specimen based upon its physical traits. The "either or" decisions of the key lead the user along a path to discover the identity of the unknown organism. We are going to use a key to Adirondack plant life to identify some of the trees found on the campus.

In any taxonomic key, there will be terms that are confusing to you. Some of the terms can be listed in couplets as they have opposite meanings. See Figure 1 for help with the definitions below.

evergreens: trees such as pines, spruces and hemlocks which are cone bearing plants. They possess needlelike or scale-like leaves which they retain throughout the winter months.
deciduous or hardwoods: trees which possess large spreading leaves that fall during the autumn.

leaves: plant organs attached to the stem of a plant by the leaf's petiole (stem). A **bud** is found where the petiole joins the stem. Leaves come in two basic varieties:

simple leaves: leaves which have a single blade.

compound leaves: leaves composed of several leaflets which are attached to the main petiole. If the leaflets are attached to the main petiole radiating from a single point the leaf is *palmately compound.* If the leaflets are arranged along the sides of the petiole the leaf is said to be *pinnately compound.*

***alternate** leaves* leaves which alternate from one side of the twig to the other as you proceed outward along the twig. The leaves are arranged singly at intervals along the twig.
***opposite** leaves* leaves that occur in pairs across from each other along the twig. Opposite leaves may be either simple or compound.

Some terms describe the edges of the leaf or the leaf margins. These terms are not necessarily opposites. They may include:

dentate: toothed along the edge. Some leaves are singularly toothed, while other leaves are doubly toothed.

lobed: indentations of the margin which proceed toward the petiole. The shape of the lobe may be used to distinguish one type of tree from another.

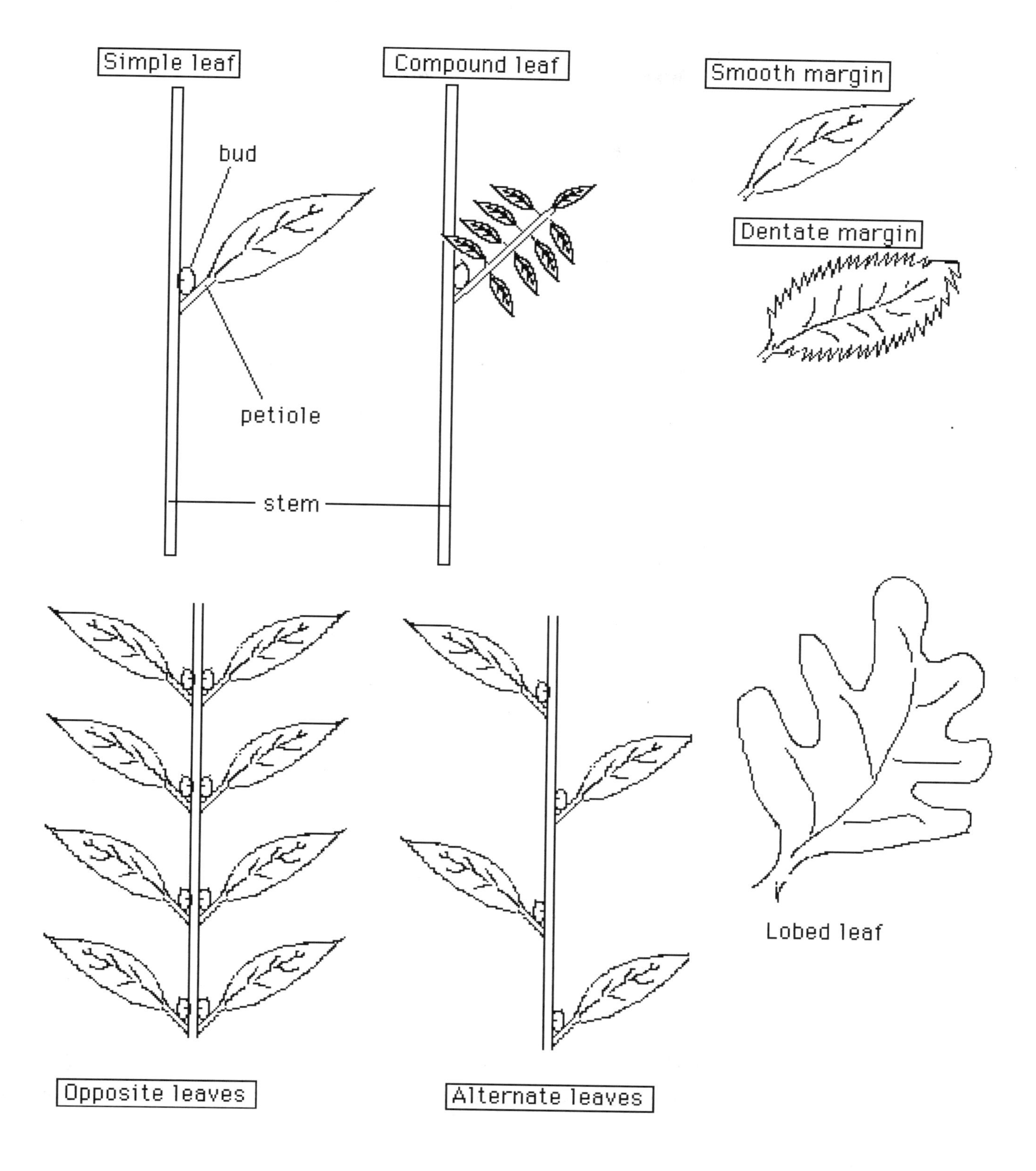

Figure 1. Leaf Terminology

Procedure

Working in groups of three or four students and using the plant key provided, follow the Par Course and identify the tagged trees you find along the way. Your instructor will demonstrate how the leaf key works on a demonstration tree before you begin. You should work on the trees for approximately one hour. If more time is needed, do not hesitate to return to the trees before the next laboratory meeting. In identifying trees, most of the categories of dichotomous selections are based on leaf characteristics. **Do not strip the leaves from the trees to observe them.** There are many General Biology students performing this exercise and the trees wish to retain some of their foliage.

There are eleven trees to identify. You only need to identify ten of them. No two are the same tree. A map of the campus with the trees is included (Figure 2). **Tree #1** is in front of Payne Hall. **Tree #2** is actually any one of three or four small coniferous trees at the beginning of the Par Course. **Trees #3 and #4** are along the Par Course before the drive into the gym. **Tree # 5** is located along the end of the tennis courts just on the other side of the drive to the gym. **Trees #6 and #7** are on the Par Course midway along the track. **Tree #8** (a grouping of three small trees with reddish leaves) is on the Par Course past the track. The leaves of tree #8 possess small teeth which terminate the side veins of the leaf. **Trees #9 and #10** are evergreen trees located on the other side of the track toward the Campus Center. They will have tags tied to their branches. **Tree #11** is located along the first base line of the baseball field. There are actually four or five specimen of this tree.

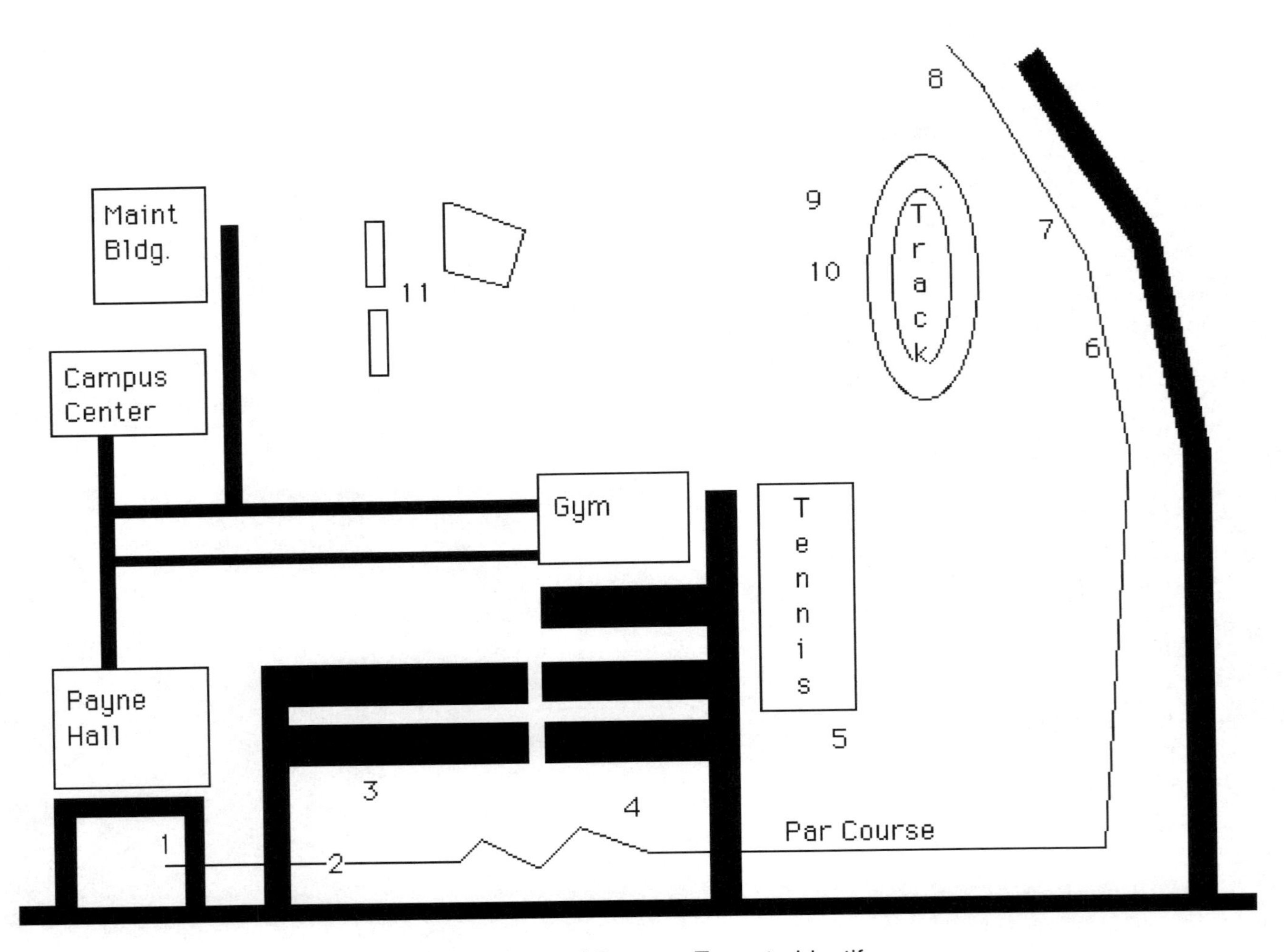

Figure 2. Map of Campus Trees to Identify

1. ______________________________

2. ______________________________

3. ______________________________

4. ______________________________

5. ______________________________

6. ______________________________

7. ______________________________

8. ______________________________

9. ______________________________

10. ______________________________

11. ______________________________

Activity 2

Multistage Classification: The Basis for a Dichotomous Key

In producing a multistage classification system, organisms are sorted based upon recognizable differences between the specimen. Characteristics which may be used to distinguish one specimen from the next might include symmetry, size, shape, color, or any other physical trait. The objective in producing a multistage classification scheme is to position each member of a larger group into its own "niche" or space in a grander scheme.

An example using the following geometric shapes will help to illustrate the procedure of producing a multistage classification system.

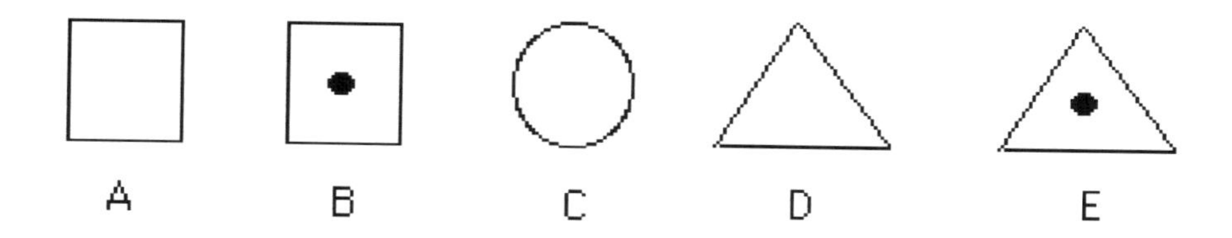

By starting with the entire collection of shapes, decide upon one physical trait that could be used to separate the group of five objects into two non-overlapping subgroups.

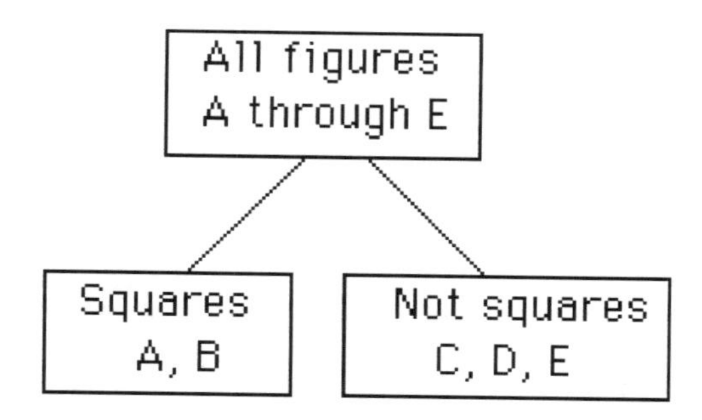

A dichotomy has just been produced. The original group of five has been separated into two non-overlapping groups, but more work needs to be done. Neither subgroup consists of only one unique specimen. The "Squares" can be distinguished based upon the presence or absence of a "navel."

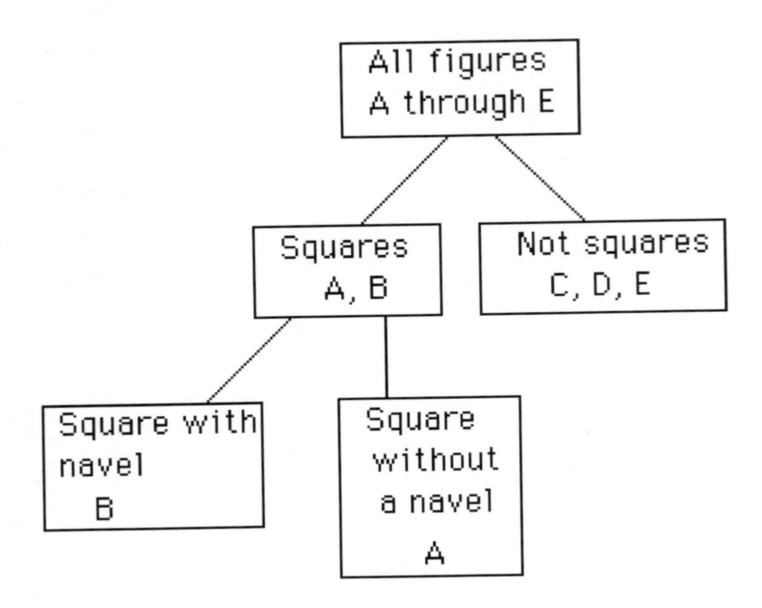

The "Squares" now each occupy their own unique place in the dichotomous branching scheme, but the right hand side still needs some work.

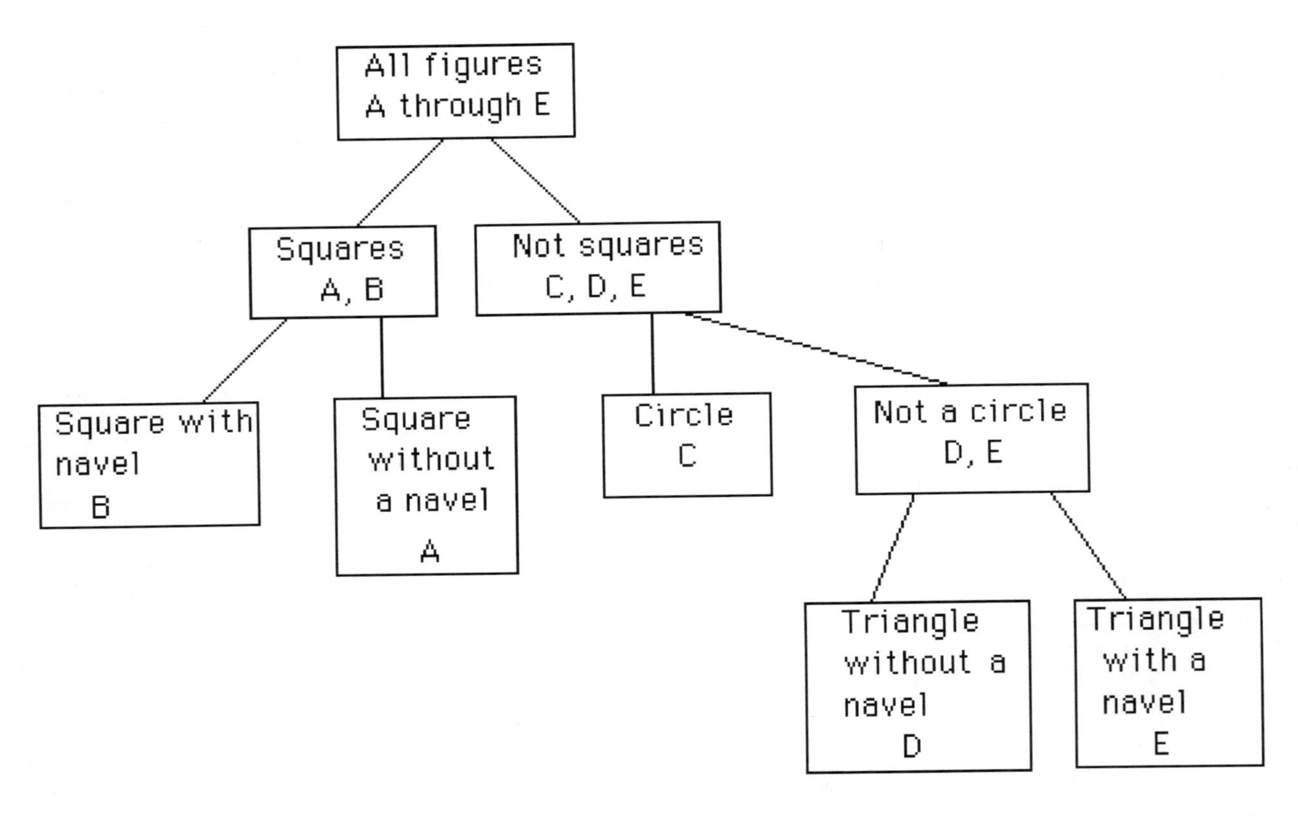

Now each object occupies its own unique position in the scheme. You should notice a few things about the dichotomous classification scheme. Only **two** branches come out of any subgroup. Once an object has been put into one branch of the classification system, it cannot reappear on the other side of the scheme. It is also a good practice to write down which objects have been put into specific subgroups so that no "one" object gets lost in the shuffle.

Procedure

Using the reasoning illustrated above, create your own multistage classification scheme using the set of "nasty" critters below.

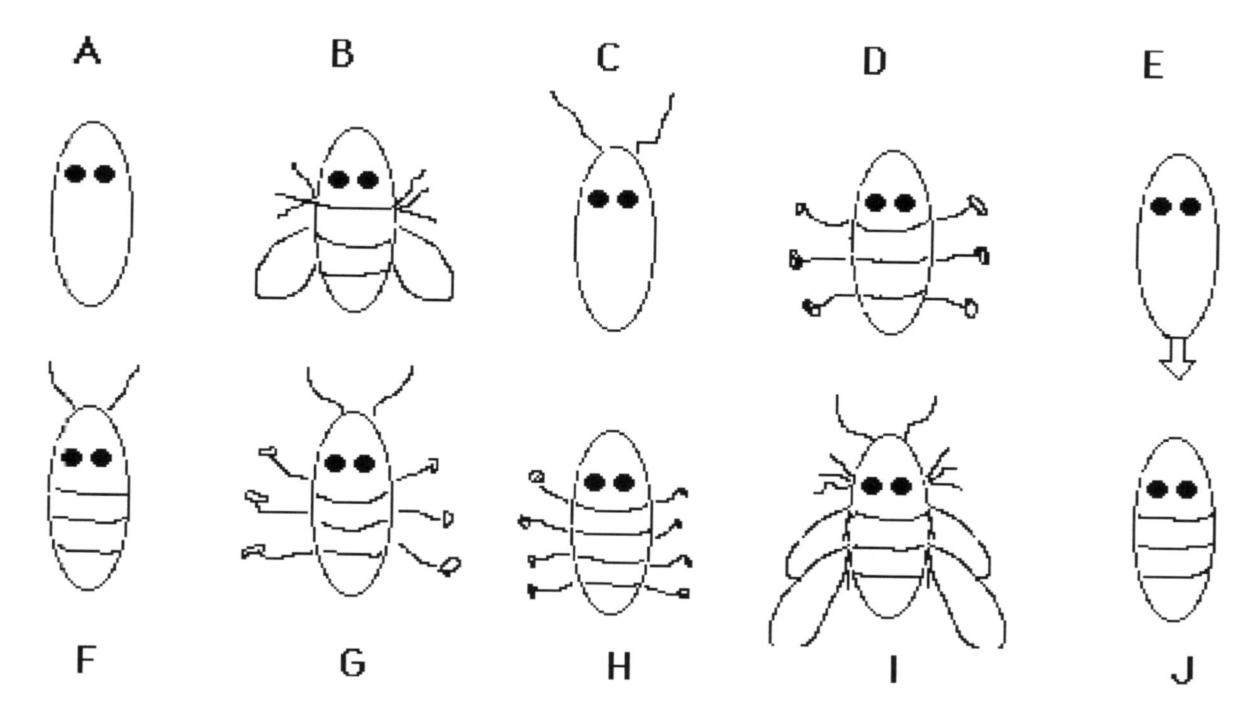

Activity 3

Producing a Dichotomous Key

A dichotomous key can be produced from a finished multistage classification scheme fairly easily. The key is a verbal description of the organization of the multistage classification system.

Each of the subcategories created in the multistage classification system receives a number with its two alternatives being labeled "a" and "b" respectively.

Once the numbering has been completed, the descriptors of the couplets are written down in ascending order. Again, this is a relatively easy step.

The third step in constructing a dichotomous key is to create the right hand column of the key. The numbers in the right hand column refer to the next appropriate couplet to be considered on the path to identifying a specimen. This can be said in another way. The number to the right of each statement in the key is the number of the dichotomous criterion appearing below the statement in the multistage classification system.

When the path terminates with one specimen in the subcategory, the specimen's name is placed in the right hand column rather than a number.

Using the previous example should make the process of writing a key more clear.

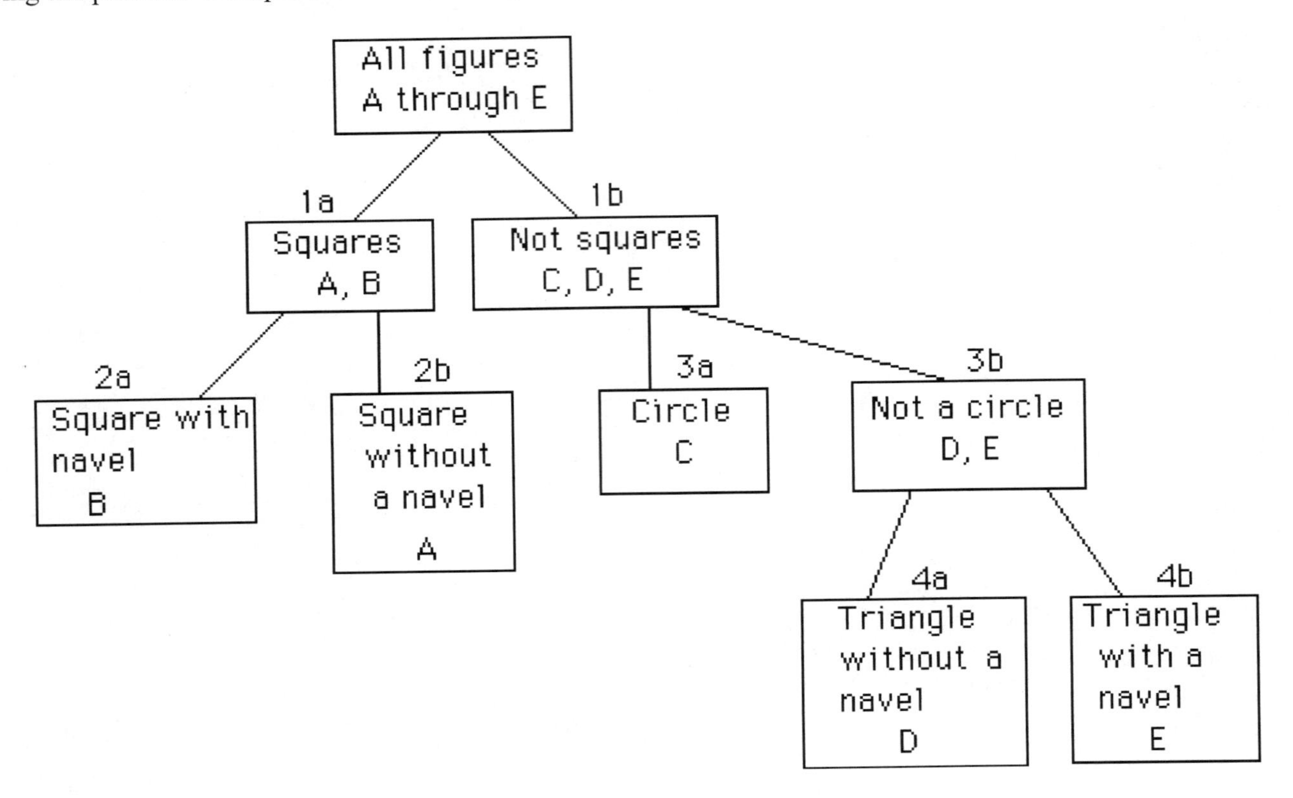

If		then go to
1a.	Squares	2
1b.	Not squares	3
2a.	Square with navel	object B
2b.	Square without navel	object A
3a.	Circle	object C
3b.	Not a circle	4
4a.	Triangle without a navel	object D
4b.	Triangle with a navel	object E

Now construct a key from your multistage classification of the "critters."

Assignment

You are to hand in:

1. a list of the trees identified on the par course.
2. a multistage classification scheme for the critters in Activity 2.
3. a key written from the classification scheme in Activity 3.
4. any of the questions below which are assigned by your laboratory instructor.

Name ______________________________

Section ______________________________

Post-lab Questions

1. What is the biological definition of the term "species"? Use your text book for this question.

2. What is a taxon in your own words?

3. What is the least restrictive taxon? **Why?**

4. What does the term "dichotomous" mean?

5. Diagram the twig of a plant which possesses alternate, palmately compound leaves. The leaflet margins are dentate.

KEY TO THE IDENTIFICATION OF TREES BY THEIR LEAVES

1a If the tree has needles 2
1b If the tree has leaves 22

TREES WITH NEEDLES

2a If the needles are long, 1/2 inch to 18 inches 3
2b if the needles are short, scale-like, overlapping 20
3a If the needles are in bundles or tufts 4
3b If the needles are borne single 13
4a If the needles are in bundles of 2 to 5 6
4b If the needles are deciduous, many in a tuft 5
5a If the branchlets are drooping, and the cones are about 1 inch long, ... EUROPEAN LARCH (Larix decidua)
5b If the branchlets are not drooping, and the cones are about 1/2 inch long, . . . AMERICAN LARCH or TAMARACK (larix laricina)
6a If there are 5 needles in a bundle,...WHITE PINE (Pinus Strobus)

6b If there are 2 or 3 needles in a bundle 7
7a If there are 2 needles in a bundle 10
7b If there are 3 needles in a bundle 8
8a If the needles are less than 5 inches long, yellow-green, twisted, the trunk and branches may be bearded with needles...PITCH PINE (Pinus rigida)

8b If the needles are 5 or more inches long 9
9a If the needles are stout, long, 5″ to 8″, not twisted (some bundles have only 2 needles)-PONDEROSA PINE or WESTERN YELLOW PINE (Pinus ponderosa)

If the needles are slender 9b
9b If the needles are 6″ to 9″, rather stiff,...LOBLOLLY PINE (Pinus Taeda)

If the needles are very long, 8″ to 18″, with a ragged sheath.... LONG-LEAF PINE (Pinus australis)

10a If the needles are short, less than 3″ 11
10b If the needles are long, 3″ to 8″ 12
11a If the needles are 1 1/2″ long, thick, spreading away from each other (divergent).......JACK PINE (Pinus Banksiana)
If the needles are 2″ to 3″ long, slender, slightly twisted slightly divergent....SCRUB PINE (Pinus virginiana)
11b If the needles are 2″ to 3″ long, slightly twisted, and the tree has branches of orange color,....SCOTCH PINE (Pinus sylvestris)
12a If the needles are slender, 3″ to 5″ long on a whitish twig, (some bundles with 3 needles).....SHORTLEAF PINE (Pinus echinata)
12b If the needles are slender, brittle; the twigs not white; the sheath long...RED PINE or NORWAY PINE (Pinus resinosa)
If the needles are stout, curved, not brittle, the sheath ragged and short......AUSTRIAN PINE (Pinus nigra)

13a If the needles are stiff, sharp, 4-sided, (can be twirled between the thumb and finger), and leave the twig rough when they fall off 14

13b If the needles are flat and pliable 17

14a If the needles are extremely sharp, and branches form a flat, horizontal spray.......COLORADO SPRUCE (Picea pungens)

14b If the needles are not very sharp, nor the branches noticeably horizontal 15

15a If the branchlets droop, and the cones are 4" to 6" long... NORWAY SPRUCE (Picea Abies)

15b If the branchlets do not droop, and the cones are shorter . . 16

16a If the needles are short, less than 1/2" or more, and the buds and twigs are hairy.....BLACK SPRUCE (Picea mariana)

16b If the needles are 1/2" or more, and the twigs and buds are hairy.....RED SPRUCE (Picea rubens)
If the needles are blue-green, and the twigs are hairless... WHITE SPRUCE (Picea glauca)

17a If the needles are 2-ranked, (like hair divided by a comb) . . 18

17b If the needles are not 2-ranked, branchlets drooping, buds red-brown, pointed,....DOUGLAS FIR (Pseudotsuga taxifolia)

18a If the needles are whitened beneath 19

18b If the needles are not whitened beneath, but of graduated lengths along the twig that is shed with them...BALD CYPRESS (Taxodium distichum)

19a If needles have broad bases, and leave twig smooth when they fall......BALSAM FIR (Abies balsamea)

19b If needles are about 1/2" long, have a narrow base, and leave the twig rough when they fall....EASTERN HEMLOCK (Tsuga canadensis)

20a If all the needles are scale-like 21

20b If part of the needles are small and scale-like and part are sharp and prickly.......RED CEDAR (Juniperus virginiana)

21a If the needles are flat, forming a flattened spray; and if there are numerous 1/2" cones; and the tree is in a swampy or limestone area........ARBOR VITAE (Thuja occidentalis)

21b If the needles are narrow scales, not in flat sprays, and if the numerous 1/4" to 1/2" cones are globular, and if the tree is in a coastal swamp.....ATLANTIC WHITE CEDAR (Chamaecyparis thyoides)

TREES WITH LEAVES

22a If the leaves or buds grow opposite 23
22b If the leaves or buds grow alternately 33
23a If the leaves are compound, composed of several leaflets (you can tell leaves from leaflets because there is no bud at the base of the leaflet) 24
23b If the leaves are simple (not composed of leaflets) 28

OPPOSITE

COMPOUND

ALTERNATE

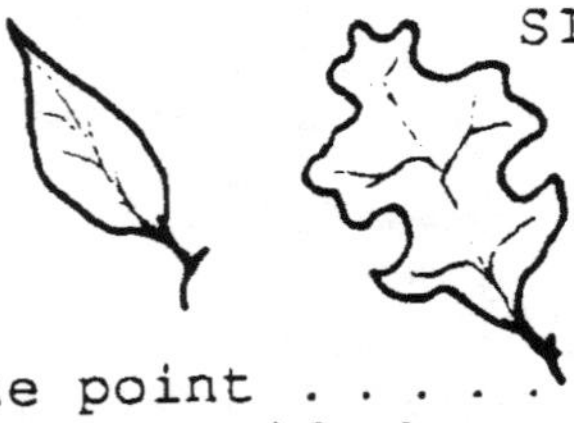

SIMPLE

24a If the 5 or more leaflets radiate from one point 25
24b If the leaflet do not radiate from one point or if there are only 3 leaflets 26
25a If there are usually 5 to 7 leaflets, doubly-toothed, and the leaflets have no stalks, and the winter buds are sticky; Leaflets widest near bluntly pointed tips.....HORSE CHESTNUT (Aesculus Hippocastanum)
25b If the leaflets are irregularly and bluntly toothed, and the end buds are keeled, and the twigs have a disagreeable smell... OHIO BUCKEYE (Aesculus glabra)
If the leaflets are regularly and finely toothed and the buds are not keeled....SWEET BUCKEYE or YELLOW BUCKEYE (Aesculus octandra)
26a If the leaflets are of different sizes and shapes...BOX ELDER (Acer Negundo)
26b If the leaflets are similar in size and shape 27
If each leaflet has a short stem 27a
27a If the leaflets have no stem...BLACK ASH (Fraxinus nigra)
If the leaflets are regularly toothed and the twig is square or with 2 long lines from leaf scars...BLUE ASH (Fraxinus quadrangulata)
If the leaflets are not regularly toothed or only toothed along the tip half of the margin and the twig is round and if the leaf stalks are hairy....RED ASH (Fraxinus pennsylvanica)
If the twigs and leaf stalks are not hairy 27b
27b If the leaflets are whitish beneath......WHITE ASH (Fraxinus americana)
If the leaflets ware green on both sides...GREEN ASH (Fraxinus pennsylvanica subintegerrima)

28a If each leaf has a single main vein with smaller side veins, and is without teeth or lobes 32

28b If each leaf has 3 to 7 main veins radiating from one point, and is lobed 29

29a If the notches between the lobes are V shaped 30

29b If the notches are U-shaped 31

30a If the leaves are distinctly 5-lobed...SILVER MAPLE (Acer saccharinum)

30b If the leaves appear 3-lobed rather than 5-lobed...RED MAPLE (acer rubrum)

31a If leaf stem shows a milky juice when broken; leaf usually wider than long; base of leaf not curving...NORWAY MAPLE (Acer platanoides)

31b If there is no milky juice; leaf about as long as wide; base of leaf curving....SUGAR MAPLE (acer saccharum) (A similar tree, but with leaves hairy beneath, 3-lobed with sides drooping).....BLACK MAPLE (Acer nigrum)

32a If the leaf tapers to both ends, and the veins curve to follow the margin....FLOWERING DOGWOOD (Cornus florida)

32b If the leaf is 6″ to 12″ long, heart-shaped...HARDY CATALPA (Catalpa speciosa)

33a If the leaves are compound, composed of several leaflets . . . 34

33b If the leaves are simple, not composed of leaflets 39

34a If the margins of the leaves are not toothed at all 35

34b If the margins of the leaves are toothed or partly toothed . . 36

35a If the leaf tips are rounded; twigs have short, paired thorns....BLACK LOCUST (Robinia Pseudo-Acacia)

35b If each leaflet is symmetrical, and the tree has some compound and some doubly-compound leaves, and the tree has large thorns, usually branched....HONEY LOCUST (Gleditsia triacanthos)

36a If the crushed leaf is aromatic and the end leaflet, if present, does not narrow gradually to an elongated, straight-sided V-base and the long section of the twig reveals layered pith 37

36b If the end leaflet narrows gradually to long, straight-sided V-shape, and the 3 end leaflets are usually distinctly larger than the basal leaflet and the husks of the nuts separate . . 38

37a If the end leaflet is small or lacking and side leaflets all taper continuously so that the sides are not parallel at any point.....BLACK WALNUT (Juglans nigra)

37b If the end leaflet is present, and if the sides of some of the leaflets are parallel along the mid-section....BUTTERNUT or WHITE WALNUT (Juglans cinerea)

38a If the leaf is 6″ to 12″ long, and the twigs are thin, without very big end bud.....PIGNUT HICKORY (Carya glabra)

38b If the leaf is small, (usually less than 12″) and the leaflets (usually 7) are slightly hairy beneath, and the buds are mustard yellow ....BITTERNUT HICKORY (Carya cordiformis)

39a If the leaf has neither teeth nor lobes 40

39b if the leaf has teeth of any kind, or a wavy margin, or lobes 41

40a If the leaf is heart-shaped, with veins branching from the base.......REDBUD (Cercis canadensis)
40b If the leaves are thin and the bark and leaves are aromatic and there are 3 forms of leaves....SASSAFRAS (Sassafras albidum) 
41a If the leaf is evergreen, tipped with stiff, sharp spines . . 42
41b If the leaf isn't evergreen 43
42a If the surface of the leaf is dull, and the edge is not wavy.....AMERICAN HOLLY (Ilex opaca)
42b If the surface of the leaf is glossy, and the edge is wavy.....ENGLISH HOLLY (Ilex aquifolium)
43a If the tree has thorns or thorn-like twigs, and is small . . . 65
43b If the tree has no thorns or thorn-like twigs 44
44a If the margin is toothed, or doubly-toothed continuously along all or almost all of its margin . . . . 45 
44b If the leaf is either deeply or shallowly lobed or waved, but not continuously saw-toothed 78
45a If the leaf is lobed as well as saw toothed, and is about as long as wide, with 3 to 5 main veins 83
45b If the leaf is not lobed 46
46a If the teeth are all of about the same size 47
46b If the margin is doubly-toothed with small teeth between the larger ones or with slightly deeper notches regularly spaced between teeth 62
47a If the teeth are of the same number as the side veins and terminate them.....AMERICAN BEECH (Fagus grandifolia)
47b If teeth are more numerous than side veins and do not terminate them 48
48a If the leaf stem is long (at least half as long as the blade) and the teeth are somewhat blunted and the blade is wide with firm texture and meshed veinlets 49
48b If the leaf does not have this combination of characteristics 52
49a If the stem of the leaf is flattened 50
49b If the stem of the leaf is not flattened..SWAMP COTTONWEED (Populus heterophylla)
50a If the leaf blade is triangular, flat at the base...... LOMBARDY POPLAR (Populus nigra variety italica)
50b If the leaf base is rounded 51
51a If the leaf is not longer than it is broad, and the teeth are many and fine....TREMBLING ASPEN (Populus tremuloides)
51b If the leaf is longer than broad, with teeth coarse and few.......BIG-TOOTHED ASPEN (Populus grandidentata)
52a If the two side veins starting from the base of the blade are longer and more conspicuous than the other side veins 53
52b If the side veins are all of about equal importance 56
53a If the base of the leaf is definitely not symmetrical . . . . . 54 
53b If the base is symmetrical or only slightly asymmetrical and the juice is milky and some leaves are lobed others unlobed 55
54a If the leaf is broad.....AMERICAN BASSWOOD (Tilia americana)
54b If the leaf is narrow long-pointed with a short stem and no teeth at the base....HACKBERRY (Celtis occidentalis)

55a If the leaf is rough above and hairy beneath, sometimes 2 or 3 lobed.....RED MULBERRY (Morus rubra)
55b If the leaves are smooth above and not hairy beneath, usually lobed.......WHITE MULBERRY (Morus alba)
56a If the leaves are long and narrow, many-veined, tapering gradually and steadily to a long point, and the twigs are slender and limber, with only one scale covering each bud . . 57
56b If the leaves and twigs are not thus, and the buds have more than one scale 58
57a If the leaf has white, silky hairs, and tapers to both ends... WHITE WILLOW (Salix alba)
57b If the leaf has no silky hairs...WEEPING WILLOW (Salix babylonica)
58a If the veins are straight, parallel, seldom branched..... SIBERIAN ELM or CHINESE ELM (Ulmus pumila)
58b If the veins are somewhat curving and branching 59
59a If the leaf stem is about 1/3 to 1/2 as long as the blade, and the base of the leaf is broadly rounded or slightly heart-shaped, and the stem and undersurface are somewhat downy......... SERVICEBERRY or JUNEBERRY (Amelanchier arborea)
59b If the leaf stem is short, and the base of the leaf is not rounded or heart-shaped, and the leaf and twig are bitter-tasting . . 60
60a If the leaf is soft, and the veinlets form a dense net-work (especially conspicuous on the under-surface) and the leaf narrows abruptly to a long, tapering tip, and the tree has thorn-like, short twigs, and shaggy bark...AMERICAN PLUM (Prunus americana)
60b If there are no thorn-like twigs, nor dense net-work of veins nor abruptly tapering point 61
61a If the teeth are somewhat incurved and the leaf is narrow.... BLACK CHERRY (Prunus serotina)
61b If the teeth are out-curved and the leaf is oval - Ridge of hairs along main vein.....CHOKE CHERRY (Prunus virginian)
62a If the base of the leaf is lop-sided 63
62b If the base of the leaf is symmetrical 71
63a If the leaf is rough beneath, as well as on the upper surface, and if a flake of bark shows layers of red...SLIPPERY ELM (Ulmus rubra)
63b If the leaf is not rough beneath 64
64a If the leaf base is only slightly lop-sided, and there are usually some twigs with corky wings....CORK ELM (Ulmus Thomasi)
64b If the leaf is distinctly lop-sided, and either sand-paper-like or smooth above.....AMERICAN ELM (Ulmus americana)
65a If the thorns are smooth, tapering, often more than an inch long 66
65b If the thorns are like stunted, pointed twigs, or stubby, blunt spurs 68
66a If the triangular leaf is hairy, soft, and with a thick, hairy stem.....DOWNY HAWTHORN (Crataegus mollis)
66b If the triangular leaf is smooth with a slender, smooth stem...67

67a If some of the leaves are deeply 3-lobed, and some unlobed, and the leaves are long-stemmed, with orange midribs....
WASHINGTON HAWTHORN (Crataegus Phaenopyrum)

67b If the triangular leaf is smooth, not deeply 3-lobed........
THICKET HAWTHORN (Crataegus pruinosa)

68a If the leaf is not lobed 70

68b If the leaf is usually somewhat lobed 69

69a If twigs, leaf stems, and undersurfaces are woolly.........
PRAIRIE CRABAPPLE (Pyrus ioensis)

69b If the leaves and twigs are not woolly....WILD CRABAPPLE (Pyrus coronaria)

70a If the leaf surface, especially the undersurface, shows an intricate network of veins, and the leaf has a long, tapering tip and the spurs are slender....AMERICAN PLUM (Prunus americana)

70b If the surface shows no intricate network, and the spurs are stout....COMMON APPLE (Pyrus Malus)

71a If the leaf has a blunt tip, fine brown hairs on underside veins.....EUROPEAN ALDER (Alnus glutinosa)

71b If the leaf has a pointed tip 72

72a If the bark is thin, (either smooth and white, or ragged and brown) with horizontal lenticels, leaves in twos (pairs) and alternate on branch 73

72b If the bark is not thin, leaves single and alternate on branch 77

73a If the trunks are white 74

73b If the trunks are yellowish to red-brown 76

74a If the leaf is triangular with a long tapering tip . . . .
GRAY BIRCH (Betula populifolia)

74b If the leaf is oval 75

75a If the buds are shiny with resin....EUROPEAN WHITE BIRCH (Betula alba)

75b If the leaf buds are not sticky and shiny, leaf width nearly 2/3 as long as the length....PAPER-WHITE BIRCH (Betula papyrifera)

76a If the base of the leaf is slightly heart-shaped with tufts of fine hairs in axils of veins...YELLOW BIRCH (Betula lutea)

76b If the base of the leaf is wedge-shaped, the bark ragged and shaggy....RED BIRCH or RIVER BIRCH (Betula nigra)

77a If the veins are unbranched, the base of the leaf is rounded....BLUE BEECH or AMERICAN HORNBEAM (Carpinus caroliniana)

77b If the veins are somewhat branched, and the base of the leaf slightly heart-shaped....IRONWOOD (Ostrya virginiana)

78a If the margin has waves, hardly indented deeply enough to be called lobes 79

78b If the leaf is lobed 82

79a If the leaf stem is long (about as long as the blade or longer) 80

79b If the leaf stem is not as long as the blade 81

80a If the leaf has a felt-like, white undersurface (some of the leaves may be lobed)....WHITE POPLAR (Populus alba)

80b If-the veins are arranged like the ribs of a fan...GINGKO (Ginkgo biloba)

81a If the waves are regular, and rounded, and the leaves are broadly oval....CHESTNUT OAK (Quercus Prinus)

81b If the waves are irregular, sometimes almost deep enough to be called lobes, and the leaves are soft and hairy beneath... SWAMP-WHITE OAK (Quercus bicolor)

82a If the leaf has only 2, 3, or 4 lobes 85

82b If the leaf has more than 4 lobes 86

83a If there are 3 main veins starting at, or almost at, the base of the leaf 84

83b If there are 5 main lobes, and the leaf is star-shaped and aromatic....SWEET GUM (Liquidambar Styraciflua)

84a If the teeth are coarse (2 to 3 to an inch) and jagged...... SYCAMORE (Plantanus occidentalis)

84b If the teeth are fine, (10 to 15) to an inch and rounded ..... MULBERRY

85a If the tree has some lobed leaves and some unlobed leaves.... SASSAFRAS (Sassafras albidum)

85b If the main vein ends in a notch, and the tip looks cut off... TULIP TREE (Liriodendron Tulipifera)

86a If the lobes are bristle-pointed 87

86b If the lobes are rounded 90

87a If the leaf is not deeply lobed, (not more than half-way to the mid-rib 88

87b If the leaf is deeply lobed, (more than half-way to the mid-rib) 89

88a If the leaf is thin, firm, smooth beneath, 5″ to 9″ long, with lobes that taper toward their tips usually more than 7-lobed ....RED OAK (Quercus rubra)

88b If the leaf is thick, leathery, usually widening toward the tip usually 7-lobed somewhat hairy beneath, (the tree usually has several different forms of leaves) and the buds are angled, lobes variable in depth, axillary tufts of hair at junctions of main veins on underside....BLACK OAK (Quercus velutina)

89a If the lobes taper toward their tips and the leaf is small, (3″ to 4″) often only 5-lobed, with a wedge-shaped base.... PIN OAK (Quercus palustris)

89b If the lobes broaden toward their tips.......SCARLET OAK (Quercus coccinea)

90a If the lobes are square cut, with the 3 end lobes much larger than the others.....POST OAK (Quercus stellata)

90b If the lobes are somewhat similar in size and shape........ WHITE OAK (Quercus alba)

THE COMPOUND MICROSCOPE

2

Laboratory Objectives

1. To identify and name the parts of the compound microscope.
2. To learn how to effectively use the microscope to study the fine detail of specimens.
3. To develop the ability to estimate the size of specimens.
4. To be able to construct a temporary wet mount of different biological specimens.

Introduction

The invention of the compound microscope opened new horizons in the study of biology. For the first time it allowed investigators to get into the world of microorganisms. Later in the semester you will have the opportunity to use the compound microscope to get a glimpse of some of the amazing structures which exist in the Protistan kingdom. The ability of the light microscope to function depends not only on its ability to magnify specimens, but also upon its ability to resolve some of the finer details of the specimen's structures.

Magnification

The compound microscope gets its name from the fact that there are two magnifying lenses placed between the observer's eye and the specimen being viewed on the stage of the microscope. The **objective** lenses are mounted on a **revolving nosepiece.** There are four different lenses which can be rotated over the specimen to vary the magnification of the optical system. Make sure that when a specific objective lens is placed into position, you hear a distinctive "click." This ensures that the lens is in the proper position. In order of power, there is a scanning lens (4X), a low power lens (10X), a high dry lens (40X), and an oil immersion lens (100X). For most of our work, we will be using the first three of the objective lenses. The oil immersion requires the use of a drop of oil placed on the surface of the microscope slide. The other magnifying lens in the system is the **ocular** lense or eyepiece. It is a 10X lens. The total magnification of the microscope can be determined by multiplying the objective and ocular lens powers together. What magnifications is your microscope capable of producing?

Resolving Power

In some murder mystery seen on television, you have probably seen a photograph magnified to some tremendous degree in an attempt to see the face of an unknown character seen in the background. The more the photograph is magnified, the more "blurred" its appearance becomes. This is because the resolving power of the camera which took the photograph was too poor to produce a clear and crisp image at the level of magnification needed to identify the "bad guy." In more technical terms, resolving power is the ability to see two objects that are very close together and still be able to recognize that they are two distinct objects. Demonstrate resolving power by having your lab partner move the two parallel lines below away from you until the two lines converge and seem to be a single line. At that distance, your eyes do not have the resolving power to distinguish the two separate units.

Two different variables affect the resolving power of the compound microscope.

$$d = \text{wavelength/numerical aperture.}$$

In the equation, d is the minimum distance which can separate two objects where the two objects can still be seen as being distinct. The shorter the wavelength of light passing through the microscope, the smaller is this distance which can still be resolved. Most of you are familiar with the colors of the rainbow: ROYGBIV. One color differs from another by its wavelength. Red light has a longer wavelength than does orange. Orange light is of longer wavelength than is yellow. Violet light possesses the shortest wavelength of the visible spectrum and is used to maximize the resolving power of research microscopes. The second variable which affects resolving power is the numerical aperture of the microscope. This number represents the microscope's ability to collect light that has passed through the specimen and to direct this light into the optical system of the microscope. The numerical aperture is affected by the position of the condenser lens and the iris diaphragm of the microscope. These two components of the microscope will be investigated below.

Activity 1

Getting to Know Your Microscope

The microscope that you will be using is an expensive and delicate instrument. Treat it gently and with respect. Follow the directions of your laboratory instructor as you identify the parts and functions of the microscope as seen in Figure 1.

1. Get a microscope from its storage cabinet. Carry the microscope back to your work area carrying it with both hands, one on its arm and the other under its base. Remove its dust cover. Always keep the microscope standing vertically. Be sure and return the microscope to the same compartment from where it came.

2. Observe the **eyepiece** (ocular lens). This lens fits into the end of the ocular tube and can be rotated within the ocular tube. This rotation will position an internal pointer to a desired location on the specimen being viewed. **Never remove the ocular lens.** Viewing the ocular lens from an oblique angle, notice that its surface may be quite dirty with oil or mascara. Now is a good opportunity to clean this lens with **lens paper** by employing a circular rubbing motion.

3. Find the **objective lenses** on the revolving nosepiece. They are of different magnifying powers. What relationship exists between power of magnification and the length of the lens? Rotate the nosepiece and observe

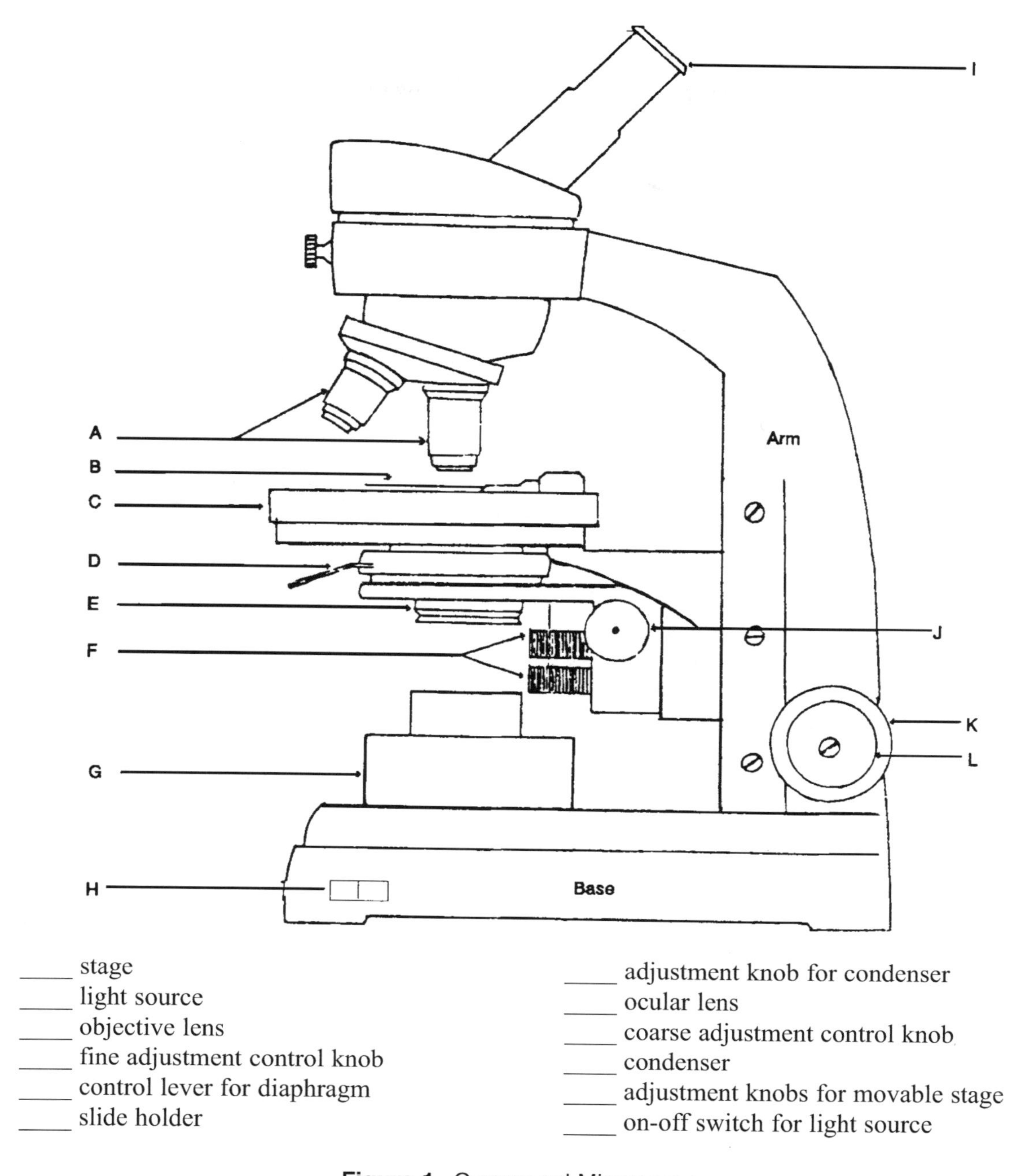

Figure 1. Compound Microscope

that there is an audible "click" when the objective lens is placed in position. Now is also a good time to clean the objective lenses with **lens paper** employing a circular rubbing motion.

4. The **stage** is the platform upon which the slide is placed. It possesses an opening which allows light to pass through the specimen being viewed. Your microscope has a mechanism for holding and maneuvering the slide. Investigate this mechanism's operation with the help of your instructor. The adjustment mechanism for the mechanical stage can be found below the stage on a vertical stem. The mechanism moves the slide north and south/east and west.

5. **Focusing adjustments.** The coarse and fine focus adjustments are located on the side of the body of the microscope. The **coarse focus** is the larger of the two knobs which produces large movements of the stage in relation to the fixed position of the objective lens. The **fine focus** control is the smaller inner knob.

6. The **condenser lens** is a third very important lens in the operation of the microscope. It is located below the stage of the microscope and is used in focusing the light from the light source onto the specimen being viewed. To demonstrate the function of the condenser lens, place the low power objective lens in place. Place a single sheet of paper on the stage over its slotted opening. Turn on your light source. Locate the **condenser lens control knob** which is located below the stage. Notice that the condenser lens travels up and down as this adjustment knob is moved. Looking at the stage from the side of the microscope (not viewing through the ocular lens), describe how the circle of light projected onto the piece of paper changes as the condenser lens is raised and lowered. In what position do you think the microscope would work most efficiently?

   Once the condenser lens is in the proper place, it is not necessary to change its position.

7. The **iris diaphragm,** associated with the condenser lens, regulates the amount of light which enters the microscope. Swing the lever adjusting the iris diaphragm back and forth. Notice the effect this has on the cone of light which is projected onto the piece of paper.

Activity 2

Focusing the Microscope

1. Clean all lenses using only the lens paper provided. Never remove any lens from the microscope for this purpose.

2. Lower the stage using the coarse focus adjustment.

3. Rotate the 4X lens into position. Make sure that you hear an audible click.

4. Place the letter "e" slide onto the mechanical stage fitting it properly in the carriage. Move the slide so that the letter "e" (not the "e" on the slide label) is positioned over the opening of the stage directly beneath the 4X objective lens.

5. Turn on the light source. You will notice that the light intensity setting possesses a scale from 1 to 10. Usually a setting of "7" or "8" will work best.

6. Using the coarse adjustment focus, raise the stage to its highest position. **Watch the stage from the side as you raise it upwards** making sure that it does not strike the objective lens.

7. While looking through the ocular lens, slowly lower the stage until something comes into focus. This will usually involve rotating the coarse adjustment knob about 1/2 turn. If you have trouble getting something into focus, ask your instructor for help **NOW.** We will be using the microscope often this semester.

8. At this point, fine tune the image that you are observing. Adjust the fine focus producing the sharpest image possible. Position the letter "e" exactly in the middle of the field. Adjust the iris diaphragm to get the level of illumination that is most comfortable for your eyes.

9. To move to a higher power, **DO NOT** touch the coarse focus. Simply rotate the nosepiece so the next higher power lens (10X) clicks into place. All of the objective lenses of the microscope should be in focus at approximately the same stage level. This property is referred to as **parfocal.**

10. Fine tune your microscope again as you did above. Remember to position the majority of the letter "e" in the middle of the field. Fine focus and adjust the light level for best illumination.

11. Repeat the above procedure to focus with the 40X (high dry lens) objective lens. **Do not touch the coarse focus for this lens change.**

12. As you use the microscope, you will find it easier to keep both of your eyes open. This will take some practice, but eye strain is reduced.

13. What is the most important thing to remember when focusing the microscope?

Activity 3

The Famous Letter "e"

1. Look at the letter "e" with your naked eyes. Examine it carefully. In the space provided, draw the letter "e" as you see it without the use of the microscope.

2. Place the letter "e" on the stage and focus with the 4X objective lens in place. Draw the letter "e" as it appears in the microscope in the space provided below. In makings sketches with the microscope, it is a good practice to include the total magnification of the system in the lower right hand corner of the sketch.

3. Change to the 10X objective lens and draw the letter "e" as it appears under the 100X total magnification.

4. Finally focus the letter "e" under the 40X objective lens drawing what you can see in the field of view.

5. After completing the above, answer the following questions:

a. Compare the orientation of the letter "e" as seen with the naked eye to that seen under the 4X objective lens. How do the two images differ?

b. What implication does this relationship have when trying to focus on a specific area of the microscope slide? When you wish to move the specimen from one area to another area?

c. As you progressed from the 4X to the 10X to the 40X objective lenses, did the "area" of the letter "e" seen get larger or smaller? As the power of the objective lens increases, the relative area of the slide seen ____________ (increases or decreases).

d. As you progressed through the objective lens series, did the image appear to get brighter or dimmer? As you increase the magnification of the objective lens, the iris diaphragm should be ____________ (opened up or closed down) for optimum viewing pleasure.

Activity 4

Estimating Sizes

The size of an object can be deceiving when looking through a microscope. One way of estimating the size is to know the diameter of the field of view. Follow the instructions below. Ask your lab instructor for help if you are confused here.

1. Focus on the stage micrometer slide using the scanning lens. Notice that it looks like a miniature metric ruler (which it is). Be very careful with this slide as it is very expensive. The micrometer scale is 1.0 mm long. The largest hash marks represent 0.1 mm lengths. The medium hash marks represent 0.05 mm, while the smallest demarcations are 0.01 mm.

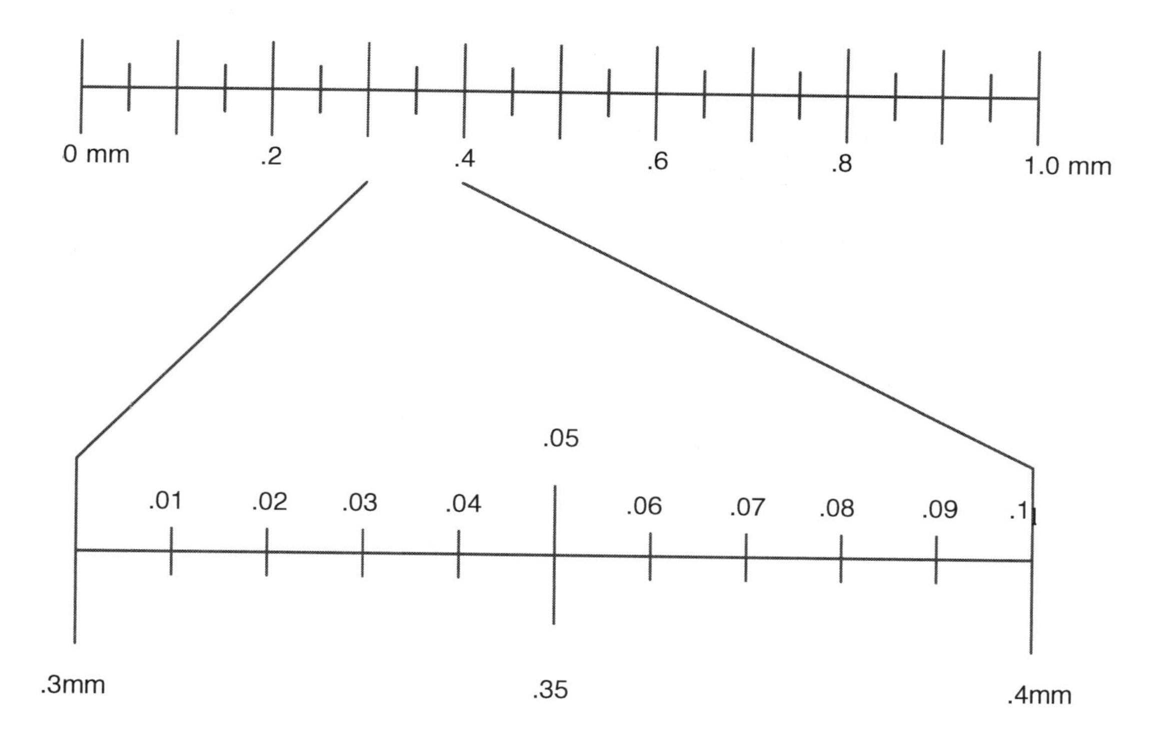

2. Move up to the 10X lens and fine focus on the micrometer. Measure the width of field of view under this **100X total magnification.** Place the left hand edge of the micrometer on the left edge of the field of view. Notice that the ruler does not reach the far right side margin of the field of view. You can mark the right end of the micrometer's position using the "pointer" contained in the ocular lens It will be necessary to reposition the ruler to measure the remaining distance to the right margin. What is the size of the field at 100X? ___________mm

3. Now focus on the micrometer using the 40X lens producing a **total magnification of 400X.** What size is the diameter of the field of view. ___________mm. This is 4X more powerful than the magnification above and the field of view is 1/4 as large.

4. As the power of the objective lens is increased, the relative size of the field of view ___________ (increases or decreases).

Activity 5

Crossed Threads and Depth of Field

Even though the specimens that we have been looking at seem as if they are only two dimensional, they are actually three dimensional. As well as possessing length and width, they also possess a depth or thickness. Depending upon the preparation, some specimen vary considerably in their thickness. Different powers of objective lenses differ in the vertical distance over which they are in focus. In other words, some objective powers are able to see more "clearly" and deeply into a specimen. We are going to use a slide possessing three colored threads which lie over one another in this exercise. The order of colors may be different on different slides, and our intent is not to cause an argument about which thread comes first, second, and third. By following the directions below, you should be able to infer which of the three objective lenses has the shallowest depth of field.

1. Focus on the crossed threads using the 4X objective lens at the point where the three threads intersect.

2. Try to determine which thread lies on top, which is the middle thread, and which thread lies on the bottom of the stack.

3. Try to determine the order of threads by focusing on the intersection of the three threads using the 10X objective lens.

4. Try to determine the order of threads by focusing on the intersection of the three threads using the 40X objective lens.

5. Answer the following questions:

   a. Under which power was it the most difficult to focus on all three threads at the same time?

   b. Under which power did all three threads appear to be in focus at the same level of the stage (at the same time)?

   c. Under which power did you first see one color clearly, then a second color as you continued to focus down through the slide, and then the third color clearly?

   d. Which objective lens has the shallowest depth of field?

   e. As the power of the objective lens increases, the depth of field of the system ________________ (increases or decreases).

Activity 6
Microscope Preparations

There are two different types of preparations which can be viewed underneath the compound microscope. A **permanent preparation** has been made so that it is preserved and lasts over time. It has usually been stained to enhance structures that might otherwise not be seen. A **temporary preparation** places living or recently living material directly on a microscope slide with no preservation process. It therefore has a relatively short shelf life. You will look at a permanent preparation of a root tip and make two temporary preparations; one of an onion skin and one of your own cheek cells.

1. **Permanent preparation of an onion root tip cell**

   Examine the onion root tip under low power magnification. The onion root was cut parallel to the long axis of the root (longitudinal section). The younger cells are closer to the tip of the root. Sketch the tip of the onion root to show the arrangement of the cells.

   Examine the root tip under the 40X objective lens to discover more about the cell's structure. You can see the **cell wall** of the plant cells very easily. The **cell membrane** is difficult to see because it is pressed up against the cell wall. The central spherical **nucleus** stands out boldly in each cell. The appearance of the nucleus may differ depending upon where in the root tip you are looking. In a nondividing plant cell, the **chromatin** in the nucleus will stain heavily in a granular fashion. A **nucleolus** (or 2 or 3 nucleoli) may be seen as smaller more densely stained spheres within the nucleus. If the cell is in process of mitosis (cell division), individual chromosomes can be seen as more string-like bodies within the nucleus. We will study the process of mitosis later in the semester. The region of the cell between the cell wall and the nuclear membrane is called **cytoplasm.** Draw a few of these cells below.

2. Temporary preparations of an onion cell

a. Slice an onion and make onion rings.

b. Separate the rings from each other.

c. Peel away the thin membrane found on the inside of an onion ring.

d. Place a one cm piece of the membrane in a drop of water placed in the center of a glass slide.

e. Cover the drop of water containing the specimen with a cover glass by lowering the left side of the cover glass to the surface of the slide. Now lower the right side of the cover glass to the surface of the slide using a dissecting needle.

f. Remove excess water with a Kim-Wipe. Be sure that no water is on the upper surface of the cover glass or the underside of the microscope slide.

g. Examine the box-like cells under low power. Note that there is little definition or contrast to the cells.

h. Place your slide on a staining rack and place a drop of methylene blue stain right next to the cover glass. It will wick under the cover glass to stain the specimen.

i. Re-examine the slide finding the **cell wall, cytoplasm, nucleus.** Make a sketch of this cell below.

3. **Temporary preparation of cheek epithelial cells**

   Animal epithelial cells are covering cells. They are easily rubbed off and will allow us to examine some of the differences between plant and animal cells.

   a Run a toothpick along the inside of your cheek scraping away some of your epithelial cells. It is not necessary to gouge the lining of your mouth.

   b. Swirl the end of the toothpick in a drop of water which has been placed in the middle of a clean slide.

   c. Over a staining tray, add a small drop of methylene blue stain.

   d. Cover the specimen with a cover glass as before.

   e. Locate the harder to find cheek epithelial cells and sketch below. Label the cell membrane, nucleus, and cytoplasm of the cheek cells.

4. List the differences which you noticed between the animal and the plant cells. You should be able to think of at least three differences.

Activity 7
Optional Preparation

A culture of living organisms may be available in the back of the room for your observation. Follow the directions of your lab instructor for the viewing of these specimen.

Lab Clean Up

The lab should be neater when you leave it than when you entered. Make sure that:

1. prepared slides have been returned to their trays in the front of the room.
2. slides which you have made have been placed in the green trays filled with soapy water.
3. microscopes have been returned to their numbered slots at the front of the room. Be sure that the dust covers have been replaced and that the cords are wrapped tightly around the base.
4. any pieces of lens paper or tissues have been thrown out in the trash can. Do not throw papers into the sinks in the lab.
5. your chairs have been pushed under the lab tables.

Name ______________________________

Section ______________________________

Pre-lab Questions

1. What does the word "compound" mean in relation to the compound microscope?

2. What are two limitations of the compound microscope? What two features have been improved to the point where they physically have reached the limit of their abilities?

3. Besides magnification, what quality is important in providing a good image in the compound microscope? How can the quality of the image be improved?

4. What is the function of the condenser lens of the microscope? What is the function of its iris diaphragm?

5. Be sure and know the parts of the microscope.

Name ______________________________

Section ______________________________

Post-lab Questions

1. What does the term parfocal mean and what is its importance in microscopy?

2. What practices do you use in order to "fine tune" the microscope for best viewing results?

3. Explain the term "limit of resolution" as it applies to a microscope.

4. What are some differences between permanent and temporary preparations of microscopic specimen?

5. What specific observations did you make relative to the differences between plant and animal cells? their similarities?

AN INTRODUCTION TO ORGANIC MOLECULES

3

Objectives

1. To appreciate the uniqueness of the carbon atom and the central role that it plays in organic molecules.
2. To use molecular models to illustrate the structure of hydrocarbon compounds.
3. To build models of some important functional groups and monomers of organic molecules.
4. To introduce the concepts of dehydration synthesis and polymerization reactions.

The Nature of the Carbon Atom

Atoms form chemical bonds to attain stability. In most cases, the way to attain this stability is to fill the valence shell (outer shell) with eight electrons. There are two different types of chemical interactions which can take place to fill outer shells to attain this stable configuration. Chemical bonds can be classified as either being **ionic** or **covalent. Ionic bonds** tend to form between elements which have either nearly empty or nearly full valence shells. In an ionic bond electrons are either gained or lost by the participating atoms as they become ions. Oppositely charged ions then attract each other. Substances held together by ionic bonds tend to dissociate (break apart) in an aqueous environment. Ionic bonds are therefore not important in the structural makeup of organic molecules. **Covalent Bonds** form between atoms which generally possess an intermediate number of valence electrons (3, 4, 5, or 6). In covalent bonds, electron pairs are shared between the participating atoms. Molecules formed in this manner are stable in aqueous (water based) solutions. Carbon, with an atomic number of 6, possesses two electrons in its first electron shell with the remaining four electrons in its valence shell. Being able to bond with four different atoms at the same time is one of the properties which makes carbon useful as the backbone for organic molecules. Carbon also tends to bond with other carbon atoms forming chains and/or rings. The chains and rings of carbon can be of varying lengths and sizes. These properties contribute to the wide diversity and complexity of organic molecules.

Activity 1
The Qualities of Carbon

Obtain a molecular model kit from the front of the room. Notice that it contains plastic pieces of various sizes and shapes. There is a key on the back of the lid identifying the various pieces.

1. Take out the piece representing the carbon atom. Each spike represents one valence electron. Therefore, an atom of carbon possesses four valence electrons and "needs" to obtain four more electrons to become stable. Carbon's ability to form four covalent bonds at one time allows it to form a great variety of molecules. What other atoms also possess the ability to form four covalent bonds? (Look at a periodic table) ______________
2. Bond a hydrogen atom to each of the four valence electrons of the carbon. Use the connecting short white tubes to represent each covalent bond of shared electrons.

a. Each bond that you have produced represents a **covalent** bond.

b. The molecule which you have produced is the simplest member of the alkane series of hydrocarbons: methane (CH_4).

Activity 2

Hydrocarbons and Isomerism

1. Hydrocarbons get their name from the fact that they only contain the elements of carbon and hydrogen. It is useful to study these molecules as they help us to understand the property of isomerism. **Isomers** are molecules which have the same kinds and numbers of atoms, but they differ is how the atoms are put together to make the molecules. Some of the hydrocarbons will also serve as radicals in the construction of more complex organic molecules. The hydrocarbons, based upon the bonding patterns between adjacent carbons (single, double, or triple bonds), are classified as being alkanes, alkenes, and the alkynes.

2. **Alkanes**

   a. Leaving your first molecule intact, build the second molecule of the alkane series: **ethane** (C_2H_6). Write the structural formula for ethane in the space provided.

   b. Build the third member of the series: propane (C_3H_8). and write down its structural formula.

   c. Build the fourth member of the series: butane (C_4H_{10}). The fourth carbon can be attached at two different and distinct points on the already existing three carbon propane. Build your model of butane showing the two possible different configurations. You are witnessing the phenomenon of **isomerism.** Isomers are molecules which have the same empirical formula, but different structural formulae. Diagram the structural formulae for the two isomers of butane below.

d. Also note that as the alkane hydrocarbons have increased in size, a predictable ratio of carbons to hydrogens has emerged. Can you come up with the relationship between carbon and hydrogen numbers?

e. The alkane series can be extended beyond butane, but we do not need to do that here. The names of the alkane hydrocarbon series will be useful in the naming of more complex organic molecules to be seen later.

1C	methane	10C	decane
2C	ethane		
3C	propane		
4C	butane		
5C	pentane		
6C	hexane		
7C	heptane		
8C	octane		
9C	nonane		

f. All of the bonds seen in the alkane series are simple and single covalent bonds. Each of the carbon atoms in the series is connected to the maximum number of hydrogens atoms that it can receive. For that reason, the molecules are said to be **saturated** with hydrogen atoms. No more hydrogens could possibly be connected to the molecule.

3. **Alkenes**

a. Carbon atoms can also bond to themselves with double bonds. Construct a molecule of ethene (C_2H_4) and write its structural formula below. The existence of the double bond (represented by two of the white plastic pieces) restricts the free rotation between the two carbon atoms as is seen in the alkane series.

b. Produce a molecule of propene (C_3H_6) and write its structural formula below.

c. These molecules are said to be **unsaturated.** If the double bond between the carbon atoms were to be broken, additional hydrogens could be added to the molecule. Where have you heard the words: **saturated versus unsaturated?** We will run across these terms later in this lab.

d. What is the ratio between the number of carbon to hydrogen atoms in the alkene series of hydrocarbons?

4. **Ring Structures** In addition to forming straight chains of carbon atoms, carbons can link together to form cyclic molecules.

a. Construct a straight chain molecule of hexane (C_6H_{14}).

b. Notice that the molecule seems to be rather flimsy. With each of the carbons able to rotate rather freely around their axis, the entire molecule seems wobbly. Rotate the carbons so that they form a circle. Remove a hydrogen from each of the end carbons that lie close to one another. Join the carbons together at the free bonding sites. You have formed **cyclohexane.** Notice how much more stable this molecule feels than the straight chain version. This phenomenon is seen in the carbohydrates to be discussed in lecture.

Activity 3

Some Functional Groups and Monomers

Even though it is carbon that produces the skeletons for the organic molecules we will study in lecture, it is the radical (functional) groups that give each of the classes of organic molecules their physical properties. These radicals (functional groups) will be seen over and over on the different classes of organic molecules.

A. **Alkyl radicals** are produced by building the alkane of the desired length and then removing one of the hydrogens. The resulting radical can then be bonded to another stem at the free bonding position. Alkyl radicals are often symbolized in general formulas by the letter **R.** The name of the radical takes its name from the number of carbon atoms making up its length with the suffix "-yl" added to the root.

1. Construct a methyl radical ($-CH_3$) and write down its structural formula.

2. Construct a propyl radical ($-C_3H_7$) and write down its structural formula.

B. **Important groups containing oxygen**

1. **Hydroxyl radical** (alcohol radical). The addition of the –OH group to an alkyl radical produces some of the organic alcohols of which you may have heard.

   a. Produce an **ethyl radical.** Attach the **hydroxyl radical** (–OH group) to the open bond position to make **ethyl alcohol.** Diagram its structural formula in the space provided.

   b. Produce **propyl alcohol** in the same manner that ethyl alcohol was made. Produce a two dimensional structural diagram of this molecule below.

2. **Carboxyl radical (acid group).** The addition of a –COOH (its diagram is seen immediately below) to an alkyl radical produces organic acids. **Acetic acid** is a **methyl radical** connected to a **carboxyl group.** Construct the molecule and produce a two dimensional structural formula for acetic acid below.

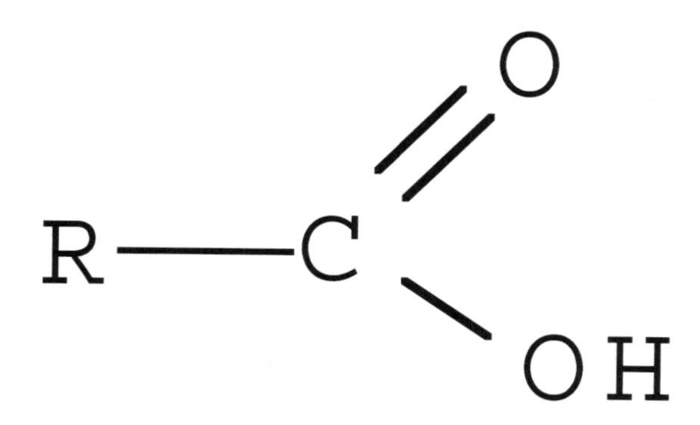

C. Amines-important radicals containing the element nitrogen

–NH2 is the amine radical. The amine can be joined to a number of different skeletons to form a variety of molecules. Construct a molecule of methyl amine. Diagram its two dimensional structural formula below.

Activity 4

Using Dehydration Synthesis in the Synthesis of Lipids and Proteins

A dehydration synthesis is an **anabolic** synthesis where two simpler molecules are joined together to form a more complex molecule. You have seen the word "anabolic" in relation to anabolic steroids which are used to increase the muscle mass of athletes. In the process of a dehydration synthesis, a water molecule is released. This occurs when two hydroxyl groups (although not always hydroxyls) lying close together are used to join the two simpler units together. An example would be S–OH + HO–B –> S–O–B + H_2O.

A. Lipid synthesis. We are going to build a model of a storage lipid called a **triglyceride.** This is an ordinary molecule of storage fat. The molecule will be composed of one glycerol molecule (a trihydric alcohol) connected to three fatty acids.

1. Construct a **glycerol** molecule. Place three carbons in a chain and place a **hydroxyl** group (–OH) on each of the carbons. Fill up the other available electron spaces with hydrogens. Orient the three alcohols so that they are all projecting from the right side of the glycerol. Produce the two dimensional structural formula below.

2. Construct three **fatty acids.** Fatty acids are made up of long hydrocarbon skeletons with a **carboxyl radical** (–COOH) attached to one end. Even though organic fatty acids are 16–20 carbon atoms long, we will produce a straight chain propyl radical and attach a carboxyl to the end. This will produce fatty acids that are four carbons long. To help minimize disorder, have the carbons of the fatty acid form a zigzag backbone. Be sure and join the carbon atoms of the fatty acids together by single covalent bonds. This will produce three **saturated** fatty acids. Produce a structural diagram of one fatty acid molecule below.

3. Line up the –OH's of the three fatty acids' carboxyl groups with the three hydroxyl groups of the glycerol.

4. Remove the –OH from the first carbon of the glycerol and the H from the first fatty acid's carboxyl group. Join these two components (–H and the –OH) together to make a molecule of water. Join the glycerol's first carbon to the open bond position on the oxygen of the carboxyl group. With this first dehydration synthesis you have produced a monoglyceride. The bond between the glycerol and the fatty acid is known as an **ester linkage.**

5. Repeat the above maneuver with the next two hydroxyl and carboxyl groups. If you are successful, you will have three water molecules floating around on the surface of the lab table with a fat molecule (triglyceride) along side. Produce a two dimensional structural diagram of the fat molecule just produced.

6. You have produced a **saturated fat.** All of the carbons in the three fatty acids are joined by single covalent bonds. Each of these carbons holds the maximum number of hydrogens that can be attached. The resulting **saturated fat** is solid at room temperature and is of animal origin. If some of the carbons in the fatty acids were connected together by double covalent bonds, the resulting fat molecule produced would be called an **unsaturated fat.** Unsaturated fats are liquid at room temperature (oils) and are produced by plants.

B. Protein synthesis via peptide bond formation

1. Amino acids are the monomers (building blocks) which are used in the construction of very complex protein molecules. Construct two generic amino acids following the diagram below. Note that all amino acids possess an amine radical (–NH_2) and a carboxyl radical (–COOH). One amino acid differs from a second amino acid by possessing different radical groups attached to the central carbon.

```
H         R        O
 \        |        ||
  N ----- C ------ C ----- O H
 /        |
H         H
```

2. Substitute a –H for the R group on one of the amino acids. This amino acid model now represents the amino acid **glycine.** Substitute a methyl group ($-CH_3$) for the second amino acid's R group. This model now represents the amino acid **alanine.** There are twenty different amino acids that are the building blocks of protein molecules. Each amino acid possesses a unique (–R) group that makes it different from the other 19 amino acids. Amino acids are said to be the **monomers** of more complex protein molecules. A monomer is essentially a building block. Protein molecules are put together by dehydration synthesis of amino acids to form long **polymers.** This type of chemical reaction can also be called a **polymerization** reaction.

3. Position the two amino acids so that the carboxyl group of the first amino acid is next to the amine group of the second amino acid. Remove the –OH group of the carboxyl radical of the first amino acid and a –H from the amine group of the second amino acid. Place the –OH together with the –H to make a water molecule. Join the first amino acid to the second amino acid through the open bond positions which have been created by the dehydration synthesis. You have just produced a dipeptide molecule through the formation of a peptide bond. Small proteins are composed of hundreds of amino acids joined in just this fashion. Larger proteins are of course composed of polypeptide chains of even greater lengths. **Polymerization** reactions combine monomers together to form longer chains of **polymers.**

Formulas

- Alkane = $C_n H_{2n+2}$
- Alkene = $C_n H_{2n}$
- Alkyne = $C_n H_{2n-2}$

AN INTRODUCTION TO THE KINGDOM PROTISTA

4

Laboratory Objectives

1. To familiarize the student with some of the life forms belonging to the Protista kingdom.

2. To estimate sizes of specimens using approximation techniques.

3. To categorize some "unknown" protistan specimens into their correct phylum.

4. To observe a protistan infusion to gain an appreciation of the living microscopic world.

Introduction

Early taxonomic schemes separated the world of living organisms into the convenient categories of plant and animal. The distinctions appeared to be clear cut and easy to see. These distinctions blur, however, in the microscopic world. Plants have chlorophyll and animals do not. Plants are immobile while animals move actively about their environment. However, some mobile chlorophyll-bearing flagellates are closely related to others lacking chlorophyll (e.g., *Euglena* and *Khawkinea*). The only apparent difference between these two genera is the presence and absence of chlorophyll. A chlorophyll bearing *Euglena* can be "converted" to the other genus simply by destroying its chlorophyll naturally (darkness or heat) or by chemical means. It does not seem appropriate to place closely related organisms into different Kingdoms based upon the presence or absence of chlorophyll. A need existed for the creation of a kingdom dedicated to one-celled eukaryotic plant and/or animal creatures.

Protists are often characterized as being simple, unicellular organisms. Both descriptors are inadequate. Organisms belonging to the Protista kingdom perform all of the functions that a multicellular organism does, but the unicellular organism does not exhibit the division of labor characteristic of a multicellular organism. The cells of "higher" organisms are often much simpler in structure because they are specialized to perform a single function. A single protistan cell performs all the functions which are necessary for life. Amoeba engulf prey items by phagocytosis. Portions of the cell flow around the item being ingested. As these "buds" meet on the far side of the "food item," a food vesicle is formed and taken into the cytoplasm for later digestion. The cytoplasm essentially swarms around the prey item. The movement of the cytoplasm is accomplished by the proteins actin and myosin. These two protein molecules are the "motor" proteins which also cause our muscles to contract. The actin and myosin of the amoeba form and breakdown as the cytoplasmic streaming continues. Our actin and myosin molecules retain a permanent home in our muscles and do not disassemble upon completion of their activity. The biochemistry of actin and myosin's action appears to be the same in the amoeba and the human muscle. It is not fair to say the cytoplasmic streaming in the amoeba is simpler than the contraction of a human muscle.

It is also unfair to say that the Protistans are unicellular, although that is what is found in most text books. When you think of a cell, you usually envision a membrane bound piece of cytoplasm whose activity is governed by one nucleus. Many of the protists have more than one nucleus per "cell." Ciliated protists like *Paramecium*, possess two different types of nuclei (macronuclei and micronuclei). One of the two is devoted to managing the activities of the cytoplasm while the other is important in reproductive efforts. A giant amoeba that you are going to view today has tens to hundreds of nuclei. Rather than categorizing the Protists as being unicellular, it might be better to think of them as not being divided into cells. Protists are acellular or noncellular.

Classification schemes of the Protista kingdom are in a state of flux. Older classification schemes separated the protistan kingdom into the plant-like protists, the animal-like protists (protozoans), and the fungus-like protists. Newer ideas of what organisms belong where are being introduced continually. Many of the new ideas are based upon similarities in biochemistry that we will study later in the semester. Many of these biochemical similarities are often not reflected in structural similarities that previously served as the basis for taxonomic work. What were once thought to be closely related forms have been put into different taxa based upon the newer biochemical criteria. It is also doubtful that the organisms presently placed in the Protistan kingdom represent a monophyletic grouping (possessing a common ancestor). Protists are very diverse in structure as well as in organization. Some taxonomists feel that organisms placed in the protist group should be reclassified into several kingdom level taxa. For simplicity, we are going to investigate three different groups of Protists which are related by patterns of locomotion (or how the organisms move about).

Mastigophora	Possess one or two long flagella. No cell wall is present May be parasitic or free-living
Ciliophora	Move about with cilia. Macro and micronuclei present Normally a cell mouth for ingesting food
Rhizopoda	Amoeboid in their locomotion with pseudopods May have shell-like structures of silica.

The purpose of this lab is to introduce you to the microscopic world of the protista.

Activity 1

Preparation of Freshwater Infusion (one week earlier)

One week prior to this lab, you should prepare an infusion of freshwater protistans. Fill a culture dish about half full of conditioned water. City water contains chlorine which will kill protistans. If tap water is allowed to stand for 24 hours the chlorine will evaporate and it will be acceptable to the organisms. To this water add 1/2 crushed infusoria tablet and a few drops of prepared culture. Be sure that the dropper contains solid material from the bottom of the culture. Protistans are usually associated with organic material and are difficult to isolate from clear water. Stack the culture dishes and cover the top one with an empty dish. Be sure to label your culture.

Activity 2

An Introduction to Mastigophora

Euglena is a typical member of this group. This green protistan is a common inhabitant of pond water. It possesses a very long flagellum (whip-like organelle) which it uses to pull itself through the water. *Euglena* possess an eyespot at the base of the flagellum which allows it to move toward light of appropriate intensity for photosynthetic purposes. Excess carbohydrate produced by photosynthesis is stored in food granules. This type of organism lacks a cell wall, but possesses strong flexible plates of protein located beneath its cell membrane (its **pellicle**). Since *Euglena* inhabit hypotonic pond water, a contractile vacuole is needed to pump out water gained by the process of osmosis. If placed in the dark, *Euglena* can live as a heterotroph by absorbing organic material from the environment.

Procedure

1. Obtain a permanent slide of *Euglena*. This is a good point to mention "composite drawings." The specimen that you are going to be observing today are in a variety of different "poses." When these prepared slides were made, some of the organisms may have either been partially contracted or fully extended. To get a good idea of what an "average" specimen looks like, it is necessary to observe many different specimen including the good points of each in a composite drawing. Draw and label the **nucleus** and the **pellicle** of the *Euglena*.

2. Prepare a wet mount of *Euglena*. Take a drop from the *Euglena* culture and add a cover slip. Examine the live culture and add the following structures to your drawing:

   a. **flagella.** The flagella may be difficult to see. It is located at the anterior of the animal. The flagellum of the *Euglena* essentially employs a "breast stroke" to pull the organism through the water Try adjusting the fine focus up and down. Also adjust the light intensity. Finally add some of the Proto-Slo to a second preparation of *Euglena* before you apply the cover slip. The Proto-Slo may slow down the action of the flagella and make them easier to see.

   b. **chloroplast.** What is the function of this organelle?

   c. **food granules** (vacuoles). These are the dark objects which may be present in single to multiple copies.

   d. **nucleus.** In the live *Euglena,* the nucleus is the same color as the organism, but by careful observation you will be able to see it as a more homogenous spherical body within the cytoplasm.

   Is the *Euglena* a plant or animal?

Activity 3
An Introduction to the Ciliophora (Ciliata)

The diverse protists of the Ciliophora group are characterized by their use of cilia to move about and feed. In contrast to flagella, cilia are much shorter, more numerous, and beat synchronously with each other. The cilia may be distributed across the entire body of some types of ciliates, while other ciliates have their cilia clustered into rows or isolated tufts.

This group is unique in its genetics. Ciliates possess two different types of nuclei. **Macronuclei** are present in multiple copies. The macronuclei contain hundreds of copies of genes which are involved in the organism's daily housekeeping chores. The **micronuclei** possessed by these organisms are involved with sexual reproduction processes that generate variability so important for the process of natural selection.

A. *Paramecia* are examples of ciliates which are covered by thousands of individual cilia. Associated with the cilia, but not visible with our microscopes, are organelles referred to as **trichocysts.** These bulb-like structures release sticky proteinaceous threads. Some experts believe that the trichocysts release their products in response to predators, while others view them as more important in stabilizing the organism as it feeds.

Paramecia feed mainly on bacteria. The bacteria are swept into the oral groove by the action of cilia. When enough bacteria have been amassed in a "bolus-like" clump, the "bolus" is taken into the cytoplasm by phagocytosis. The food vacuole is then acted upon by lysosomes for digestive purposes. Just as with the *Euglena* mentioned above, paramecia possess contractile vacuoles to expel water absorbed from their hypotonic environment.

Procedure

1. Obtain slides of *Paramecium caudatum* and *P. bursaria*. *P. caudatum* is the example most often shown in text books, but at times it does not show everything that you need to locate. *P. bursaria* is unique in that it contains symbiotic green algae which carry out photosynthesis. Use both slides to draw a composite example of a paramecium labelling the **micronucleus, macronucleus, cilia, oral groove, food vacuoles and contractile vacuoles.** You will need to look at several specimen to find the proper orientation to locate the above structures.

2. How large are the paramecia? In the microscopy lab you learned how to estimate the diameter of the microscopic field. Refer to that lab to recall the actual dimensions of the field of view. ________ You can now determine the size of an organism by estimating how many Paramecia it takes to span the field by placing them end to end.

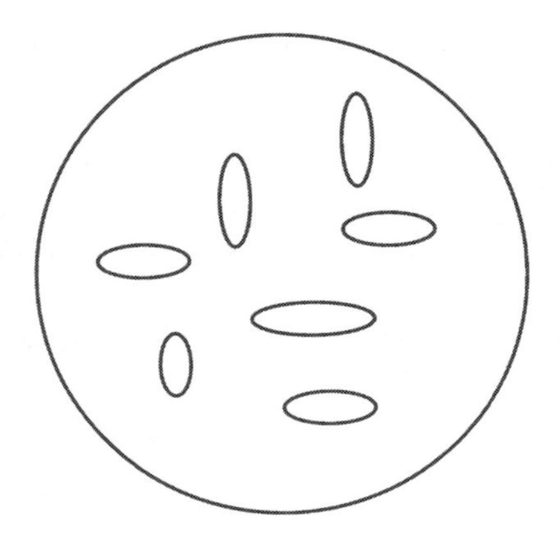

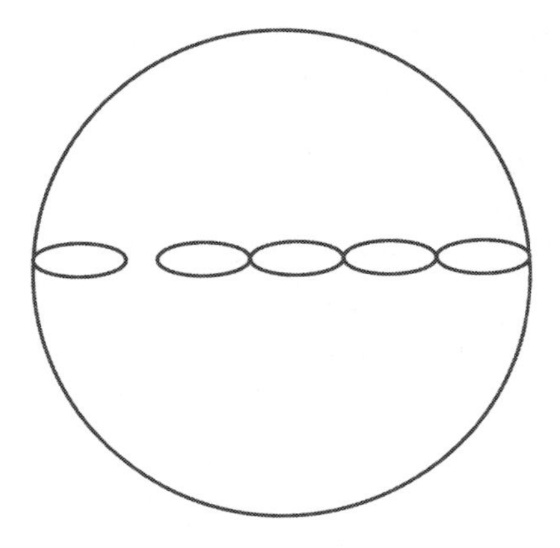

What you actually see looks like the left hand field. What you should visualize looks more like the right hand field. If the diameter of the field is 0.5 mm and you estimate that five paramecium would fit across the field, then the size of each paramecium is 0.5 mm/5 or .01 mm each.

3. Prepare a wet mount from the protista infusoria started the week before. Search the infusoria for living examples of paramecium.

   a. Observe how they move about?

   b. Is there a pattern to their movement? Do they swim about in the open spaces of your slide or do they tend to cluster near the organic matter floating about on your slide? What activity may they be engaged in?

B. *Stentor* is another common example of a ciliate. This organism moves about its environment by individual cilia along its sides and membrane-like folds which spiral around its broader funnel-like anterior end. The membrane-like folds rhythmically undulate helping to propel the organism. In its feeding mode, *Stentor* attaches to the substratum with its narrower posterior end and, with its membrane-like fins, creates a cyclonic flow of water at its broader anterior bringing food particles toward its cell "mouth."

Procedure

1. Obtain a prepared slide of *Stentor*. The organisms will be in many different positions and orientations. You will have to look at several different specimen to find the best views. In its full glory, *Stentor* should resemble a bugle with a flaring anterior end. When disturbed by some outside force, this organism is capable of contracting itself into a protective ball. The sketch you do should be a composite of several organisms. Illustrate the organisms general shape labelling its anterior and posterior ends. Label the **macronuclei** which will appear as light beaded strands running the length of the organism. Also label the cell mouth.

2. Using the wet mount prepared above (or a new wet mount if the first is dried up), search for specimen that resemble the *Stentor* seen in the prepared slide above.

Activity 4

An Introduction to the Rhizopoda (Amoeba and related organisms)

No members of this phylum possess flagella at any stage in their life histories. Instead, amoeba move about their environment using cellular extensions called pseudopodia (false feet). A pseudopod represents a "polyp" that can bulge out of any portion of the cell. The end of the pseudopod is then anchored to the substratum with the rest of the cytoplasm flowing into the pseudopod. The microtubules of the cell along with the microfilaments functions to produce this amoeboid activity. Amoeba use this type of locomotion to swarm around prey items which they then package into phagocytic vesicles. You can find amoeba in salt water and fresh water environments along with those that live in different soil types. This group of organisms exhibits no meiosis or sexual reproduction, but reproduce by binary fission.

Procedure

1. Draw several examples of *Amoeba proteus* from the slide provided. The organism is surrounded by a plasma membrane. Immediately inside of this membrane is a thin, clear, nongranular layer of protoplasm called **ectoplasm.** Beneath this layer is the main body mass of granular **endoplasm.** Within the endoplasm is a large disc-like **nucleus.** Food vacuoles may also be seen. Irregularly shaped extensions of the amoeba's body are called **pseudopods.** Label the structures indicated above. Amoeba also possess contractile vacuoles which help to expel excess water. These organelles may be very hard to see in preserved specimen. Why would an amoeba tend to gain water from its environment?

2. Estimate the size of the amoeba using the procedure outlined above.

3. Prepare another wet mount from the protistan infusoria sampling mainly from the "scum" near the junction of the fluid and the side of the container. Look for amoeba using low light levels.

Activity 5
Identification of Unknowns

For each of the two unknown organisms below, make a simple sketch and identify the group to which you think the organism belongs. Tell me the basis for your decision.

A. *Chaos chaos (carolinensis)*

B. *Chlamydomonas sp.* This organism tends to cluster. Select solitary examples to sketch.

Questions for Further Thought

1. What functions do cilia, flagella, and pseudopods have in common?
2. What factors may account for the wide distribution and great structural diversity seen among the unicellular organisms?
3. In what sense are protists "primitive"? In what sense are they "advanced"?

MEMBRANES, DIFFUSION, AND OSMOSIS

5

Laboratory Objectives

1. To understand the meaning of "kinetic molecular activity."
2. To be able to differentiate between diffusion, osmosis, and dialysis.
3. To be able to predict which direction a substance will diffuse.
4. To learn the terms of osmotic pressure and the effect osmotic pressure has on the directionality and rate of osmosis.
5. To learn how to interpret data and to write a laboratory report.

Introduction

Despite change in its external environment, the cell must maintain its internal environment to preserve its integrity and carry on its life functions. Maintenance of the internal environment in a steady state is a dynamic equilibrium process in which the cell membrane plays a crucial role. It is the cell membrane which controls movement of materials into and out of the cell. Some substances such as water pass freely through membranes, while others are controlled and still others are excluded. For this reason, cell membranes are described as being **semipermeable.** The term "semipermeable" means that some substances are able to cross the membrane into or out of the cell quite easily. Other materials cannot cross the cell's outer boundary at all. Still other materials, by the process of active transport, cross the cell membrane through the expenditure of energy. We are going to study some of the processes involved in the **passive transport** (no energy expenditure) of materials across the cell membrane. These processes depend upon the concept known as the kinetic molecular theory.

The kinetic molecular theory states that all matter, whether it exists as a gas, liquid, or solid, is in a state of constant motion. Molecules of a gas are constantly moving about independently of one another. The molecules of a liquid slide past one another as they take the shape of their container. The molecules of a solid, although locked into a specific locus with reference to their neighbors, vibrate in their set location. At a given temperature, all particles possess a given amount of kinetic energy (energy associated with motion) moving at temperature dependent intensities. As temperature increases, the thermal motion and collision rates of these particles increase. A lower temperature would of course have the opposite influence. Hypothetically all molecular movement ceases at absolute zero (−273°C). Molecular motion, with the subsequent collisions between molecules, favors a random distribution of molecules. Your instructor will now demonstrate kinetic molecular motion with a piece of apparatus at the front of the room.

The kinetic molecular theory, along with differing concentration gradients of materials, produces the phenomenon of diffusion. **Diffusion** is defined as the movement of particles away from a region of their higher concentration to a region of their lower concentration (given a uniform temperature and pressure). This movement is caused by the greater frequency of collisions between particles located in the more concentrated region. Molecules from the more concentrated areas spread outward toward less populated spaces. Can you cite some examples of diffusion which you have experienced in the last day?

Diffusion occurs in and through all states of matter. As the laboratory proceeds, your instructor will demonstrate everal different examples of solids, liquids, and gases diffusing into each other.

Osmosis is a special case of diffusion involving the diffusion of solvent (water in biological applications) through a semipermeable membrane. Remembering that diffusion is movement of material from an area of its high to low concentration, osmosis becomes the movement of water from an area where there is relatively more water to an area where there is relatively less water.

In talking about osmosis some terms of solution chemistry are needed:

A **solution** is produced by the homogeneous intermingling of the molecules of a mixture.
The **solvent** is the substance with the greater number of molecules in the solution.
A **solute** is a substance with a lesser number of molecules in the solution (mixture).

Normally when we think of a solution, a sugar and water mixture comes to mind. The sugar is the solute and the water is the solvent. Air is another example of a solution. Air is a mixture of the gases: nitrogen, oxygen, carbon dioxide, carbon monoxide, and others. As nitrogen is the most common gas in air, it is by definition the solvent of the solution. The other gases, being less common, are defined as being the solutes of the solution.

In a solution, the relative concentration of solvent is influenced by the amount of solute dissolved. As the amount of solute in the mixture increases, the relative amount of solvent decreases. The purpose of the following activity is to investigate the process of osmosis and to determine the effect of solute concentration on the rate of osmosis.

Activity 1

Experimental Design

Osmosis is a special case of diffusion involving the passage of water molecules through a differentially permeable membrane. This movement of water molecules will lessen the concentration gradient of water on opposite sides of the membrane separating the two solutions.

You will construct artificial cells from dialysis tubing. Different concentrations of sugar solutions will be placed in the artificial cells. Sugar cannot diffuse through the dialysis tubing, but water can. The artificial cells will be placed in different environments for sixty minutes. The weight of the artificial cells will be determined at ten minute intervals to follow the flow of water into or out of the cells.

Procedure

1. Prepare five artificial cells

    a. Using the presoaked dialysis tubing, **fold, twist, and tie** one end of the tube. The "tubing" looks like a piece of plastic tape. It is a tube that needs to be opened.

    b. Add 15 ml of a specified solution to one of five cells as detailed below. Squeeze gently to remove as much of the remaining air in the cell as possible. **Fold, twist, and tie** the open end of the cell. If you notice any leakage at this time, you have to start over with the production of a new cell.

c. The cells are to receive the following material:
   Cell 1: 15 ml of tap water
   Cell 2: 15 ml of 20% sucrose
   Cell 3: 15 ml of 40% sucrose
   Cell 4 15 ml of 60% sucrose
   Cell 5 15 ml of tap water

d. Each of the above cells should be rinsed off, patted dry, and **weighed** to the initial 0.1g on the electronic balance. Record this weight in Table 1 at time zero.

2. Place each cell in the correspondingly numbered beaker.
   Beaker 1 200 ml of tap water
   Beaker 2 200 ml of tap water
   Beaker 3 200 ml of tap water
   Beaker 4 200 ml of tap water
   Beaker 5 200 ml of 60% sucrose solution

3. At ten minute intervals, remove the cells, carefully blot off all excess water, and weigh each cell to the nearest 0.1 grams. Record the weights in Table 1.

Table 1. Osmosis Data

	Weight (grams)				
Time (min)	Cell 1	Cell 2	Cell 3	Cell 4	Cell 5
0					
10					
20					
30					
40					
50					
60					

Activity 2

Lab Report

Follow the format outlined in Appendix A of your laboratory manual in writing this report. Below are some specific suggestions for inclusion in this lab report.

Data Analysis

1. Prepare a data table showing the change in weight from the initial weight for each cell for each time interval.
2. Graph the change in weight of each cell against time. All five cells will be placed on one graph. All lines should originate at the same point on the graph (0 ml of weight gain at 0 sec of time).
3. Calculate the rate of weight change (gr/min) for each time interval for cell #4 and record the results in Table 2. The trends shown by this cell should resemble the trends shown by the other cells. You are essentially calculating the slope of the curve for the particular time interval that you are interested in. See below for an explanation.

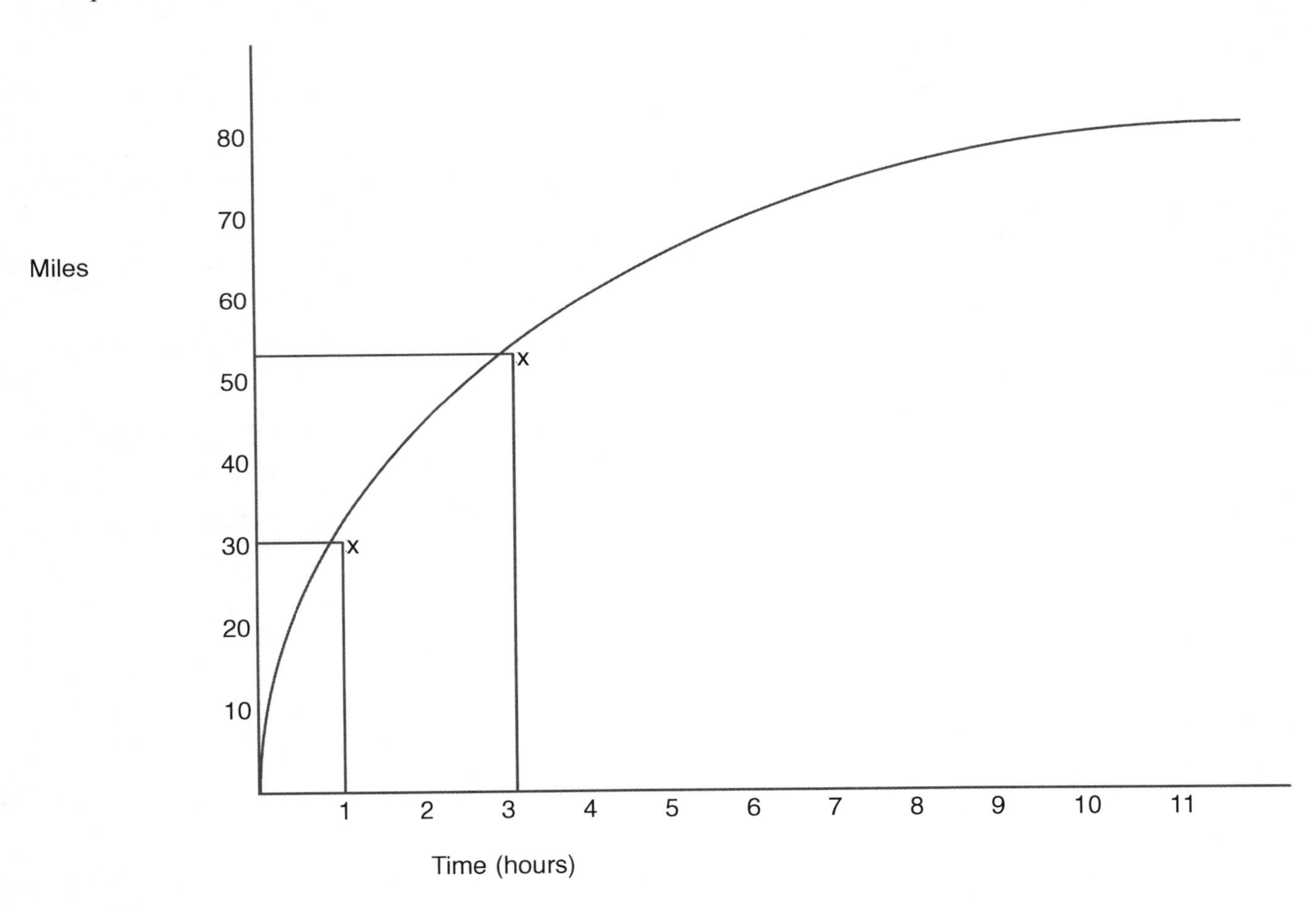

To determine the rate of movement between the first and the third hour, you divide the change in the Y value (miles) by the change in the X value (time) Rate = (52-30) miles/(3-1) hours = 22 miles/2 hours = 11 mph.

Table 2	Time Intervals (min)					
	Initial (0–10)	10–20	20–30	30–40	40–50	50–60
Rates*						

*grams/min

Data Interpretation

The following points should be addressed in the **Discussion Section** of your lab report. You do not need to address the points in the order listed nor should you number your replies to these questions. The questions asked below are there to help you develop your discussion.

1. Which cell gained the most weight and why?
2. What relationship can be inferred between concentrations of sucrose and rates of osmosis? Cite specific examples as evidence.
3. Using the terms of tonicity found in your text book (hypertonic, isotonic, and hypotonic) explain the direction of water flow in each of the artificial cells produced.
4. Was there a control in the experimental design? Why?
5. Did the rate of water gain change in each cell as time progressed as shown by Cell #4? Why or why not? Would each cell continually gain weight forever or would an equilibrium be reached? Why or why not?
6. How do you know that the movement of water is responsible for the changing weights of the cells and not the movement of sugar? Explain your reasoning.
7. Explain any discrepancies seen in your data.

Lab Cleanup

The lab will look cleaner after you are done than before you started. To insure this result complete the following steps:

1. At the end of the sixty minute time period, place the artificial cells in the beaker at the front of the room.
2. Rinse all of the glassware which you used and place it in the soap-filled tubs on the bench partition in front of your work station. Make sure any tape placed on the glassware has been removed.

3. Throw any and all paper products used during the lab into the waste basket.

4. Clean the lab surface upon which your group was working. Do not just wipe the surface. Wash the table tops with the sponges and soap provided.

5. After turning the balances off, clean their surfaces in the same manner as you did the lab table tops.

6. Push your chairs under the desk.

Optional Demonstrations (time permitting)

A. Dialysis

Dialysis involves the differential movement of **solute** through a membrane either based on size or solubility characteristics. In other words, dialysis is the separation of one class of solute particle from a second class due to differing abilities to cross a semipermeable membrane. To demonstrate dialysis:

1. Construct an artificial cell filling it three-fourths full with a starch-salt solution. Rinse the "cell" under tap water. Blot off the excess water and weigh the cell to the nearest 0.1 gram.

2. Immerse the cell in a 250 ml beaker of distilled water. Let it stand for twenty minutes, then reweigh the cell.

3. Test the water in the beaker for starch and salt.

4. Starch test:

   a. Place ten drops of the beaker fluid to be tested in a clean test tube.

   b. Add two drops of iodine solution.

   c. A blue/black color indicates the presence of starch.

5. Sodium chloride test

   a. Place 1 ml of the fluid from the beaker in a clean test tube.

   b. Add several drops of silver nitrate ($AgNO_3$)

   c. Formation of a white precipitate indicates the presence of chloride ions.

6. Interpretation of the demonstration:

   1. Which material was able to permeate the artificial cell's membrane?

   2. What characteristic or quality of the solute is responsible for this differential ability to pass through the membrane?

B. Diffusion of Gas into Air

The apparatus for this demonstration consists of a length of large-bore glass tubing clamped to a ring stand. Two stoppers, each with a ball of absorbent cotton attached, will be placed in either end. One of the stoppers will be saturated with a 1N ammonium hydroxide (NH_4OH) solution. The other stopper will be saturated with 1N hydrochloric acid solution (HCl). Both stoppers will be inserted simultaneously into the glass tubing. The two solutions will dissociate and vaporize. The ions will begin to diffuse toward the middle of the tube where they will meet. At their point of contact, a salt (NH_4Cl) and water droplets will form.

Interpretation of the demonstration:

1. What happens when a base is combined with an acid? How will this reaction show up inside of the large bore glass tube?

2. Does the reaction occur in the middle of the tube, closer to one end, or the other?

3. Given the molecular weights of the ammonium chloride (NH_4Cl) and hydrochloric acid (HCl) as 49 and 37 amu's respectively, suggest a relationship between molecular weight (or particle size) and the rate of diffusion.

C. Gas into a Liquid

Several drops of phenolphthalein solution are added to a large water-filled test tube. Phenolphthalein is an indicator of pH change and will act to mark the progression of the basic gas into the liquid of the test tube. The end of the test tube is closed with a piece of moist dialysis membrane held in place with a rubber band. The tube is inverted and clamped to a ring stand over a beaker of ammonium hydroxide.

Interpretation of the demonstration:

1. Describe what happens.

2. Account in molecular terms for what is happening.

3. Is this a demonstration of the semipermeability of dialysis tubing? Why or why not?

D. Diffusion of a Solid into a Gel (Solid)

A petri dish filled with a 2% agar gel is used for this demonstration. A crystal of potassium permanganate ($KMnO_4$) is placed on one side of the petri dish, and a similar sized crystal of methylene blue is placed on the opposite side. The two crystals will be left in place for the duration of the lab period. The molecular weights of the potassium permanganate and methylene blue are 158 and 320 amu's respectively.

Interpretation of results:

1. Measure the sizes of the two crystals from the original site of deposition to the edge of the arc produced.

2. Which has diffused faster? Why?

THE USE OF VOLUMETRIC GLASSWARE

6

Laboratory Objectives

1. To be able to use a pipette to draw and deliver accurately small volumes of fluid.

2. To be able to determine the relative concentration of hydrogen peroxide in samples of varying concentrations using titration.

Introduction

This laboratory is intended to allow you to practice and master two techniques which you will need in the Enzyme Catalysis lab. The use of the pipette will be very important in transferring safely small quantities of fluid from one container to another. The buret will be used to measure the concentration of hydrogen peroxide (H_2O_2) in five stock solutions by delivering potassium permanganate drop by drop until the hydrogen peroxide has been consumed in the chemical reaction taking place. Hydrogen peroxide will be the substrate digested by enzyme activity in next week's laboratory.

Operation of the Pipette

Pipettes are an inexpensive tool used to measure and transfer relatively small quantities of liquid in the laboratory.

Before operating the pipette, carefully observe the calibrations on its side. Calibrations differ between styles of pipettes. One style shows the entire scale etched onto the side of the pipette. The other style does not include the final value. **See Figure 1.**

In using the pipette on the left, you will empty the entire column of fluid to deliver 10 ml of the solution. With the variety of pipette shown on the right, allow the fluid to drain from the pipette until the meniscus reaches the demarcation of "10." Note that there is still a considerable volume of solution left in the pipette which will not be delivered.

A pipette is filled by immersing its tip into a solution and drawing the solution up into the pipette by suction. Materials should never be drawn up by using your mouth. In this lab we will be using a valved bulb **(Figure 2)** to power the pipette. The bulb has a number of valves, which when pinched sequentially, direct the flow of material in the desired direction.

To operate the valved bulb:

1. Place the end of the pipette into the base of the bulb. **Do not insert the pipette more than one half of an inch into the bulb** to avoid damage to the valve at the base of the bulb's stem.

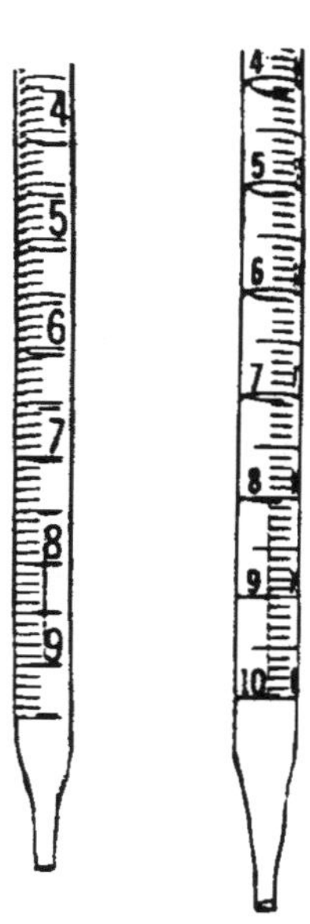

Figure 1

2. Depress the first pressure point (Indicated by the number "1" or letter "A") and squeeze the bulb to force air out. This action produces the "suction force" which will be used to pull fluid into the pipette.

3. Place the tip of the pipette into the solution being sampled and depress the second valve (indicated by the number "2" or the letter "S"). Draw the fluid into the pipette a bit beyond the "0" mark. **Be sure and keep the tip of the pipette in the solution being drawn up or you will suck air into the apparatus.**

4. Wipe the outside of the pipette with a Kimwipe to remove any excess liquid.

5. Holding the pipette vertically, slowly depress the third pressure point (indicated by the number "3" or the letter "E"). Allow the fluid to flow out of the pipette until its meniscus touches the zero line. Depress the third pressure point slowly and evenly so the flow out of the pipette does not leave droplets of the solution of the pipette's inner walls.

6. Touch the tip of the pipette to the side of the vessel that you are depositing the solution to insure complete delivery of the desired amount of solution. This procedure removes any droplets from the pipette tip.

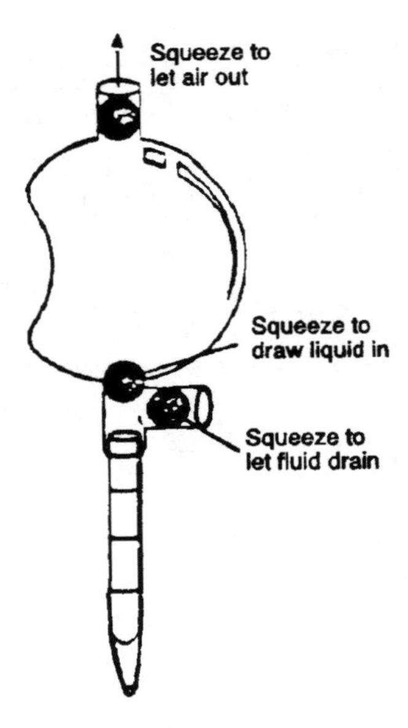

Figure 2

Activity 1

Complete the Following Procedures Using Water

1. Weigh a small beaker carefully on an electronic balance to the nearest tenth of a gram. _________ g

2. Deliver 5 ml of water to the previously weighed beaker.

3. Reweigh the beaker to the nearest tenth of a gram. _________ g

4. What relationship exists between the volume of water delivered and its weight? One milliliter of water should weigh 1 g.

5. Continue practicing until you are **very comfortable** with the pipette's operation. Make sure that everyone in your group practices and becomes proficient with the procedure. Your grade will depend upon it.

Activity 2

Operation of a Burette

In this activity, you are going to use a burette to determine how much hydrogen peroxide is found in five different stock solutions located at the front of the room. The 50 ml burrette is calibrated from 0 to 50 ml in 0.1 ml increments. All volumes delivered from the burette should be between the calibration marks. Do not estimate above the 0 ml mark or below the 50 ml mark. Figure 3 is an illustration of a common setup with a stopcock burette.

The amount of hydrogen peroxide in a given volume of stock solution can be (assayed) measured by a chemical titration. The basis for this determination lies in the following reaction:

$$H_2O_2 + KMnO_4 + H_2SO_4 = KSO_4 + MnSO_4 + H_2O + O_2.$$

Hydrogen peroxide reacts with potassium permanganate in the presence of sulfuric acid to form the products listed on the right. Potassium permanganate possesses a very deep purple color. It will be added drop by drop to a known volume of stock solution containing the hydrogen peroxide. The first drop of $KMnO_4$ will react with the hydrogen peroxide and be consumed in the reaction. The deep purple color of the potassium permanganate will be lost and the solution will remain clear. Potassium permanganate will continually be consumed in the reaction (drop by drop) as long as there is hydrogen peroxide in the reaction mixture. At the precise moment when all of the hydrogen peroxide has been consumed, the addition of the next drop of potassium permanganate will turn the solution a **very light persistent shade of pink.** The pink color is produced by the dilution of the potassium permanganate that persists. At this point there is no more hydrogen peroxide present with which the potassium permanganate can react. The amount of potassium permanganate added to the stock solution is directly proportional to the amount of hydrogen peroxide present. **For purposes of this experiment, we will assume that a 1:1 relationship exists between the hydrogen peroxide present in each stock solution and the potassium permanganate delivered by the burette.** If you recorded the initial level of potassium permanganate present in the burette and the final level after the titration is complete, the difference is directly proportional to the amount of hydrogen peroxide in the original sample

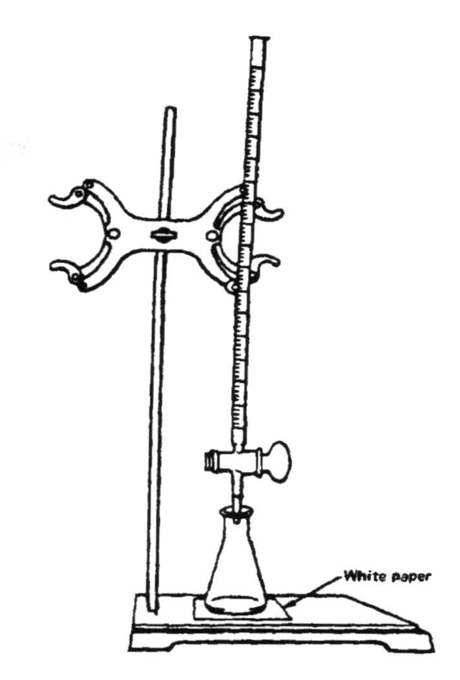

Figure 3. Setup with Stopcock Buret

Procedure for Activity 2

1. Put 10 ml of a stock hydrogen peroxide solution into a 50 ml beaker.

2. Add 10 ml of 2 N sulfuric acid carefully to the same 50 ml beaker using the premeasured dispenser at the front of the room. Use care. **Be sure that all the members of your team are wearing safety goggles.** Mix carefully by swirling.

3. Remove **5 ml of the 20 ml** and assay for the amount of hydrogen peroxide present using a burette with the potassium permanganate.

    a. Fill the burette with potassium permanganate to a point below the "0" mark and record. Make sure that the nipple at the burette's end is completely filled with the potassium permanganate solution.

    b. To deliver the solution from the burette, turn the stopcock with the forefinger and the thumb of your left hand to allow the solution to enter the flask **(Figure 4).** You want to deliver the solution in a continuous drip fashion rather than in gushes (a flood and then nothing). This procedure leaves your right hand free to swirl the solution in the flask during the titration. With a little practice, you can control the flow so as to deliver as little as 1 drop of the potassium permanganate.

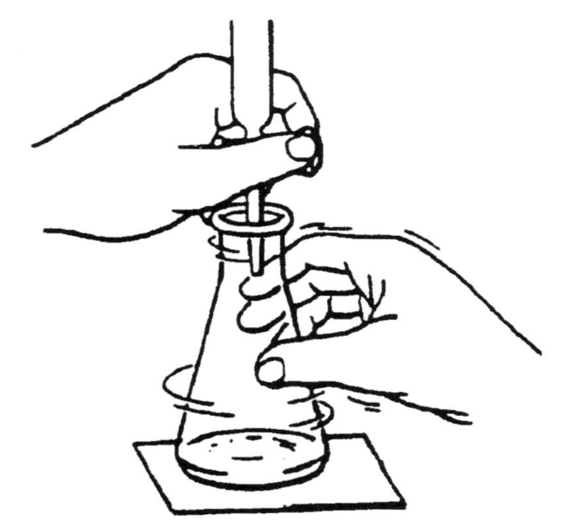

Figure 4. Titration Technique

c. Continue adding the potassium permanganate until a persistent light pink color is obtained. If a brown color is produced, you have added too much potassium permanganate and gone past the end point. When a persistent color change has been obtained in the titration process, immediately **dump** out the contents of the beaker into the sink (with a steady flow of water running through), **rinse** the beaker with water, and **wipe** the beaker dry with Kimwipes. **Remember to DUMP, RINSE, and WIPE your glassware after each titration.**

d. Determine the amount of potassium permanganate added. You know where the column of potassium permanganate was at the beginning of the titration. Read its level at end of the titration. The difference between these two numbers is the amount of potassium permanganate added. When reading the burette, be sure that your line of sight is level with the bottom of the meniscus in order to avoid parallax errors. (see **Figure 5**). Record your results in Table 1.

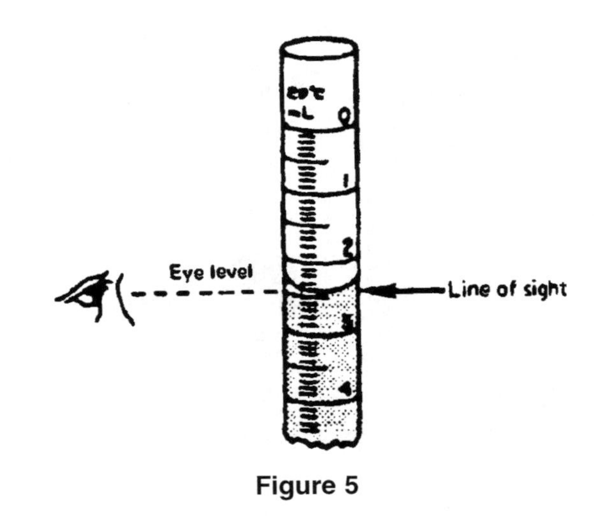

Figure 5

e. Repeat the titration to verify your first determination. You have plenty of solution left for this second titration. From the original volume of 20 ml, you should have 15 ml remaining. If the second titration verifies your first determination, proceed on to the next stock solution. If the two results are very different, do a third titration using the 10 ml of solution which still remains. Calculate an average of the recordings.

Stock Solution Number		Final Reading	Initital Reading	Relative Amount of Hydrogen Peroxide
1	Trial 1			
	Trial 2			
	Trial 3			
2	Trial 1			
	Trial 2			
	Trial 3			
3	Trial 1			
	Trial 2			
	Trial 3			
4	Trial 1			
	Trial 2			
	Trial 3			
5	Trial 1			
	Trial 2			
	Trial 3			

Table 1. Amounts of Potassium Permanganate Delivered

6. The amount of potassium permanganate added is proportional to the amount of hydrogen peroxide present in each stock solution. In looking at your results, what relationship do you think exists between the different stock solutions?

Lab Clean Up Procedures

The laboratory activity just completed produces a messy lab. In the clean up process, be sure and do the following:

1. **Dump** into the sink, **rinse, and wipe** all titration vessels after you have seen a persistent color change.

2. Place all dirty glassware (beakers, pipettes, and graduated cylinders) in the glassware discard bin located on the divider in front of your work station.

3. Drain all burettes into the large beaker located at the front of the room. Rinse the burettes with water allowing a fairly large portion to run out the bottom of the burette. Return the burettes to their stands, but store them upside down with the stopcocks open.

4. Return the pipette bulbs, protective eye wear, and anything else that I have forgotten to the green trays at your work station.

5. Your work station should look cleaner and more organized than it did when you started your experiment.

Thank you for your help in keeping the lab environment clean and safe.

ENZYME CATALYSIS

7

Laboratory Objectives

1. To become familiar with enzymes as biocatalysts
2. To be able to use a variety of laboratory apparatus to determine amounts of substrate consumed over time.
3. To be able to determine the rate of a chemical reaction and analyze how that rate changes over time.
4. To produce a standard lab report

Introduction

It seems difficult to imagine, but the very organized state called "life" is dependent upon random molecular collisions producing the chemical reactions needed to sustain life. Earlier in the semester we studied the process of osmosis and diffusion and used the concept of the **kinetic molecular theory** to explain why diffusion and osmosis work the way they do. Kinetic molecular theory also helps to explain reaction chemistry. Reactant molecules continually move randomly through a reaction mixture. If two reactant molecules are moving fast enough and collide with enough vigor and proper orientation (a productive collision), a chemical reaction may occur. To speed up chemical reactions, chemists usually try to increase the number of random collisions taking place. Heat can be added to increase the molecular movement of the substrate molecules, and therefore the number of successful collisions leading to a chemical reaction. A higher concentration of reactants also increases the chances of producing more random productive collisions. The chemist might stir the reaction mixture to further agitate the reactant molecules. Some of the chemical "tricks" listed above for increasing reaction rates are not conducive to life. I have tried for years to light a fire under some of my more lethargic students, but it doesn't seem to work. Organic catalysts called **enzymes** produce more order in chemical reactions reducing some of the randomness described above.

Enzymes

Enzymes are protein molecules which function as biocatalysts. A catalyst is any material which increases the rate of a chemical reaction without being consumed in the process. Enzymes appear to function by "attracting" reactant molecules to their surface. A complimentary fit exists between an enzyme's reactive site and the specific substrates involved in the chemical reaction. The term "complimentary fit" refers to the relationship which exists between a "hand and a glove" or a "lock and a key." When reactants attach to enzymes, a more "orderly" state is produced. A number of things may occur in this situation facilitating the chemical reaction. For example, before new compounds can form, substrate molecules must be broken apart. Existing internal bonds in the reacting molecules may be stressed when the substrate molecule attaches to the enzyme. Secondly, the reactive sites of the substrates may become properly aligned when attached to an enzyme. Thirdly, the local environment of the enzyme binding site may be different than the surrounding medium. This local difference may be conducive for the chemical reaction. Do not overlook, however, that the attachment of the substrate to the enzyme is still due to random molecular collisions. Whatever the method, enzymes facilitate chemical reactions by lowering the **energy of activation** needed to start the chemical reaction. The energy of activation is the energy which must be invested to make the substrates reactive.

Catalase is an enzyme found within almost all cells and is housed within the peroxisomes of cells. Catalase degrades hydrogen peroxide which can accumulate in cell cytoplasm as a by-product of metabolic processes. Hydrogen peroxide at high levels is toxic to the cell and its rapid breakdown is necessary for the health of the cell. Hydrogen peroxide (H_2O_2) will naturally decompose to O_2 and H_2O, but it does so very slowly. Catalase speeds up the reaction considerably. The purpose of this experiment is to determine the reaction rate of catalase and to see how this reaction rate changes over time.

Reaction Rates

Reaction rates can be measured as either the amount of product formed or the amount of initial substrate degraded over a period of time. In this exercise, the amount of hydrogen peroxide consumed over different time intervals will be determined. Reaction rates will be calculated by plotting the amount of hydrogen peroxide digested against the amount of time the reaction is allowed to proceed.

Experimental Design

It is difficult to measure the amount of hydrogen peroxide consumed directly during a given time interval. We will measure this quantity with an indirect method. First, the amount of hydrogen peroxide present in 5 ml. of a 3% hydrogen peroxide solution will be determined. This quantity will be referred to as the **baseline.** We will then digest a given quantity of hydrogen peroxide with a constant volume of catalase for different lengths of time. At the end of each digestion, the amount of hydrogen peroxide remaining will be determined. The amount of hydrogen peroxide consumed in the reaction is the difference between the amount present at the beginning of the digestion (baseline) and the amount remaining at the end of the digestion. An analogy might be helpful. If a person starts the day with five dollars in her pocket (baseline) and finds that $2.25 remains at the end of the day, she knows that she has spent $2.75. She knows how much money has been consumed without keeping track of each individual transaction.

The amount of hydrogen peroxide remaining in a given volume of reaction mixture will be determined by a chemical titration. The basis for this determination lies in the following reaction:

$$H_2O_2 + KMnO_4 + H_2SO_4 = KSO_4 + MnSO_4 + H_2O + O_2.$$

This equation is not balanced chemically, but shows the reactants and products. Hydrogen peroxide reacts with potassium permanganate in the presence of sulfuric acid to form the products on the right. Potassium permanganate possesses a very deep purple color. It will be added drop by drop to a known volume of reaction mixture containing hydrogen peroxide. The first drop of $KMnO_4$ will react with the hydrogen peroxide and be consumed in the reaction. The deep purple color of the potassium permanganate will be lost and the solution will remain clear. At the point when all of the hydrogen peroxide has been consumed, the reaction mixture will turn a very light persistent shade of pink. The pink color persists because the diluted dark purple drop of potassium permanganate remains. There was no hydrogen peroxide present with which it could have reacted. The amount of potassium permanganate added to the reaction mixture to achieve this color change is directly proportional to the amount of hydrogen peroxide remaining in the reaction mixture. For purposes of this experiment, we will assume that a 1:1 relationship exists between the hydrogen peroxide and potassium permanganate.

The basic sequence of the experiment:

1. A one milliliter purified catalase extract (enzyme) is added to the hydrogen peroxide (substrate). The enzyme catalyzes the breakdown of hydrogen peroxide to water and oxygen (liberated as a gas). **If you do not see oxygen bubbles being given off here, you should immediately stop and determine what has gone wrong with your procedure.**

2. The reaction will be stopped after different specified periods of time by mixing sulfuric acid to the reaction mixture. The acid changes the pH of the reaction mixture denaturing the enzyme, **catalase.** Probably there has been a loss of complementary fit between enzyme and substrate.

3. The amount of substrate remaining (hydrogen peroxide) will be determined by the titration method discussed above.

Activity 1

The Baseline
(See Figure 1 for protocol)

To determine the amount of hydrogen peroxide initially present in a 5% solution, we will perform the steps of the process described above without adding the catalase (enzyme) to the reaction mixture. This will allow us to determine how much hydrogen peroxide is present at the beginning of the experiment (the amount of money that the person started the day with in the analogy described above). To do this:

1. Put 10 ml. of 5% hydrogen peroxide in a clean beaker using a **dedicated** ten ml pipette. This pipette should only be used to transfer hydrogen peroxide from the stock bottle to the reaction flask. **Contamination of this pipette with acid will ruin your experiment.**

2. Add 1 ml of water to the beaker. This takes the place of the enzyme added in a later procedure.

3. Add 10 ml of 2N H_2SO_4. Use care in handling the acid and be sure to wear your **protective eye goggles.**

4. Mix well but carefully.

5. Remove **five** milliliters of the mixture and titrate for the amount of H_2O_2 remaining. Use a burette to add $KMnO_4$ to the solution until a persistent pink color is obtained. If you add too much of the potassium permanganate, the result will be a brownish color. **Check to be sure that you understand the calibrations on the burette.** Record your results in Table 1.

   Repeat the titration to verify your first baseline determination. You have enough reaction mixture left for a second and third titration. If the second titration verifies your first determination, proceed to the next step. If the two results are very different, do a third titration on the remaining reaction mixture. Calculate an average of the two closest recordings.

   When a persistent color change has been obtained in the titration process, immediately **dump** out the contents of the beaker, **rinse** the beaker with water, and **wipe** the beaker dry with Kimwipes. **Remember to DUMP, RINSE, and WIPE your glassware after each titration.**

Table 1. Baseline Determination of Hydrogen Peroxide Levels

	Final Reading	Initial Reading	Amount of Hydrogen Peroxide
First titration			
Second titration			
Third titration			
Average	X	X	

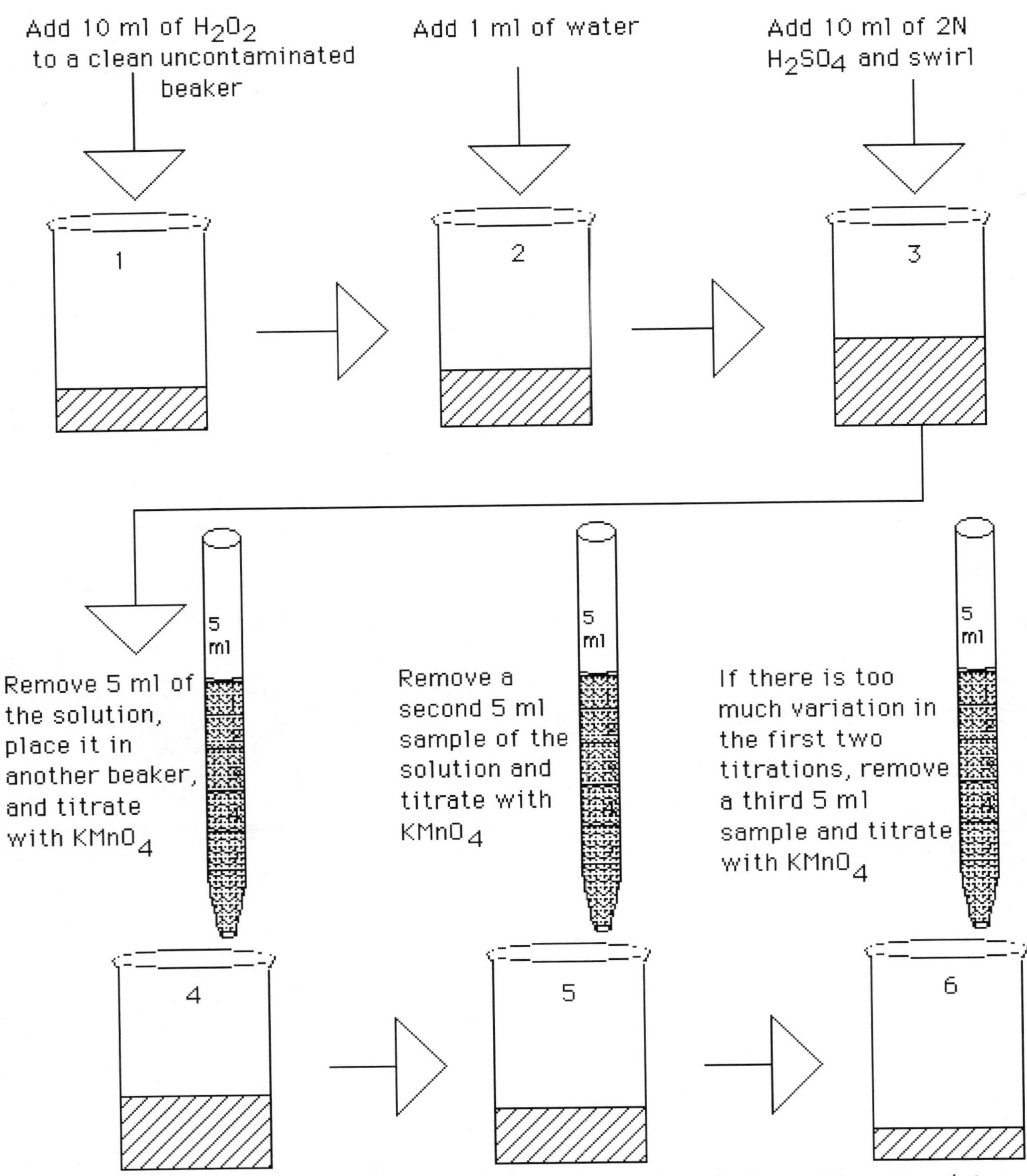

Figure 1. Baseline Protocol

Activity 2

The Enzyme-Catalyzed Rate of H_2O_2 Digestion (See Figure 2 for experimental protocol)

In this experiment, you will determine the rate at which a 5% hydrogen peroxide solution decomposes when catalyzed by the purified catalase extract. To do this, you must determine how much hydrogen peroxide has been consumed after digestions of 10, 30, 60, 120, 180, and 360 seconds. **Start with the 360 second time interval and work backward toward the shorter length digestions.** This will allow time for your team to formulate an assembly line approach for the procedures. For each time interval:

1. Put 10 ml of a 5% hydrogen peroxide solution in a 50 ml clean glass beaker using a **dedicated pipette.** This 10 ml pipette should not be used for anything except transferring the 10 ml of hydrogen peroxide.

2. Add 1 ml of catalase (enzyme) extract.

3. Swirl gently for the indicated time period. Notice the vigorous release of oxygen bubbles from the reaction mixture. If no bubbles are being released, **stop immediately** and determine what has gone wrong.

4. At the end of the indicated time period, add 10 ml of 2N sulfuric acid to the reaction mixture. Each team should have a dedicated acid person. She should have the 10 ml of acid premeasured in a separate dedicated beaker ready to pour into the catalase reaction mixture at the end of the digestion period.

5. Remove 5 ml of the resulting solution and assay for the amount of hydrogen peroxide remaining. Follow the same procedures for the assay done above performing a minimum of two titrations on each reaction solution. Record your results in Table 2 below.

Time (sec)		Final Reading	Initital Reading	Amount of Hydrogen Peroxide Remaining	Amount of Hydrogen Preroxide Used*
10	Trial 1				
	Trial 2				
	Trial 3				
30	Trial 1				
	Trial 2				
	Trial 3				
60	Trial 1				
	Trial 2				
	Trial 3				
120	Trial 1				
	Trial 2				
	Trial 3				
180	Trial 1				
	Trial 2				
	Trial 3				
360	Trial 1				
	Trial 2				
	Trial 3				

*To determine hydrogen peroxide used, subtract the amount of hydrogen peroxide remaining from the baseline measure previously determined.

Table 2. Time Course Determination of Hydrogen Peroxide Digestion

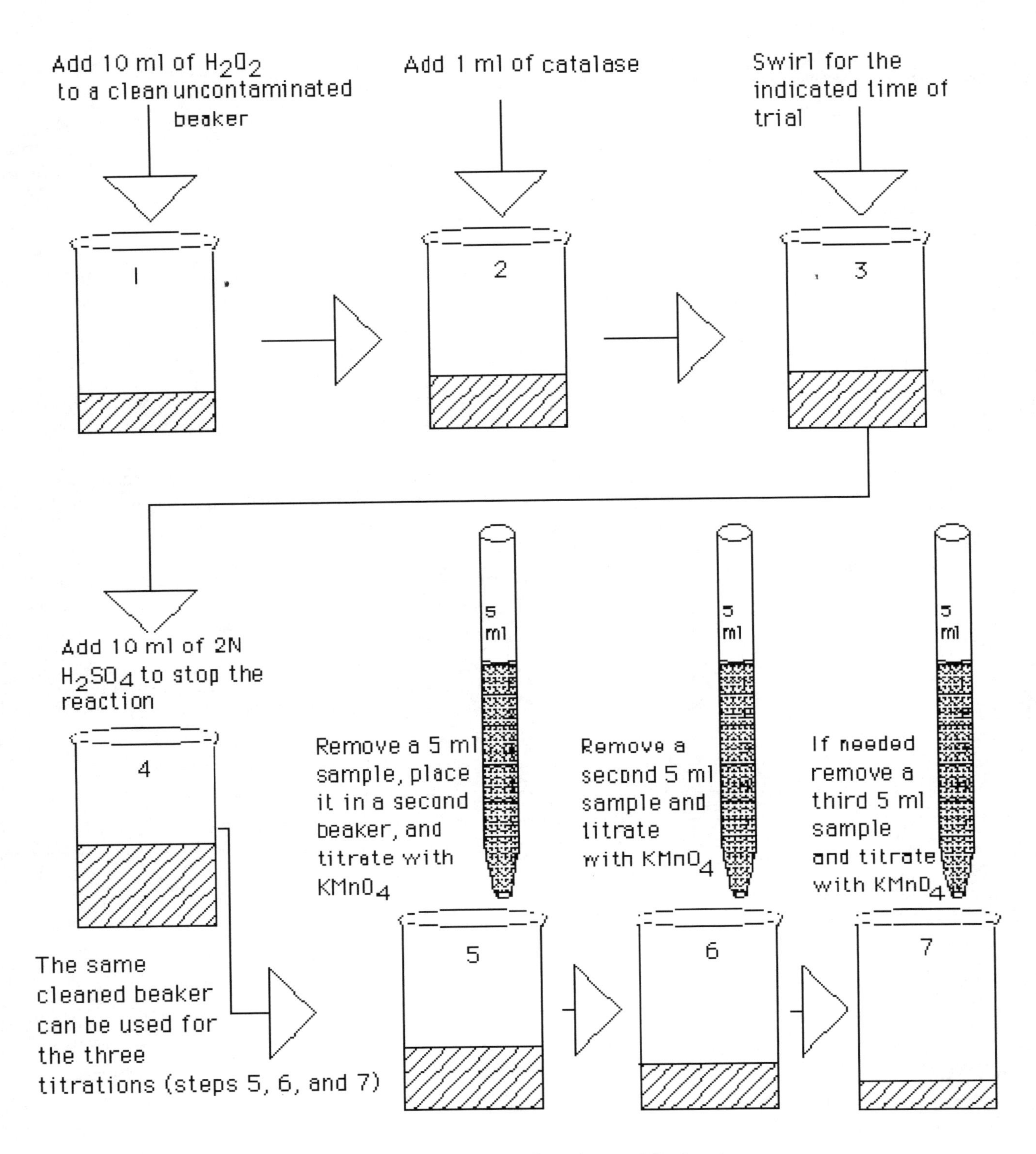

Figure 2. Experimental Protocol

Data Manipulation

A. On a graph, show catalase activity as a function of time by plotting the amount of hydrogen peroxide used on the y-axis and time in seconds on the x-axis. Include a title for your graph and use graph paper. **Be sure to have equal time intervals represented on the x-axis.** In drawing the curve for catalase activity, do not play "connect-the-dots" on the graph. Try to produce a smooth curve with as many data points falling below as fall above your line. You are visually trying to produce the line of best fit for your data.

B. From your graph, determine the initial rate of the reaction, and the rates of the reaction for each of the time intervals thereafter. Record the rates in Table 3. When you determine the reaction rates for each time interval, you are really determining the slope of the curve over that time interval. The slope of a line geometrically is equivalent to the rise over the run. Another way to define the slope of a line is the change in Y value over the change in X value. Since your Y value is recorded in ml and the X value is in seconds, your reaction rates will possess the units of ml/sec.

For example:

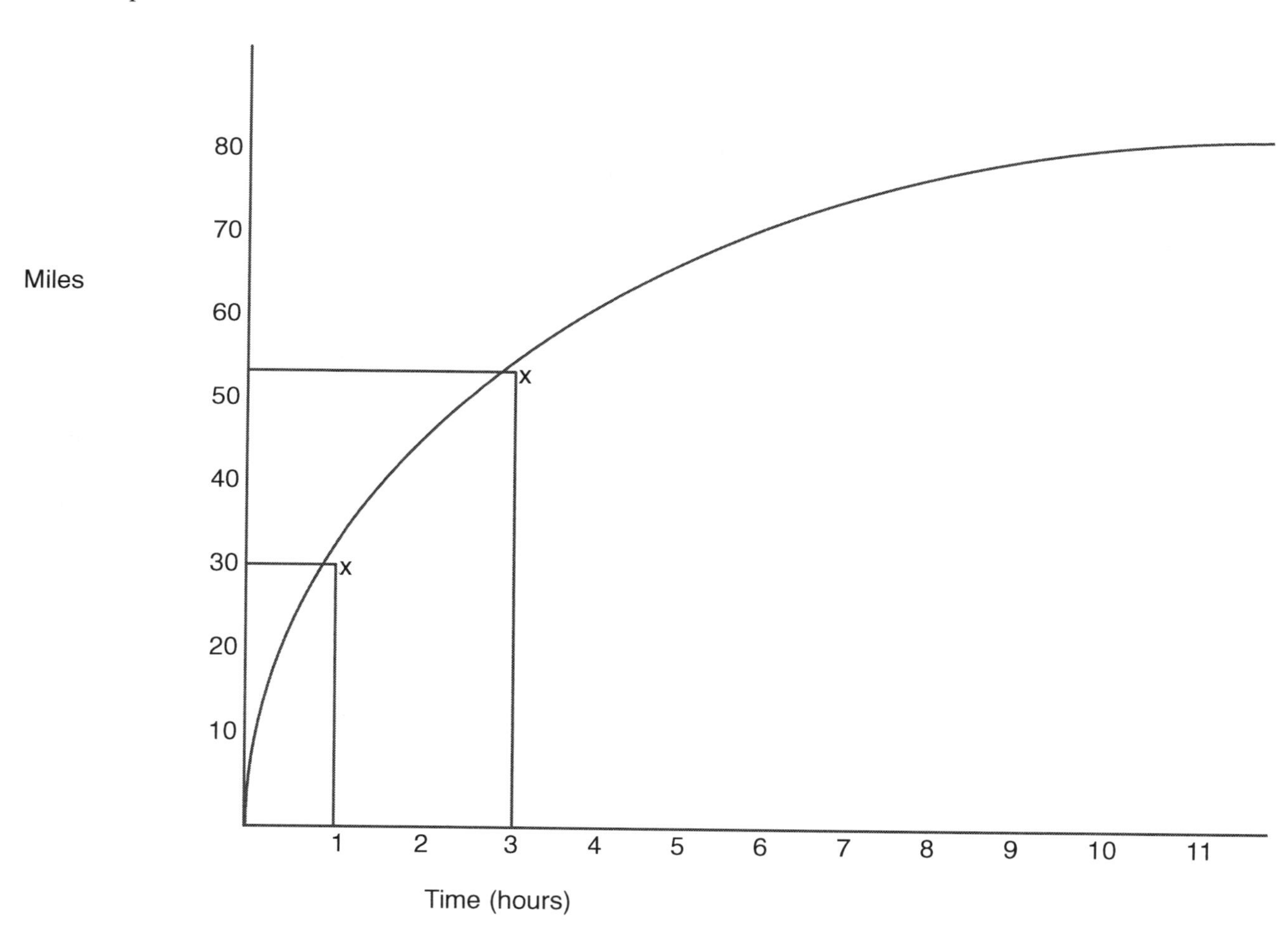

To determine the rate of movement between the first and the third hour, divide the change in the Y value (miles) by the change in the X value (time). The rate between hours 1 and 3 = (52–30) miles/(3–1) hours = 22 miles/2 hours = 11 mph.

Table 3. Reaction Rates						
	Time Intervals (secs)					
	0: 10	10: 30	30: 60	60: 120	120: 180	180: 360
Rates*						

*reaction rate (ml H_2O_2/sec)

Activity 3
Lab Report

Follow the report format outlined in Appendix A of your lab manual. Below are some questions which might be addressed in the Discussion section of your lab report.

When is the rate of the reaction highest? Why? If you have not calculated your reaction rates correctly, the answer to this question may not be so obvious. To answer the above, you have to relate reaction rate to what is happening at the molecular level (kinetic molecular theory). When is the concentration of substrate molecule the highest? How would this effect reaction rate?

When is the reaction rate lowest? Why?

Explain the inhibiting action of sulfuric acid on catalase activity.

Hypothesize about the effect of lowering temperatures on the rates of enzyme activity? Explain your reasoning.

Did your results come out as you expected? If not, why not? Was there one data point that looked out of line? In discussing possible sources of error, not only identify the error source, but explain how your results were affected by this error.

Include any other points that you think are relevant in this section.

Lab Clean Up Procedures

The laboratory activity just completed produces a messy lab. In the clean up process, be sure and do the following:

1. **Dump, rinse, and wipe** all titration vessels after you have seen the persistent color change.
2. Place all dirty glassware (beakers, pipettes, and graduated cylinders) in the glassware discard bin located on the divider running down the length of the laboratory table.

3. Drain all burettes into the large beaker located at the front of the room. Rinse the burettes with water allowing a large portion to run out the bottom of the burette. Return the burettes to their stands, but store them upside down with the stopcocks open.

4. Return the pipette bulbs, protective eye wear, and anything else that I have forgotten to the green trays at your work station.

5. Your work station should look cleaner and more organized than it did when you started your experiment.

Thank you for your help in keeping the lab environment clean and safe.

AN INTRODUCTION TO PLANT TISSUES

8

Laboratory Objectives

1. To introduce the student to the different cell and tissue types seen in plant specimens.
2. To expose the student to the structures of three primary plant organs; the leaf, stem, and root.
3. To distinguish between the organization of monocot and dicot plants.
4. To produce detailed and labelled diagrams of the various plant organs described.
5. To compare the structure of the gymnosperm leaf to a dicot leaf.

Instructor Expectations

1. Your lab instructor expects you to produce large and detailed sketches of the following plant structures on separate pieces of paper. Do your sketches in pencil. You do not need to be good artists to produce accurate sketches, but you do need to take time and care in their production. You may use your textbook as a resource, but do not copy their illustrations. Provide labels for the specific structures **(bold faced type)** described in the following exercises. Your sketches will serve as your evaluation of this material, therefore your grade is in your own hands. Ask your lab instructor or fellow students for help when you are confused. If you do not have enough time to complete your work, come into lab during other times to finish your assignments. Be neat and organize your sketches following the order of presentation in the lab handout.
2. At the end of each work session, the slides will be put back in their proper order listed on the top of the slide box. Microscopes will be returned to their proper shelf in the storage cabinet.

Introduction

In order to understand the anatomy of plant organs, an understanding of their component tissues is important. Below is a listing of plant tissues with some identifying characteristics. After being introduced to their general organization, different plant organs will be surveyed to view these tissues in plant organs.

I. Meristem tissues show continual mitotic (cell division) activity producing new plant growth

A. **Primary meristems** produce primary plant tissues which result in plant elongation. Primary meristems are located at the shoot or root tips of the plant (apical meristems) or in lateral buds (axillary meristems).

B. **Secondary meristems** produce a growth in the width (girth) of the plant. This growth is referred to as secondary growth because it occurs after the plant has grown in length. There are two major types of secondary meristems.

1. **Vascular cambium** produces secondary xylem and phloem in the stem. This increases the diameter of the stems of perennial plants.

2. **Cork cambium** produces the periderm (bark) of stems which have lost their original epidermal covering.

3. **Pericycle** produces lateral roots growing from tap roots of dicotyledonous plants. This may not be the place in the organization to put this tissue, but I am not sure where else it fits.

II. Permanent tissues are for the most part nondividing. They can be subdivided based upon the roles which they play in the biology of the plant.

A. **Dermal tissues** cover the surfaces of plant organs regulating what does and what does not enter the plant organ.

1. **Epidermis** is a primary tissue (produced by a primary meristem) which covers young plant organs. The epidermis is modified in a structural/functional way to reflect the needs of the organ which is being covered. Leaf epidermal cells are closely packed together and produce a fatty wax-like material (cutin). The deposited **cutin** makes up the cuticle of the leaf. The more cutin deposited, the greater is the retention of water by the leaf. Specialized guard cells are interspersed between the epidermal cells of the leaf reflecting the need for gas exchange between the leaf and the atmosphere. The epidermis of roots lacks cutin, but possesses thin cytoplasmic extensions which permeate the surrounding soil particles. These root hairs increase the absorptive surface area available for water uptake.

2. **Periderm** is a secondary plant tissue produced by the secondary meristem, cork cambium. Periderm is usually several cell layers thick. As the original epidermal layer is sloughed off due to secondary growth, the periderm is essentially a "second set of clothes" for the plant.

3. **Endodermis** is an internal "dermal" tissue which surrounds the stele (vascular tissues) of roots. It regulates what will and what will not enter the stele of the root.

B. **Vascular tissues** transport materials throughout the plant.

1. **Xylem** tissues are concerned with the transport of water and inorganic nutrients usually upward from the root to the shoot. The cells of xylem tissues are dead at functional maturity. No living cytoplasm is to be found, but the highly re-enforced secondary cell walls left behind form the conduction pathways for water transport. Two different types of xylem cells exist. **Tracheids** are the more primitive of the two being found in greater abundance in the cone bearing plants. Tracheids are closed and tapered at both ends. They do not directly abut the next cell in the path of water flow. The tapered ends of the tracheids overlap with other tracheids. At the point of overlap, pit pairs are found between adjacent tracheids. The pits allow for the passage of water from one cell to the next. **Vessel elements** are more highly advanced predominantly occurring in the angiosperms or flowering plants. Vessel elements possess open ends and are lined up consecutively providing water a more direct path from the root to the shoot. The organization vessel elements is similar to placing straws end to end linking one location to a second.

2. **Phloem tissues** transport sugar and other organic products of the plant. Phloem tissues are alive possessing cytoplasm. Two cell types functioning together move organic compounds both up and down the plant body. **Sieve tube cells** provide the actual conduction pathway. Sieve tube cells, laid end to end, form the sieve tubes. Unlike vessel elements of xylem which have open ends, the sieve tube cells possess porous **sieve plates** which allow cytoplasm to extend from one cell to the next. Sieve tube cells lack nuclei to direct their cytoplasmic activity. Adjacent companion cells provide the needed instructions to the sieve tube cells which is necessary for the active transport of food through the sieve tubes.

C. **Ground tissues** are difficult to describe as they are a heterogeneous grouping. They are essentially like the connective tissues of animals in that their main role in plant architecture is to provide structural support and storage.

1. **Parenchyma cells** are the most general of the ground tissues. They form the "bulk" of most of the organs which you are going to see. Parenchyma cells are large, possess living cytoplasm, and have large central vacuoles used for storage of food, water, or various other chemicals. Parenchyma cells become modified when located in the different organs of the plant. Their cytoplasm may be loaded with chloroplasts if in the leaf or

starch may be stored when observing parenchyma in a root. Spaces are commonly found between the parenchyma cells.

2. **Collenchyma cells,** like parenchyma, have living cytoplasm and may remain alive for a long time. The collenchyma differ in having unevenly thickened cell walls at the intersection of three or more collenchyma cells. These cells are typically longer than they are wide. Found just under the epidermal layer in many herbaceous stalks, their pliable yet strong cell wall provide flexible support for the stem of the plant.

3. **Sclerenchyma cells** do not have living cytoplasm when seen in the mature plant. Before the death of the cell, the secondary cell walls are impregnated with a very tough and strong compound, lignin. Their chief function is in support. There are two different kinds of sclerenchyma cell. **Sclereids** (stone cells) are as long as they are wide with the lignin in their secondary cell walls being distributed in a web-like distribution across the surface of the cell. The appearance of stone cells in the pear fruit produces its gritty texture. Sclerenchyma fibers (as their name implies) are much longer than they are wide. Commonly seen around vascular tissue, they are easily identified in cross section due to their very thick cell walls enclosing a tiny cavity or lumen which once housed the cell's cytoplasm. Sclerenchyma fibers are used in the production of textile goods, rope, twine, canvas, and other products.

During this exercise, you are going to be viewing plant tissues from two major groups within the Angiosperm (flowering plant) division. The monocots differ from the dicots in many ways. Some of these differences will show up in our discussion of the plant organs. These differences will be pointed out as they arise.

You are going to make sketches on separate pieces of paper of the following slides labelling the indicated structures. Each sketch that you make should be large enough to show details of cell structure. A good practice is to draw larger secondary sketches of areas of concern so that more detail can be included. Please emphasize the word *large* in the phrase "large sketches."

Activity 1

Coleus Stem Tip Longitudinal Section

A. Meristem tissues. Observe the slide of the *Coleus* stem under low power to orient yourself to the slide. The profile of this stem tip resembles bat man. At its top (between the two "ears," you will see a densely staining patch of cells that represent the **apical meristem.** This primary meristem is the source of all of the other cells which you see on your slide.

Further down the stem, in the angle between the attachment point of young leaves to the stem, find the **axillary meristem.** Axillary meristems are pockets of primary meristem which have been left behind for the growth of lateral shoots from the main axis of the plant.

Meristem cells in general are small, tightly packed cells with relatively large nuclei. They exhibit no structural specializations as these will develop later.

B. Dermal tissue is the outer covering tissue of a plant. The *Coleus* stem is covered by epidermis. It is a primary tissue as it was produced by the apical meristem. The epidermis immediately above apical meristem is very young. As you proceed downward along the stem, the epidermal cells become progressively older. Can you detect any changes in the appearance of the epidermal cells as you progress down the stem? Make sure that these structural changes are reflected in your sketches. The extensions which you observe from some of the older epidermal cells may be two to three cells in length. They are referred to as **trichomes.** The fuzzy feel of a tomato stem is due to the presence of trichomes. Notice that there are different shapes of trichomes. Some may contain irritating chemicals. What might their function be?

C. Vascular Tissues include both xylem (water transport) and phloem (sugar transport).

1. **Xylem** tissue can be fairly easily seen running up and down the stem of the Coleus. The thickened spiralling secondary cell walls of the xylem can be seen running longitudinally in from the epidermal layer. What is the significance of the spiral thickening of the secondary cell walls of these cells.

2. **Phloem** tissue may be seen just to the outside of the xylem. Distinctions between sieve tube cells and companion cells will not be observable in this preparation.

D. Ground Tissue. The only observable type of ground tissue seen in the *Coleus* slide is the parenchyma in the center of the *Coleus* stem (pith). Describe their general appearance below.

The central vacuoles of the parenchyma cells contain a hypertonic fluid. What might be the functional significance of this structural arrangement.

Activity 2

Leaf Structure

Examine a prepared slide of the *Ligustrum* (privet) leaf. Begin with low power to get an overall view of the leaf's structure. Remember that in making a sketch of this organ, you should use a separate piece of paper to hold a large sketch. As you study this leaf, list the characteristics which are adaptations promoting its photosynthetic function? What characteristics are needed to make a leaf an efficient solar collector, yet also an efficient conserver of water?

A. **Dermal tissues.**

1. Locate the upper and lower layers of cells on the leaf's surface; these are the upper and lower **epidermis** respectively. There is a distinct noncellular outer coat of **cutin** making up the leaf cuticle. What is the function of this hydrophobic layer? Where is the cuticle thicker, on the upper or lower surface of the leaf? Do you see any chloroplasts in the cytoplasm of the epidermal cells? **Stomata** are openings through the lower epidermal layer of the leaf which allow for gas exchange. Why are these openings restricted to the lower surface of the Ligustrum leaf? (Think about temperature!!) Adjacent to the leaf's stomata are smaller cells called **guard cells.** These guard cells control the opening and closing of the stomates.

2. **(This may be an instructor demonstration.)** Obtain a leaf from a Wandering Jew plant located someplace in the lab. Strip off a portion of the lower epidermis and prepare a temporary slide for observation using tap water. Cut off any thick portion of the leaf which may still be attached to the lower epidermis.

Make a sketch of the lower epidermis showing a few stomata and guard cells.

Flood the slide with a 10% NaCl solution by introducing a drop of the solution to one side of the cover slip. Allow the salt solution to wick under and across the temporary mount. Observe and sketch any differences which appear in the guard cells of the stomates.

B. **Ground Tissue.** Return to the slide of the *Ligustrum* leaf. Notice that the leaf resembles a sandwich with the lower and upper epidermis being the bread. The middle layer of the leaf is referred to as the **mesophyll** area and is mainly filled with **parenchyma cells.** The parenchyma cells are arranged in two distinct patterns. The upper layer of parenchyma cells is referred to as the **palisade parenchyma.** The cells are all closely packed together with the cells resembling columns of a building. Note the position of the chloroplasts within the cytoplasm of these cells. Are they located centrally within the cytoplasm or peripherally? What is the significance of their distribution? Where would the sunlight be brightest? Below the palisade parenchyma is the **spongy parenchyma** layer. These cells also possess chloroplasts, but the cells are much more loosely arranged with numerous air spaces between the cells. What is the functional importance of this structural arrangement?

C. Vascular tissue of the leaf includes both observable **xylem** and **phloem.** The xylem and phloem of the leaf travel together in the veins (vascular bundles) of the leaf. Locate the main vein of the Ligustrum leaf. Find the xylem (larger cells with thicker cells walls normally stained red) located in the upper portion of the vein. The phloem cells are visible below the larger xylem vessels. Sieve tube cells and companion cells of the phloem are not yet visible.

The veins of the *Ligustrum* leaf show the branching net-like vein pattern characteristic of the dicot plants of the Angiosperm division. Therefore it is possible in the cross section of the *Ligustrum* leaf to see side veins running parallel to the upper and lower epidermis of the leaf. Within the side veins, the spiral secondary cell walls of the xylem are visible as they were in the Coleus stem tip.

D. **Comparison between the structures of monocot and dicot leaves.** (Optional: check with your instructor). Working with neighboring students, set up three microscopes side by side. On one microscope place the slide of the Ligustrum leaf. This is the example of a dicot leaf. On the second microscope, place a corn *(Zea)* leaf. On the third microscope, place the leaf of blue grass *(Poa)*. The second and third microscopes contain examples of the monocot group of angiosperms (lilies, orchids and grasses) which characteristically have parallel veins.

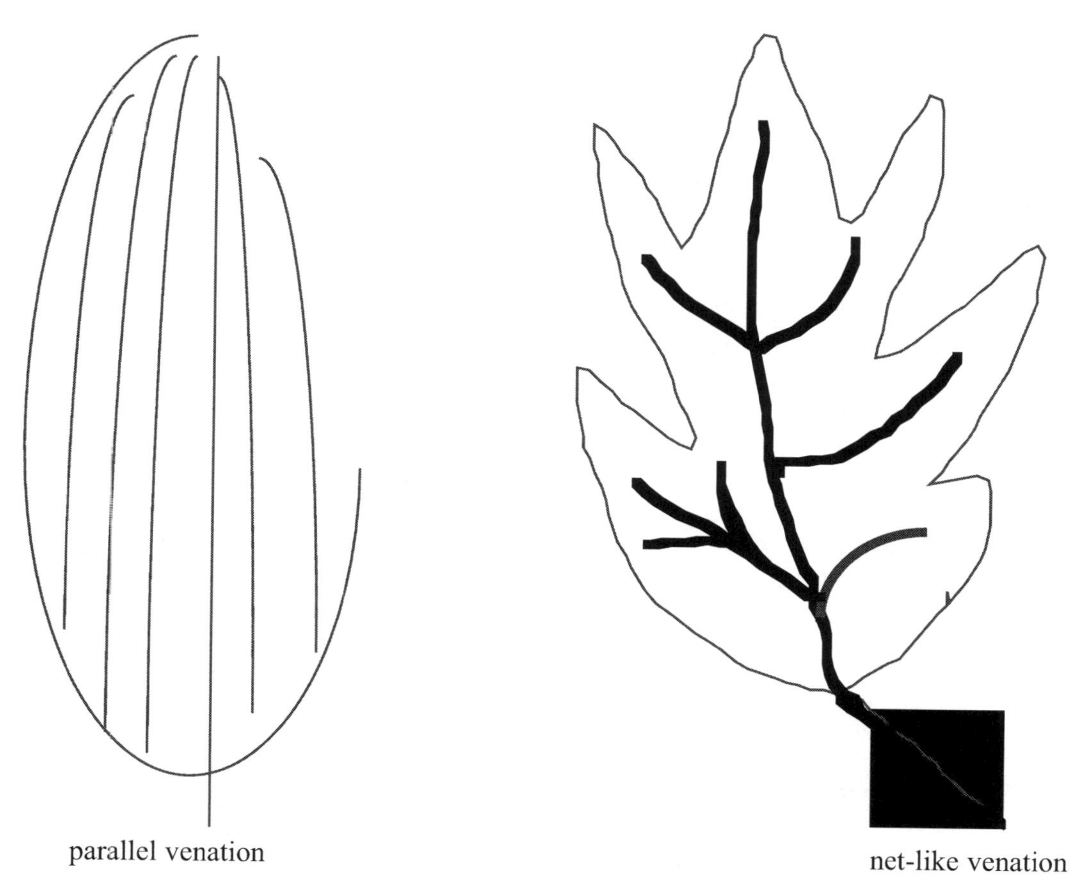

1. How does the appearance of net-like and parallel venation show up structurally on the slides you are observing?

2. Notice the well developed **bundle cells** surrounding the vascular bundles of the monocot leaves. The corn leaf *(Zea)* exhibits Kranz anatomy characteristic of plants displaying C_4 photosynthesis. As you may see later in lecture, C_4 photosynthetic pathways allow the corn to continue to produce sugar under dry and hot conditions. What differences do you notice between the bundle sheath cells of *Zea* and *Poa?*

3. Both the leaves of *Zea* and *Poa* grow upright (vertically) rather than horizontally like the Ligustrum leaf. What features of the Zea and Poa leaf reflect this difference in growth pattern? Look for stomates.

Activity 3

Stem Structure

Place the slide of the *Medicago* (alfalfa) stem on your microscope's stage. *Medicago* is an herbaceous (non-weedy) dicot. The stem does not show any woody secondary growth in girth. The **pith** found in the center of the stem with the vascular tissue arrayed around the periphery of the stem is characteristic of dicot stems. Make a detailed sketch of this stem labelling the structures indicated below.

A. Dermal tissue, as before, covers the surface of the stem. It is a single layer of heavily cutinized primary **epi-dermal cells.** The cells within the cell walls have shriveled somewhat revealing the cell membranes within the lumen of the cell. At fairly regular intervals you should see **"stomates"** with guard cells within the epidermis. These structures are more appropriately called **lenticels** when found on the surface of the stem. What is their function?

B. Vascular tissues are seen in bundles arranged peripherally around the central pith of the stem. Locate one vascular bundle and focus on it with the 40X objective lens.

1. The **xylem** cells are quite large with their characteristically red staining cell wall.

2. The **phloem** tissue is seem immediately to the outside of the xylem being four or five cell layers deep. The sieve tube and companion cells of the phloem tissue cannot be distinguished in this stem.

C. Ground tissues are easily seen in this slide.

1. In the center of the stem, the **pith** houses many large, thin walled **parenchyma** cells. Notice that these cells are not as closely packed as the epidermal cells. These cells possesses a hypertonic sap filling their central vacuoles. What is the function of this hypertonic cell sap? Just below the epidermis, a second category of parenchyma cells (four or five cell layers deep) can be seen. This area houses the **cortical parenchyma cells.** Observe their cytoplasm. What organelles are contained within that are not found in the pith's parenchyma cells? (Hint: what is the color of an herbaceous stem?)

2. Around the periphery of the stem under low power, notice the seven or eight prominent ribs (ridges) which protrude from the surface of the stem. These ribs are analogous to the ribs seen on celery stalks. Focus with higher power to observe **collenchyma cells.** Their cell walls are much thicker than the cell walls of the parenchyma. Notice that the cell walls are characteristically thicker at the intersection where three or four cells come together.

3. Return to low power to find one of the vascular bundles. **Sclerenchyma** fibers are best seen to the out-side of the phloem tissue observed above. The sclerenchyma fibers are small in diameter with very heavily stain-ing secondary cell walls which help to support the vascular bundle. The most visible sclerenchyma is seen 5–10 cell layers below the epidermis just to the outside of the vascular bundles.

D. Organization of a woody dicot stem. Place the older *Tilia* stem under your microscope with the 4X lens in place. This stem differs from the herbaceous stem just seen as it has accumulated secondary tissue resulting in a growth in diameter. This secondary growth (diameter) is the result of the activity of two secondary meristems (mitotic tissue) to be described below.

1. **Vascular tissue.**

a. The rings of **xylem** are the most conspicuous feature of this plant organ. The rings of xylem are produced by the cyclical change in the diameters of these cells. Notice that within one ring, the xylem cells closest to the pith are largest. Their diameters decrease in size as you move away from the pith until the next growth ring begins. This pattern of larger cells produced in the spring with smaller cells produced through the summer and into fall months reflects water availability during the growth season. With plentiful water in the spring, high internal pressures stretch the newly produced xylem cells to their limit. As the season progresses and water becomes more limiting, the cells decrease in size. Notice that some of the rings are larger in size than others. What implications do these differences in size of growth rings have (about the growing conditions) of that particular summer? How old is the specimen of Tilia that you are looking at? Also notice the **vascular rays** which are seen radiating outward from the pith through the xylem like the spokes of a wheel? The rays are made up of parenchyma cells and they function for lateral transport within the stem.

b. **Phloem** tissue is seen to the outside of the xylem. Under low power the phloem appears to be arranged in darker red staining triangular patches. The apices of the triangles point toward the edge of the stem. Neither the sieve tube cells nor the companion cells can be distinguished.

2. **Ground tissues.** The most prominent ground tissue seen in this slide is the **parenchyma** tissue of the central **pith.** The prominent central pith is the trademark of the dicot stem.

3. **Dermal tissues.** The surface of the perennial stem is covered with **cork** (bark). The cork is approximately 5–7 cell layers thick. This secondary tissue has replaced the original epidermis that has been sloughed off as the tree stem has increased in diameter through secondary growth. Recall that epidermal tissue, being a permanent tissue, does not exhibit mitotic activity. Since the epidermis cannot grow as the stem increases in diameter, it is sloughed off.

4. **Secondary meristems**

a. The **vascular cambium** is a single ring of cells located between the stem's outermost xylem layer and the stem's innermost phloem layer. Through mitotic activity, the vascular cambium produces new xylem vessels to the inside and new phloem cells to the outside.

b. The **cork cambium** is located six to seven cells below the outer perimeter of the stem. By mitotic activity it replaces the cork (bark) which is sloughed off as the tree stem increases in diameter.

E. Organization of a monocot herbaceous stem (Zea)

Survey this stem under low power. Note that there is no distinct pith area of the stem. Scattered **vascular bundles** are interspersed between the large accumulation of **parenchyma cells.** Find a relatively large vascular bundle (sometimes said to look like a monkey face) under low power. Move to high power. Sketch the vascular bundle (monkey face) and label the following structures:

1. The **xylem** of the vascular bundles look like the eyes, nose, and mouth of the monkey face. The xylem cells are again characteristically large with prominent red staining cell walls.

2. The **phloem** tissue is seen above the eyes of the monkey in what would be the monkey's forehead region (staining a light green). For the first time, the specific cells of the phloem tissue are visible.

a. The conducting **sieve tube cells** are the larger diameter cells.

b. The significantly smaller cells located between the sieve tube cells are the **companion cells.** These cells help to pump sugar into the conductive sieve tubes.

c. **Sieve plates** can be found in some of the sieve tube cells. Remember that the sieve plate is found at the junction between one sieve tube cell and the next cell in the conduction pathway. Their intersection is marked by the sieve plate. As its name implies, the sieve plate possesses small pores through which cytoplasmic strands may pass from one cell to the next. The plate is not visible in all of the cells because not all of the cells are cut at their terminal end. By fine focusing and adjustment of the iris diaphragm, you can see the small perforations passing through the sieve plate.

3. Carrying our analogy to the fullest, the chin and hair areas of the monkey face are occupied by **sclerenchyma fibers.** These supportive cells possess extremely thickened secondary cell walls which nearly fill the lumen (space) once occupied by the cytoplasm of these cells.

Activity 4
Root Structure

The distinguishing feature of a root is the location of the vascular tissue in the center of the root rather than on the periphery as seen in the stem. The vascular portion of the root is referred to as the **stele.** The area between the stele and the **epidermal** tissue is the **cortex** of the root. As in other plant organs, the root itself is covered by dermal tissue.

A. Dicot Mature Root: *Ranunculus* (Buttercup)

1. The **epidermis** of the root is different than the epidermis covering the aerial parts of the plant. Rather than retarding water loss with the possession of a cuticle, the epidermis of the root lacks a cuticle which promotes water absorption. If your specimen were cut in the appropriate location, cell extensions called root hairs would extend from the epidermal layer to intermingle between soil particles increasing the surface area for the absorption of water.

2. The **cortex** of the root (between the epidermis and the stele of the root) is occupied by **parenchyma** cells. Notice that these cells are thin-walled and not packed as tightly as other plant tissues. In the **Ranunculus** root you can see the presence of starch granules inside the cytoplasm of the parenchymal cells. Carbohydrate is stored in the plant as starch. The innermost cell layer of the cortex is the **endodermis** which directly borders the stele of the root. Using a higher power objective lens, notice that the cell walls of the endodermis are much thicker than the other cells of the cortical layer. The heavily staining material composes the **Casparian strip.** This material is essentially like the mortar which holds together the bricks of a wall. The Casparian strip insures that material entering the stele passes through the cells of the endodermis rather than between the cells.

3. The vascular tissue dominates the region known as the **stele.**

a. **Xylem** is distributed in a very characteristic X-shaped pattern through the middle of the stele. The star-shaped pattern of the xylem is another distinctive characteristic of the dicot world. The cell walls of xylem are again thicker than surrounding cell walls and are dyed their characteristic pink/red color.

b. The **phloem** tissue of the root is located between the arms of the X-shaped xylem. It is difficult to distinguish between the sieve tube cells and the companion cells.

4. The **pericycle** of the stele is the ring of cells directly to the inside of the endodermis. This is a meristematic ring of cells that produce the lateral roots of the main tap root.

B. Monocot Mature Root: *Zea* (corn)

This root has the same regions that the dicot root possesses, although it possesses some distinct differences in structure. Note the following differences in the sketch that you make of this root.

1. The outermost epidermis shows more extensive development of root hairs.
2. The **parenchyma** cells in the root cortex lack the stored starch granules that were seen in the Ranunculus root. Considering the structure of corn, why is starch absent from its roots.
3. The **endodermis** encircles a relatively larger stele than was seen in the *Ranunculus*.

4. The large **xylem** vessels are distributed around the periphery of the stele. This distribution pattern is characteristic of monocot roots. The much smaller green-tinted phloem cells are located between the xylem and the endodermis.

5. The central portion of the stele resembles the pith of the stem in possessing large thin walled **parenchyma** cells.

Activity 5

The Pine Leaf— An Example of Gymnosperm Adaptation

Up to this point, we have dealt with Angiosperm structure. The pine needle of a typical gymnosperm (evergreen) allows us to do some comparative study. Pine trees retain their leaves (needles) all winter unlike the deciduous angiosperms. Winter in the temperate zone is a particularly difficult season to live through. In addition to temperatures being so very cold, water (in its frozen state) is not readily available for plant use. It would be difficult to transport ice cubes through the xylem of a plant. Angiosperms have handled the problems winter presents by losing their leaves and closing up shop. They basically "hibernate" through this difficult growth season. The conifers have evolved structural solutions to handle these difficult conditions. Make a detailed sketch of the pine needle labelling as many structures as you can from your previous angiosperm experience. You should be able to include: **epidermis, stomates, xylem, phloem, parenchyma cells with chloroplasts,** and an **"endodermis"** (like seen in the root) around the vascular tissue. There are also **resin ducts** which carry the "pitch" to the leaf's surface. Below your sketch list at least four structural modifications this leaf possesses which would help it to reduce its water loss.

CELLULAR RESPIRATION 9

Laboratory Objectives

1. To understand the role of O_2 and CO_2 in the respiratory process.
2. To be able to set up the laboratory apparatus used to measure respiration, understanding what the apparatus is designed to do.
3. To be able to collect and record data from an experimental and control situation understanding the significance of each.
4. To determine the effect of temperature on respiration rate.
5. To prepare a laboratory report following a standard format.

Introduction

In looking up the definition of respiration in the dictionary, the first entry you find is "the act or process of inhaling and exhaling; breathing." This definition applies to the organismal level of biology. This process of respiration is necessary to support the second definition which describes respiration as "the metabolic process by which an organism assimilates oxygen and releases carbon dioxide and other products of oxidation." This second process, investigated in today's lab, occurs at the cellular level of organization in the eukaryotic cell. The dictionary's definition of cellular respiration leaves much to be desired. The process of cellular respiration can better be represented by the equation below:

$$C_6H_{12}O_6 + 6O_2 = 6CO_2 + 6H_2O + 36\ ATP$$

The dictionary's definition ignores one of the most important products of cellular respiration, and that is ATP. ATP is sometimes referred to as the cell's "energy currency." With the "burning" of glucose in the cell's mitochondria, some of the released energy is temporarily stored in the chemical bonds of ATP. Later, if a muscle is going to contract or a flagella is going to wiggle, ATP must be present to power the process.

As you have probably discussed in lecture, the Second Law of Thermodynamics states the amount of disorder in the universe is on the increase. Every time energy changes form, some energy is lost in the form of waste heat. In such a situation, entropy (disorder) is always on the increase. Living cells combat this tendency toward disorder through the expenditure of energy to repair, grow, and reproduce more cells like themselves. Denied a source of energy, cells cannot carry out the chemistry of life and they die. Cellular respiration is the chemical process which makes battle against entropy possible. The chemical energy stored in the chemical bonds of glucose can be harvested and stored in the ATP molecule to pay for the war against increasing disorder.

Glucose oxidized in the presence of oxygen yields carbon dioxide, water, and a large quantity of ATP. As you will study in lecture, the process is much more complicated than the summary equation seen above. In eukaryotic cells, the complete oxidation of a glucose molecule involves the processes of glycolysis, the Kreb's citric acid cycle, and the electron transport chain. In reality, respiration represents a controlled "burning" of glucose allowing the energy present in the bonds of glucose to be successfully captured by the cell. Most of you realize that oxygen is consumed in the process of burning something, and in the absence of oxygen the material is not burned very efficiently. A wood stove that does not receive an adequate supply of oxygen will burn at a very low temperature releasing a lot of smoke through its chimney. In a similar manner, a cell deprived of an oxygen supply will continue to "eke-out" some energy from the glucose, but a lot of the original energy present in the glucose molecule is lost.

The purpose of today's lab is to determine the effect of temperature on the rate of respiration. Recall that as temperature increases, the speed of randomly moving molecules increases. As molecules move faster and faster, the number of random collisions between molecules rises. What effect should this have on the rate of a chemical reaction?

When a rate is determined, you either measure how fast something is being consumed or how fast something else is being created. It is theoretically possible to measure the rate of glucose utilization or ATP production, but the procedures are beyond the scope of our facilities. The rate of respiration can be determined more easily by measuring either the production of CO_2 or the consumption of O_2. The method most often used depends on the measurement of changes in gas volumes or pressures within a closed system in which an organism is respiring. Any change in the volume or pressure represents the net difference between oxygen consumption and carbon dioxide production. The consumption of oxygen in a closed system would decrease the volume of gases while the production of carbon dioxide would increase the volume of gases in the closed system. If the carbon dioxide released by the organism can be absorbed, any changes in volume are directly attributed to the consumption of oxygen.

Activity 1

Measuring Respiration Rate

A. Experimental Design

In this exercise, a simple respirometer will be used to detect changes in gas volume. Respiring material will be placed in a closed test tube along with a carbon dioxide absorbing agent (soda lime). Changes in volume will be measured over time at five different temperatures for later comparison. There will be one group of students responsible for each of the temperatures investigated. Because gas volumes can be influenced by outside factors such as temperature and air pressure, a second chamber (just like the first but lacking the living material) will be placed alongside the experimental chamber. Any changes in the control chamber (hopefully small) will be taken into consideration when evaluating the changes in volume of the experimental chamber.

Your experimental organism is the common garden pea. The peas should be actively engaged in the process of germination with their radicles emerging from the seed coat. Glucose, stored in the pea as a fuel source, is being consumed in the presence of oxygen to fuel the germination process. Remember that plants have mitochondria too. Just because your experimental organism is green does not mean that it is incapable of respiration. Smell the germinating peas to insure that the process of decomposition has not begun. Healthy germinating peas possess little to no odor. If the peas are dying rather than germinating, the distinctly unpleasant odor of decomposition will be present.

B. Procedure

1. Prepare two large test tubes as seen in Figure 1. One of these tubes will contain respiring material (germinating peas) and the second tube will contain plastic beads. The second tube will serve as the compensation chamber used to detect any changes in gas pressure caused by temperature and/or atmospheric pressure changes. Any changes in the volume of the compensation chamber will be applied to the volume changes noted in the experimental chamber.

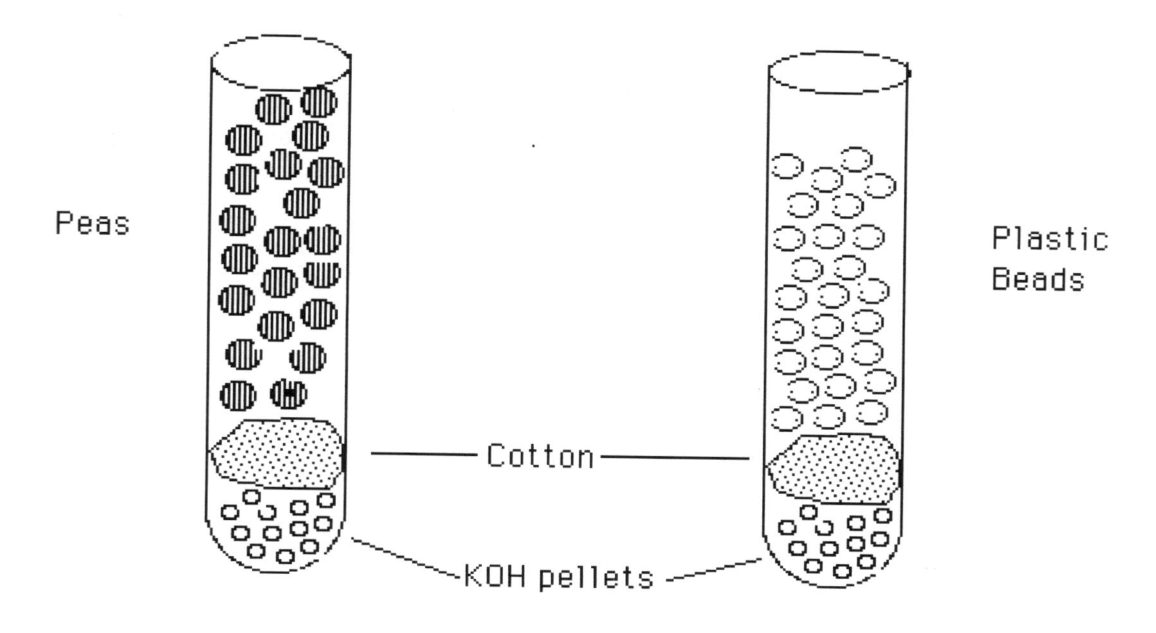

Figure 1. Test tube apparatus

   a. Place an inch of KOH pellets (soda lime) in the bottom of each tube. The soda lime is caustic so handle it with the spatula provided. Immediately pick up any pellets that fall on the floor as they will eat away the linoleum.
   b. Place a medium plug of cotton into each test tube making sure that it is neither too big and nor too small.
   c. Fill the experimental test tube almost to the top with peas. After filling, empty the peas back out to **weigh them to the nearest 0.1 grams.** Return the peas to the experimental chamber.
   d. Fill the compensation chamber with plastic beads to approximately the same height as the peas in the experimental chamber.

2. Firmly insert a rubber stopper with attached capillary tubing into each test tube. Be firm but not overly aggressive. It is possible to break the test tubes by being too rough.

3. Clamp the tubes vertically in a ring stand and place the entire assembly in a water bath for a 1/2 hour equilibration period. Make sure that the majority of both test tubes are submerged in the water bath.

4. The capillary tubes from each chamber (experimental and compensation) should be connected to the horizontal arm of the "U-shaped" manometer supported by on a wooden block. See Figure 2.

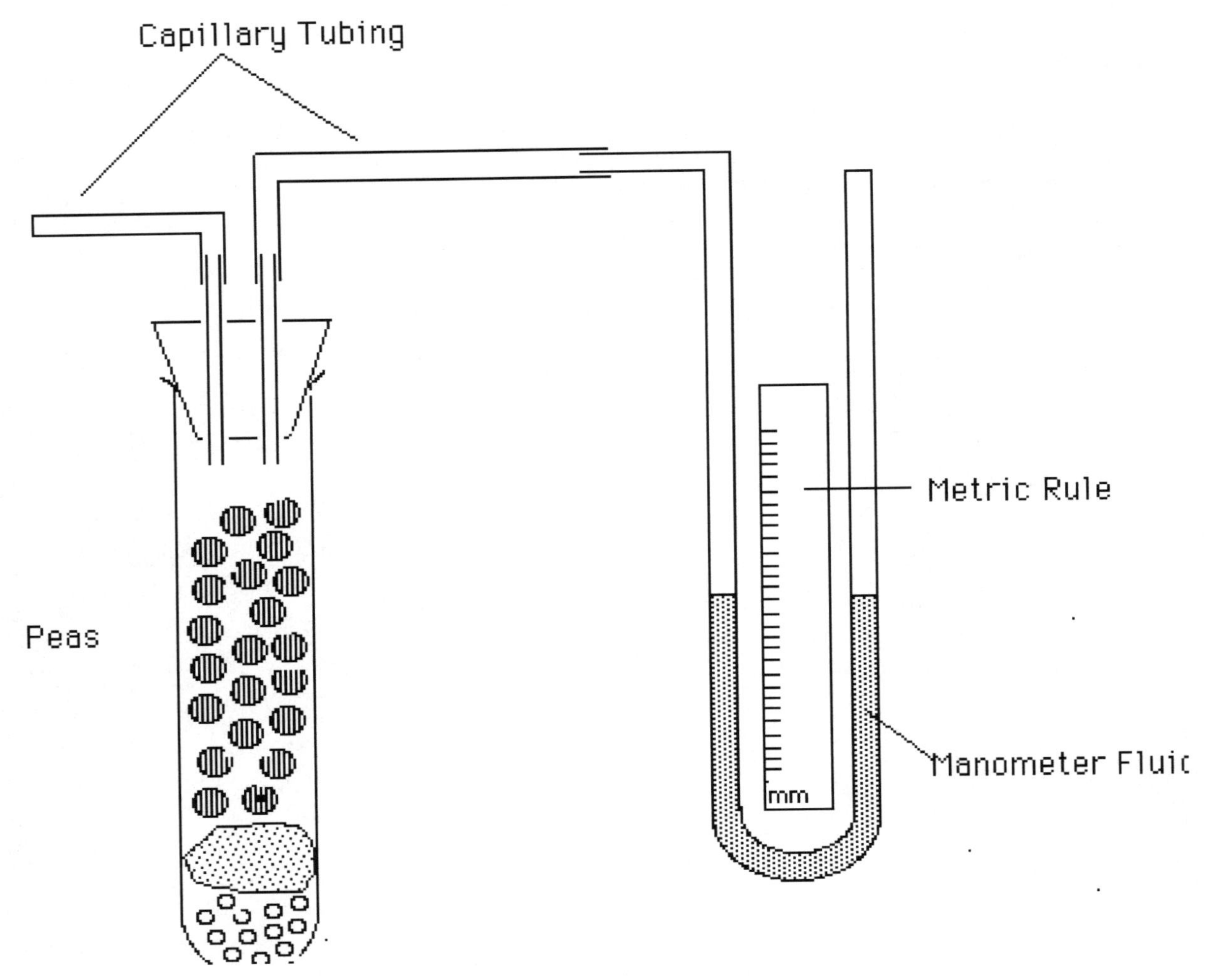

Figure 2. Respirometer Connected to the Manometer

5. Add dye to the manometer tube so that it reaches the middle of the metric rule.
6. Following the equilibration period of 1/2 hour, fold over the rubber tubing on the end of each of the test tubes and attach pinch clamps.
7. When the manometer fluid begins to move, note the position of the dye column on the metric rule in millimeters (mm). Record this value as your initial reading at time zero in Table 1. Also record the position of the compensation tube and the temperature of your water bath.
8. Take readings of the location of the dye columns at one minute intervals for the next fifteen minutes. Record your data in Table 1. Make a note of the temperature of the water bath every other minute.
9. In reading the compensation chamber, make a note as to whether the column of dye moved toward or away from the test tube. This will make a difference in your evaluation of data. Record your data as illustrated in Table 2.

Table 1. Germinating Peas at ____˚ C

Time (min)	Respiration Chamber (mm)	Change in Respiration Chamber (mm)	Compensation Chamber (mm)	Change in Compensation Chamber (mm)	Corrected Reading (mm)
0					
1					
2					
3					
4					
5					
6					
7					
8					
9					
10					
11					
12					
13					
14					
15					

C. Data Manipulation (determination of a respiratory rate)

1. Complete filling in Table 1's "Corrected Readings" column. See Table 2 for an explanation of the calculations.

Table 2. Sample Table Calculation

Time (min)	Respiration Chamber (mm)	Change in Respiration Chamber (mm)	Compensation Chamber (mm)	Change in Compensation Chamber (mm)	Corrected Reading (mm)
0	56		34		
1	58	2	35	0	2
2	61	3	36 (toward)	–1	2
3	62	1	37 (toward)	–1	0
4	64	2	36 (away)	+1	3
5					

a. In the first minute, the respiration chamber moved 2 mm while the compensation chamber remained stationary. The Corrected Reading is simply 2 mm.

b. During the second minute, the compensation tube moved 1 mm toward the compensation chamber. Since the same force was responsible for movement in the experimental chamber, the 3 mm total movement shown by the experimental chamber must be reduced by 1 mm for a corrected value of 2 mm.

c. During the fourth minute, the dye in the compensation tube moved 1 mm away from the compensation chamber. Since the experimental tube was subjected to the same forces moving the dye away, this number must be added to the actual movement of the experimental tube for a corrected reading of 3 mm.

2. Add up the total corrected movement for the fifteen minute period and enter that number here.
   ____________ Total dye movement in 15 minutes

3. Divide the total dye movement by 15 minutes to give the rate of dye movement. Enter that number here.
   ____________ mm/min

4. Since the mass of peas would affect respiration rate, the rate determined in #3 above must be adjusted to reflect the different masses of peas used by each group. Divide the rate determined in #3 by the weight of the peas used in your experiment and enter the number here.
   ____________ mm/min/gram

5. Place your determination of weight-adjusted rate along with the other rates determined in your lab section in Table 3 below and on the board in the front of the room.

Table 3. Class Data of Respiration Rates

Temperature	Respiration Rate (mm/min/gr)

Activity 2

Laboratory Report

Prepare a laboratory report following Appendix A of your laboratory manual. Be sure to label the various sections of the report. Some specific points (although not an exhaustive list) which should be included in this laboratory report are listed below. They are included in a random order. By now, you should know where these various points belong in your report.

1. Prepare two graphs of your data. One graph should have the movement of dye for your apparatus plotted against time? What does this graph look like and what does it tell you about the consumption of oxygen in your experiment? The second graph should be of the class data showing respiration rate as a function of temperature?

2. What was the manipulated variable in this experiment? What were some of the controlled variables?

3. Why did the dye move toward the respiration chamber? Under what circumstances might the dye have moved away from the respiration chamber?

4. How does the compensation chamber serve as a control for this experiment?

5. How is oxygen involved in the respiratory process and why is it used up?

6. You determined respiration rate in terms of change in the length of a column of fluid over time (mm/min). How can you convert this to a volume of oxygen consumed (ie mm^3/min)?

7. What is the relationship between temperature and the rate of respiration? What is the basis for this relationship?

8. Does the basic relationship discussed above hold for the entire temperature range investigated? What might happen to respiration rate if the temperature were to rise or fall beyond the ranges investigated? Support your answer.

Laboratory Clean Up Procedures

1. **Remove the tubing from the manometer before disassembling the rest of the apparatus.** This will avoid breakage of manometer U tube.

2. Place the used peas in the designated containers.

3. Return the plastic beads to the prescription bottles.

4. Remove the cotton plugs from the experimental and compensation chambers. Use a long wooden applicator stick and insert it into the tube until it reaches the cotton. Twirl the stick while withdrawing it from the tube. The cotton will adhere to the stick and be drawn out of the tube.

5. Place the used cotton plug in the trash can.

6. Place the KOH pellets (soda lime) in the dessicator at the front of the room. The dessicator's lid slides to the side allowing the pellets to be put inside. Handle the KOH with care. If any pellets are spilled on the floor, clean them up as the KOH eats away at the linoleum over time.

7. Rinse out the test tubes and place them in the designated containers.

8. Put your chairs back under the lab desks and throw away any miscellaneous papers found in the laboratory.

9. Thank you for your help.

MITOSIS

10

Laboratory Objectives

1. To learn that the process of cell division is only a portion of a cell's dynamic life cycle.
2. To learn the nuclear and cellular events which occur in the different stages of cell division.
3. To find, recognize, and sketch the different stages of mitosis in both plant and animal cell preparations.

Terms of Mitosis

The following terms will be useful in your study of mitosis. They will be used in the description of the cellular and nuclear events which are important in the creation of two identical daughter cells.

chromatin: the aggregrate mass of dispersed genetic material composed of DNA observed between periods of cell division in eukaryotic cells. Chromatin can be compared to strands of pasta that have just been cooked for ten minutes.

chromosome: rod-like bodies of tightly coiled chromatin visible during cell division. A chromosome would be analogous to the pasta mentioned above coiled around a fork. When the genetic material is in this state, it is not sending out signals directing cytoplasmic activity. Chromosomes will be seen in two distinct configurations. Chromosomes possessing a single strand of DNA **(monads)** must be replicated before a cell divides. Double-stranded chromosomes consist of two strands **(sister chromatids)** joined together at a centromere. Sometimes chromosomes in this configuration are called **dyads.**

centromere: the region of the chromosome where two sister chromatids are joined. The **kinetochore** is a specialized region of the centromere where microtubules of the mitotic spindle apparatus connect to the chromatids.

Introduction

Cell division is the process by which one cell produces two identical daughter cells. This process is very important for both the growth of an individual from a single fertilized egg cell and the replacement and repair of existing adult tissues. Cell division is really composed of two separate processes which usually occur at approximately the same time. **Mitosis** is the division of the nucleus producing two identical daughter nuclei. Mitosis is the dance of the chromosomes as they sort them selves into two equal portions. The second act of cell division, **cytokinesis,** involves the partitioning of cytoplasm into the two forming daughter cells. Usually cytokinesis occurs toward the end of mitosis, but in some organisms cytokinesis may be absent. In this case, multinucleated cells are produced.

Interphase

Before a mitotic division can occur, a lot of work must be done as cells grow. During the G_1 phase of the cell life cycle (see Figure 1) molecules, such as fats, proteins, and nucleic acids, are synthesized from food molecules using energy derived from respiration. Cellular organelles must also be synthesized. New cell membrane, thousands of ribosomes, hundreds of mitochondria (and hundreds of chloroplasts if a plant cell), must be produced. Healthy cells double their components and then divide these components in half during mitosis to produce two identical daughter cells. During the **S**(synthesis) stage of the cell life cycle, the complete store of genetic information is replicated so that each daughter cell receives an identical blue print for its own development. This is no easy task. If the information contained in the chromosomes of one of your cells were printed using this size font, it would fill about 900 books as thick as your biology text book (1200 pages each). A chromosome enters this stage of the cycle singly stranded, but by the time the S phase is complete, each chromosome is composed of two sister chromatids attached by a centromere. After the genetic material has been replicated, the cell still requires a second period of normal metabolic activity (G_2) to get ready for cell division. To better place the mitotic process in perspective, examine the cell cycle shown in Figure 1.

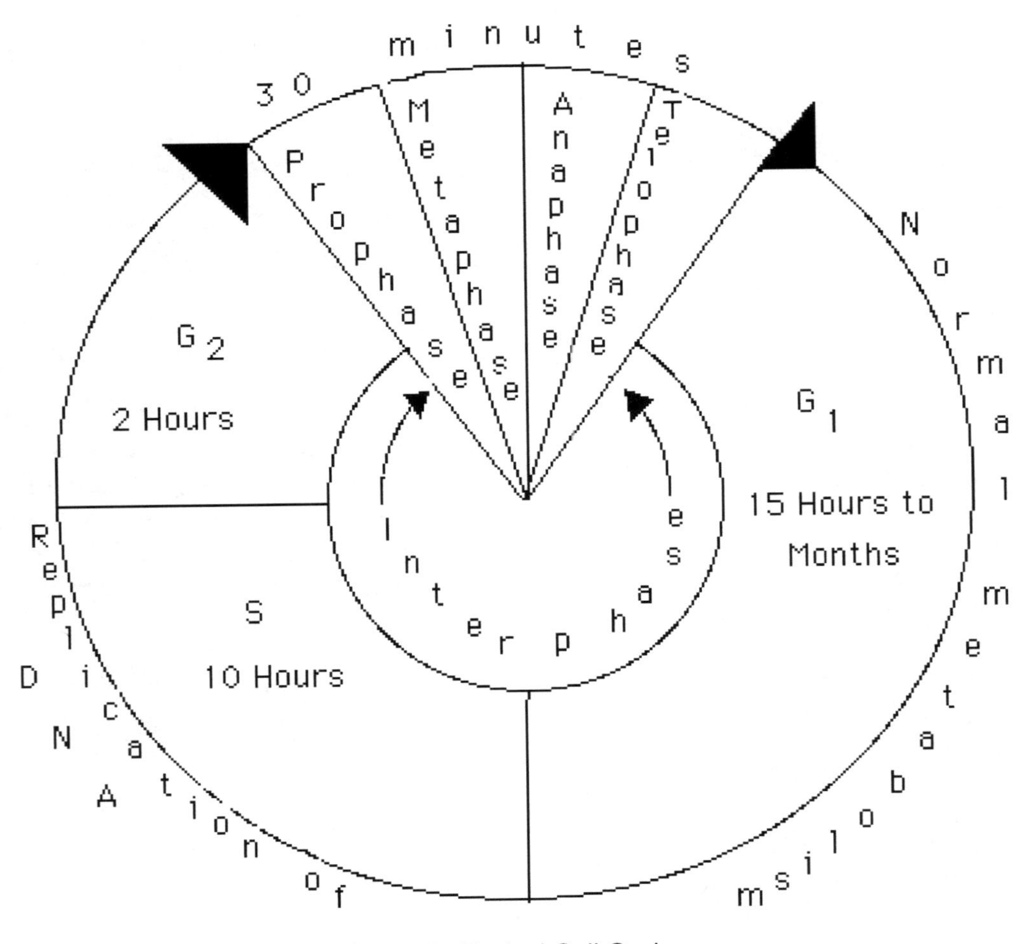

Figure 1. Typical Cell Cycle

No synthesis of DNA or RNA occurs during the second gap (G_2). During the entire interphase period (G_1, S, and G_2) there is a clearly defined nucleus present with a network of chromatin threads spread throughout. A darker staining nucleolus, or possibly several nucleoli, may also be observed within the nucleus. The nucleolus is a center of RNA synthesis (particularly the RNA associated with ribosome structure).

Two words of caution. Never think of interphase as a resting stage in the cell cycle. Secondly, the word "typical" in the phrase "typical cell cycle" is a dangerous word to use because all cells are different. Some cells, like mammalian muscle and nerve cells, stop dividing when they are mature. Some cells will go through the G_1 phase of the cell cycle in 15 hours while others take months to complete this phase. As you study mitosis, also keep in the back of your mind that each individual mitotic phase is a part of a larger dynamic process.

Mitosis

Even though a dynamic process, mitosis is arbitrarily divided up into different stages for ease of study. You will be looking for examples of each of the stages to be described below and it is easy to think of these phases as being static events. In reality, mitosis is a very dynamic process as the videotape will illustrate.

1. **Prophase** can be thought of as the phase of **p**reparation.
   a. Chromatin is transformed into long slender threads, which will further condense and coil into the chromosomes of eukaryotes. At this stage, each chromosome consists of two longitudinal halves (sister chromatids) joined together at a centromere.
   b. The nucleoli (if present) fade from view.
   c. The nuclear membrane slowly dissolves and has completely disappeared by late prophase.
   d. The mitotic spindle has begun to form during this stage. The spindle apparatus is formed by the centricles in animals. A pair of centrioles, located near the periphery of the nuclear membrane, replicate and begin to migrate toward opposite poles of the nucleus. As the centrioles migrate, they "spin" the microtubules of the spindle apparatus which will later "pull" sister chromatids apart. The mechanism of spindle fiber formation in plant cells is not well understood.

   The events of prophase help to organize the genetic material of the cell so that it can be successfully halved and transported into the two new daughter cells. Chromatin can be thought of as pasta on a plate. If you are going to eat the pasta politely, you twirl the strands of pasta around your fork to more efficiently move it from the plate to your mouth. Similarly, the chromatin of the nucleus has to be coiled into the more easily managed chromosomes.

2. **Metaphase** is the phase of orientation where the doubly stranded chromosomes align along the **m**iddle of the cell's equator. The centromeres of the chromosomes are located centrally along the equatorial plate (mitotic plate) and are attached to the fibers of the now completed spindle apparatus.

   Imagine you are in charge of a kindergarten class of 15 identical pairs of twins. Your job as teacher is to separate the members of each pair so that they end up in different class rooms. An efficient approach to solving this problem would be to have the twins line up in the center of the classroom while holding hands. This will help to organize the separation which is about to commence.

3. **Anaphase** is the period during which sister chromosomes move **a**way from one another. The centromeres split allowing the spindle fibers to move the once "sister chromatids" toward opposite sides of the cell. Note that as soon as the sister chromatids are released by the centromeres, they are more properly referred to as sister **chromosomes.**

4. **Telophase** is the fourth and last stage of the mitotic cycle. It is also the phase of reconstruction, being the reverse of prophase. Chromosomes unravel to form chromatin. New nuclear membranes assemble around the chromatin. The nucleoli reappear as the new cell gears up for another round of protein synthesis. Cytokinesis, to be described below, also commences. Mitosis is completed with the two daughter cells produced possessing equal genetic potentials.

Activity 1
Videotape-Mitosis in Endosperm

You will view the videotape *Mitosis in Endosperm*. This is an excerpt from a longer research film produced by Andrew Bajer and his wife, J. Mole-Bajer, in 1957 at the Jagellonian University, Cracow, Poland.

In the film, *plant endosperm,* the stored food material of the seed, is observed dividing. Most of the sequences are of the African Blood Lily *(Haemanthus katherinae).*

Note especially that mitosis is a continuous dynamic process, each phase setting the stage for and moving into the next phase in a definite, highly choreographed dance. The chromosomes appear, move about, separate, and disappear in a regular pattern.

These preparations are viewed with a phase contrast microscope. This form of microscopy detects variations in structure by means of slight changes in the rate of transmission of light waves as they pass through the object. It accomplishes, in effect, "optical staining" of living material making fine details more pronounced.

The cells were photographed at varying time intervals (time lapse) on movie film and processed for projection at standard speed. The preparations were illuminated only during exposure. To maintain the cells in viable condition, they were maintained in a moist chamber during the filming period.

Activity 2
Mitotic Activity in the Onion (Allium) *Root Tip (l.s.)*

You are going to look for cells arrested in different stages of mitosis when this plant tissue was fixed and stained. The tissue is taken from the tip of the onion root.

A. Observe the onion root tip slide with the 4X objective lens. There may be two or three root tips on your slide depending upon the firm that processed the material. There are four different regions or zones comprising the length of this small root. The **root cap** is at the very tip of the root. It covers the tip much like a thimble would cover the end of a finger. The root cap protects the fragile cells lying underneath as the root is pushed further into the soil. Immediately behind the root cap is the root's **primary meristem.** It is here that cells in different phases of mitotic division will be found. Above the root apical meristem is the **zone of elongation.** The newly formed cells in this region are actively growing in length. This growth zone is the motive force propelling the root tip further into the soil. Above the zone of elongation is the **zone of maturation.** In a functioning rootlet, root hairs would develop from epidermal tissue in this zone to increase the surface area for absorptive purposes.

B. Focus on the root meristem with the 40X objective lens. Find each stage of mitosis and draw it in the appropriate location in Table 1. It is not necessary to draw the cells in sequential order. If you locate a good mitotic figure, draw it in the appropriate box. If not, you may have trouble re-locating it.

Cytokinesis begins during telophase of mitosis. Daughter cells of animal mitosis essentially are formed by a "pinching" process (referred to as the cleavage furrow) which separates the cytoplasm of the original cell into two relatively equal masses. Plant mitosis accomplishes the division of cytoplasm by building a new cell wall between the two daughter cells. The beginning of this new cell wall is seen during telophase when the **"cell plate"** (the beginning of the new cell wall) begins to form between the two daughter cells. The presence of a rigid cell wall prohibits the cytokinetic approach utilized by animal cells.

Table 1. Sketches of Plant and Animal Mitosis

	Plant Mitosis	Animal Mitosis
Interphase		
Prophase		
Metaphase		
Anaphase		
Telophase		

C. Mitosis lasts for about **90 minutes** in onion root tip cells. Each of the four phases takes a different amount of time. The phase lasting the longest will be the most commonly observed. Working with a partner, look at 100 ***dividing*** cells, and tally the frequency of occurrence of the phases in Table 2. Because frequency of occurrence is directly proportional to the length of a phase, multiply the percentage of the cells in a phase times the length of mitosis to obtain an estimate of the length of a phase in mitosis. You may have to look at several root tip slides to find 100 mitotic figures. As you search a root tip, be careful not to cover the same ground avoiding cells already included in your count. Search the slide as shown below.

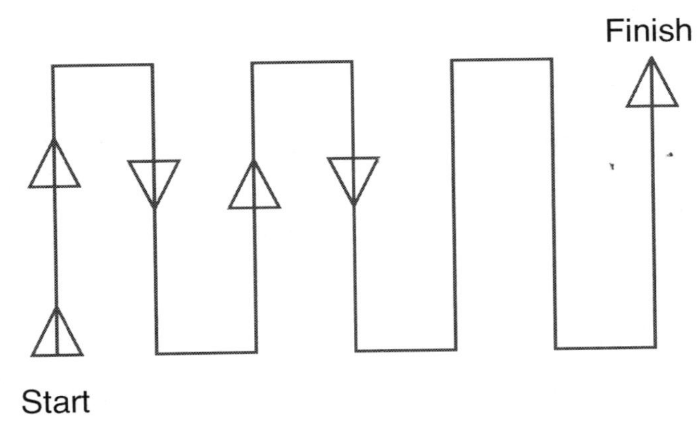

Table 2. Determining Length of Mitotic Phases

Phase	Number seen	% of total	Length in minutes
Prophase			
Metaphase			
Anaphase			
Telophase			
Total observed			

Activity 3

Whitefish (Coregonus) *Blastula (c.s.)*

Mitosis is also easily observed in growing, embryonic animal tissue. The whitefish blastula is an early embryonic stage in the development of the whitefish from a fertilized egg. At this stage, the embryo is essentially a disc of dividing cells on the top of a globe of yolk.

A. Obtain a slide of a whitefish blastula and observe it under the 4X objective lens. You will notice approximately 12 masses of tissue. Each mass is one section (cut) through the blastula stage and contains many cells caught in different mitotic stages.

B. Move to the 10X objective lens to survey one of the microscopic sections of the blastula. At this power, some of the cells will appear to have darker staining material in their center, the material being the chromosomes of cells engaged in mitosis. Other cells will appear to be empty. Remember that this slide is a slice through a three dimensional mass of cells. Some cells will be cut through their centers revealing nuclear events. If the very anterior or posterior portion of a cell were cut when the slide was prepared, the nuclear region may not be included in the section. It will appear as if the cell completely lacks a nucleus. This appearance is only an artifact of the process by which the cell was made.

C. Move to the 40X objective lens to study and sketch (Table 1) individual cells showing mitotic figures. Note the following differences from the onion:

1. During prophase, centrioles (absent from plants) divide and separate. The mitotic spindle forms between the centrioles.

2. In metaphase, the centrioles have migrated to opposite ends of the cell and the spindle fibers range between them. Each of the centrioles is at the center of a radial array of short fibers. This is called an **aster.**

3. Cytokinesis is accomplished by formation of a **cleavage furrow** which pinches off the two daughter cells from one another. This process begins in telophase.

Name ______________________________

Section ______________________________

Pre-lab Questions

1. List in sequence the stages of the life cycle of the cell and indicate what happens during the G_1, S, and the G_2 stages.

2. What is interphase, and why is it important in the cell cycle?

3. How do mitosis and cytokinesis compare? How do they differ?

4. What is the relationship between chromatin, a chromosome, and sister chromatids?

6. List in proper order the stages of mitosis.

7. How does the phase of reconstruction differ from the phase of orientation?

Name ______________________

Section ______________________

Post-lab Questions

1. Compare and contrast plant and animal mitosis.

2. What is the zone of elongation and where would you find it?

3. What do the cell plate and the cleavage furrow have in common?

4. Draw an animal cell, possessing six chromosomes, as it would appear during anaphase of mitosis. Label your diagram completely.

MEIOSIS

11

Laboratory Objectives

1. To be able to recognize and describe the events of meiosis.
2. To understand the role of meiosis in sexual reproduction.
3. To distinguish the various stages of spermatogenesis in microscopic slides.
4. To understand how crossing over works and the importance of crossing over in producing genetic variability.
5. To be able to compare and contrast the processes of mitosis and meiosis.

Introduction

Meiosis is a special form of cell division associated with the production of sex cells (gametes) in eukaryotes. Sexual reproduction is accomplished by the fusion of gametes to produce a new zygote. Each gamete contains DNA, so the resulting zygote has twice as much DNA as each of the gametes. Can you see a potential problem?

Mitosis produces cells that have exactly the same number of chromosomes as the parent cell. If cells produced by mitosis were to function as gametes, a species' chromosome number would double each generation. Human somatic (nongametic) cells possess 46 chromosomes. If gametes possessed 46 chromosomes also, a human zygote would possess 92 chromosomes. A third generation produced in this manner would have 184 chromosomes. Clearly, this is not the way sexual reproduction works. Constancy of chromosome number from generation to generation is the rule of biological systems. The role of meiosis in reproduction is to reduce the number of chromosomes in gametes by one-half maintaining the species' chromosome number.

The physical appearance of a cell's chromosomes becomes evident when chromatin supercoils during mitotic or meiotic divisions. Chromosomes in this tightly coiled state differ in physical appearance. Some chromosomes are longer than others, and the positions of the chromosomes' centromeres varies. When a karyotype is done (analysis of a cell's chromosomes), it becomes evident that a **diploid** cell possesses pairs of **homologous** chromosomes that are identical in size, shape, and genes carried. The term **"diploid"** or **"2N"** refers to the fact that two sets of chromosomes are present. A human diploid cell contains 46 chromosomes. Upon closer inspection there appear to be 23 distinct types of chromosomes. One chromosome of each type came from the sperm which fertilized the original egg cell. The sperm cell, possessing one complete set of chromosomes, is said to be **haploid** or **"1N."** The other complete set of chromosomes was already present in the haploid egg awaiting fertilization. Meiosis is the process by which these cells became haploid.

A word of caution is needed here. It was said above that homologous chromosomes carry the same genes. A **gene** is a site on a chromosome where information is stored. The actual information present at the site is referred to as an **allele.** A pair of homologous chromosomes both may carry a gene controlling eye color. The information stored at the gene (the allele) may result in the production of blue eyes or nonblue eyes. Homologous chromosomes carry the same genes affecting the same traits. The specific information stored at the genes may differ.

A general overview of the meiotic process will make the following exercises more easily understood. Meiosis employs two separate cell divisions to produce haploid cells from the original diploid parent cell. The first meiotic division (meiosis I) reduces the chromosome number by one-half (reduction division). The second division (meiosis II) separates sister chromatids from one another. Meiosis II, therefore, closely resembles a mitotic division.

Meiosis, just like mitosis, is preceded by an interphase. Recall that chromosomal replication occurs during the S phase of interphase, so the chromosomes of a cell entering the meiotic process exist in the **dyad** condition. The diploid cell, which already contained two sets of genetic instructions (one maternal set and one paternal set) now contains "four sets" of genetic instructions. One of each of these fours sets of instructions is going to be partitioned into four haploid daughter cells. This complicated manuever requires two nuclear divisions rather than the single nuclear event seen in mitosis.

Each chromosome **(dyad)** entering prophase I possesses two identical sister chromatids joined by a centromere (crossing over has yet occurred). During meiosis I, homologous pairs of chromosomes come together forming **synaptic pairs.** The arms of two dyads line up side-by-side and wind around each other becoming intertwined. The chromatids may physically connect at points called **chiasmata** (spot welds). A synapsed homologous pair of chromosomes is referred to as a **tetrad** (two dyads = four chromatids = four sets of genetic instructions).

The reduction in chromosome number that occurs during Meiosis I does not occur haphazardly. The daughter cells produced from Meiosis I contain only one member of each of the homologous pairs of chromosomes present in the original diploid cell. Even though haploid cells are produced by meiosis I, the production of gametes is not yet complete. The chromosomes still exist in the dyad condition. It is the job of meiosis II to separate sister chromatids from one another.

Refer to Figure 1 below to help clarify some of the chromosomal terms used above. Chromosomes A and B represent one homologous pair of chromosomes in the monad state. During the S phase of interphase, each of

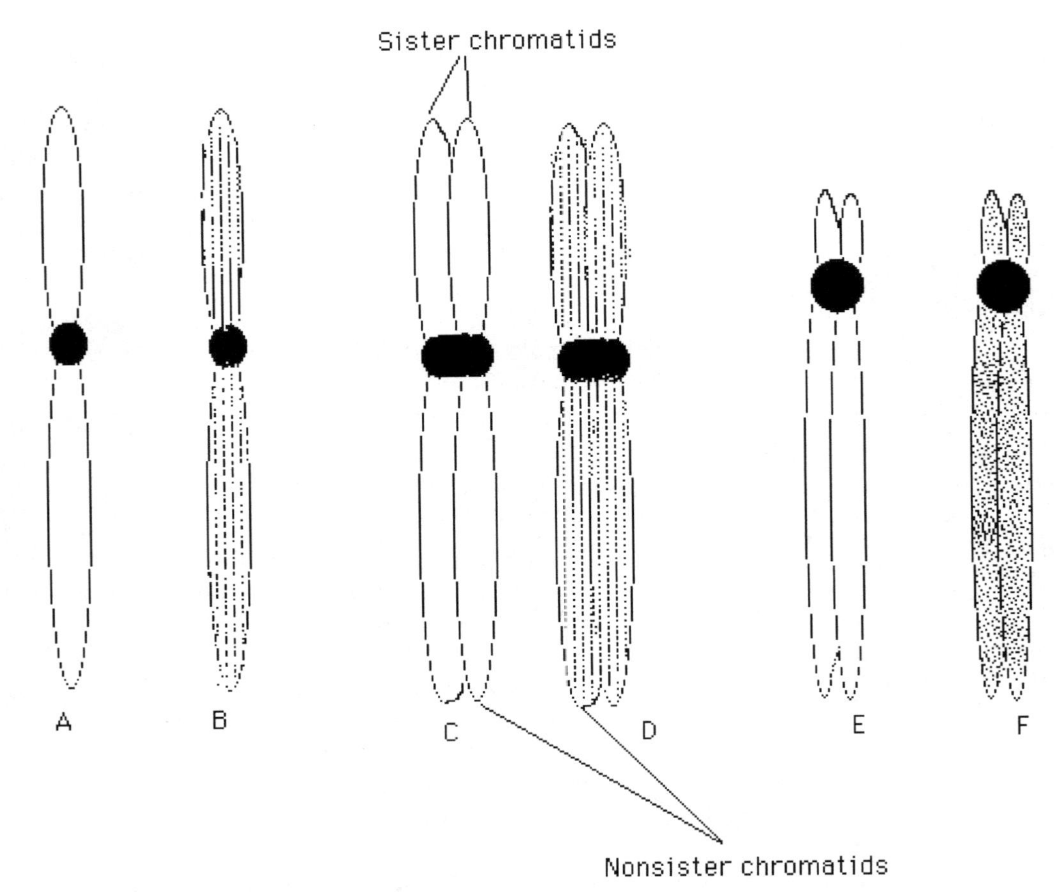

Figure 1. Different Forms of Chromosomes

the monads is replicated to produce the dyad state as shown by C and D. The identical chromatids connected by the same centromere are sister chromatids. They are identical in information which they carry (unless a crossing over event has occurred). Chromatids of homologous chromosomes not connected by a centromere are referred to as nonsister chromatids. They may or may not carry the same genetic information. E and F represent a second homologous pair of chromosomes in the dyad state.

Questions

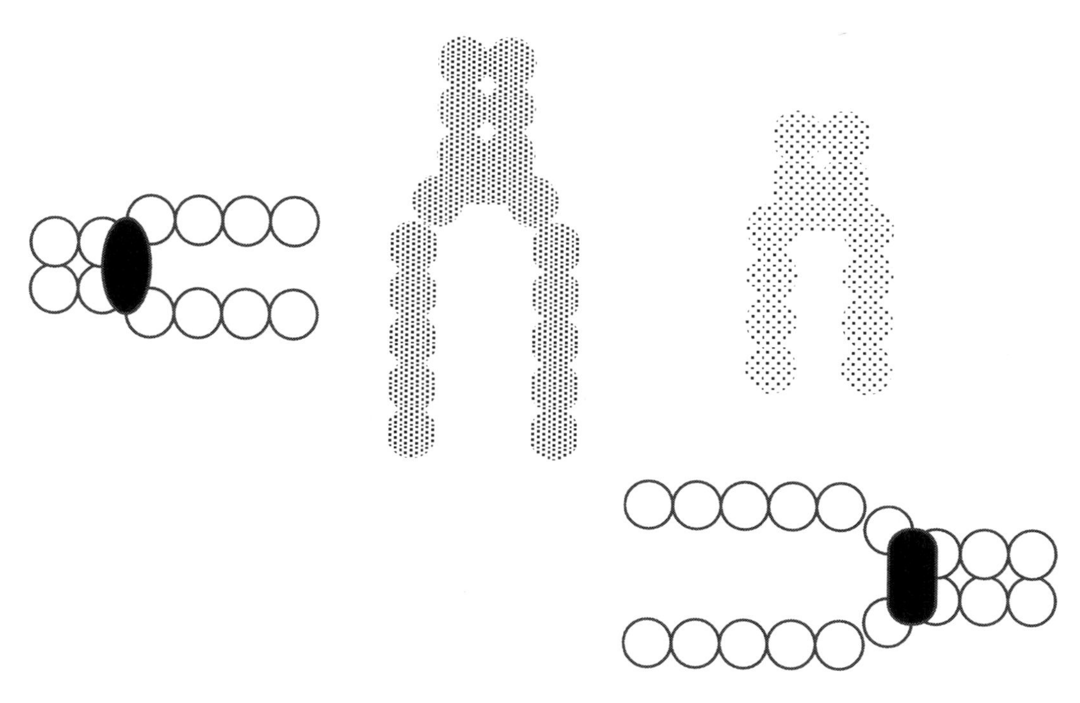

1. How many chromosomes are present in the nucleus above?
2. How many homologous pairs of chromosomes are there?
3. How many chromatids are present in this nucleus?
4. Is the nucleus depicted above in the haploid or diploid state?

A general outline of meiosis follows. Each of the meiotic divisions, designated as Meiosis I and Meiosis II, can be further divided into phases similar to those of mitosis. There are, however, significant differences in the events of each.

Meiosis I

Prophase I of meiosis is much slower and more complex than mitotic prophase. Chromosomes shorten and thicken, appearing as dyads with sister chromatids joined at the centromere. Spindle fibers form and the nuclear membrane disappears. A unique feature of Prophase I is the synapsis of homologous chromosomes bringing dyads together to form tetrads. The sister chromatids of each homologue are still attached to their centromere. At points along their length, nonsister chromatids are held together at points of attachment called chiasmata (sing., chiasma). When tetrads break apart later in the process of meiosis I, nonsister chromatids may exchange segments. This significant event, called crossing over, reshuffles alleles into new combinations.

Metaphase I occurs when homologous pairs align along the equatorial plate. The homologous pairs of chromosomes straddle the equatorial zone. It is important to note here that the order in which the homologous chromosomes align along the plate and the side of the plate that each chromosome moves to is completely random. This characteristic forms the underpinnings of the concept of independent assortment so important to the work of Mendel.

Anaphase I has the homologous chromosome pairs separating and moving toward opposite poles. Note that there is no splitting of the centromeres at this stage and sister chromatids remain attached to each other. The dyads remain intact.

Telophase I is marked by the disappearance of the spindle fibers and the reappearance of nuclear membranes. Cytokinesis follows producing two haploid cells. Each cell possesses one chromosome of each homologous pair present in the original diploid cell. The chromosomes are in the dyad condition.

Interkinesis

Interkinesis is the gap between Meiosis I and Meiosis II. The length of this time gap is quite variable among different species. In some, neither telophase I nor interkinesis intervenes between Meiosis I and II. The cell goes directly from Anaphase I to Prophase II. Other species possess an interkinesis that may never lead to Meiosis II unless fertilization occurs. These individual differences are not important at this time, but it is crucial to remember that there is no additional synthesis of DNA during this stage.

Meiosis II

The events of the second meiotic division are similar to a mitotic division. During prophase II, chromosomes are present as dyads. The centromeres of the dyads attach to the spindle and orient at the metaphase plate during metaphase II. Division of the centromeres initiates anaphase II leading to the separation of the chromatids. During telophase II, the formation of four new cells is completed. The resulting cells contain one-half (haploid condition) the number of chromosomes of the parent cell and these chromosomes exist as monads.

What Is the Significance of Meiosis?

The reduction of chromosome number from the diploid to the haploid state sets the stage for the union of two gametes, providing a means by which the traits of two different parents can be combined. Meiosis also provides for variation within the gamete pool produced by each parent. Random assortment of maternal and paternal homologues during the reduction division and crossing over of various portions of maternal and paternal homologues shuffles the alleles present in gametes. New combinations of physical traits produce the variation so important to the process of evolution and natural selection.

It is necessary that meiosis occur sometime between fertilization and gamete formation in sexually reproducing forms. In animals, meiosis leads directly to the production of gametes. In plants, meiosis produces spores, and only later are gametes derived from the tissue arising from the spores.

Activity 1
The Meiosis Model

You are going to simulate the process of meiosis by following two pairs of homologous chromosomes as they proceed through meiosis I and II. You will create a pictorial history of the events by drawing the meiotic phases and the physical appearance of your chromosomes during these phases on a long piece of paper toweling stretched along the top of the lab bench. Draw circles on the piece of paper toweling representing cells in the different stages of meiosis.

A. With "pop" beads, form two homologous pairs of chromosomes. Make the chromosomes of one homologous pair nine beads long. The chromosomes of the other homologous pair will be six beads long. Symbolize maternal and paternal chromosome sets using different colors. One chromosome of each homologous pair will be blue (paternal in origin), and the other chromosome of the pair will be green (maternal in origin). Each chromosome will be in the dyad condition. Sister chromatids will be attached to each other with a "twistem." The possession of a centromere (twistem) is the defining characteristic of a chromosome. Remember that sister chromatids are identical to each other as they were produced during the S phase of interphase. The sister chromatids of each chromosome will therefore be of the same color.

Carefully observe the chromosomes that you have made. How many chromosomes should you have in front of you? Remember that you are to make two homologous pairs (2 X 2 = 4). The centromere (twistem) is the defining characteristic of a chromosome.

Questions

How many twistems have you used in the construction of your model chromosomes?

How many dyads are in front of you?

How many chromatids did you make?

After becoming frustrated with the above process, check to see if your chromosomes look like those in Figure 4 at the end of the lab.

B. Place your four replicated chromosomes at one end of a long sheet of paper towel. Move your chromosomes through the following stages diagramming their physical appearance as you go.

Prophase I

Move the homologous chromosomes side by side, matching their centromeres and beads along their length. Intertwine the arms of the chromosomes to form a **tetrad.** The resulting union represents synapsis of homologous chromosomes. **Centrioles** are migrating to opposite sides of the nucleus spinning **spindle fibers.** The nuclear membrane is disintegrating. At the end of prophase I, crossing over may occur, but this will be considered later in another activity.

Metaphase I

Move the synaptic pairs to the metaphase plate. The centromeres of homologous chromosomes are on opposite sides of the plate with homologous chromosomes parallel to one another.

It is important to note that the side "chosen" by each dyad of a homologous pair occurs completely at random. This random event forms the basis of the independent assortment of chromosomes during meiosis. Look carefully at the arrangement of your chromosomes at this point? Are any other combinations possible?

Anaphase I

Separate the homologues by pulling on the strings connected to their centromeres. Notice that the centromeres remain intact. The dyads stay together as they move toward opposite poles.

Telophase I

Reformation of two haploid nuclei occurs. How many chromosomes are there in each of the two daughter cells produced? How many chromatids do you see?

Interkinesis is a transient stage that may be absent in some cases. No DNA synthesis occurs making this stage very different from interphase. Our models of chromosomes do not allow for the formation of chromatin to be illustrated at this point.

Prophase II

Thickening of chromatin to form chromosomes occurs once more. Sister chromatids are present and remain attached to their centromeres. Spindle fiber formation has begun as centrioles again migrate toward opposite sides of the nucleus.

Metaphase II

Move the chromosomes to the metaphase plate. Attach strings to the centromeres present. The strings representing the spindle fibers should lead back to the opposite poles of the meiotic spindle.

Anaphase II

The centromeres divide (remove twistems making sure that each former chromatid gets a new twistem), and pull the newly formed chromosomes toward opposite poles.

Telophase II

Reconstruction of nuclei now housing chromosomes in the monad condition occurs. Note that four cells are now present, each with half of the number of chromosomes present in the original interphase cell.

Before leaving this activity, have your instructor check your work.

Activity 2
Meiosis with Crossing Over

A very important event occurs in prophase I that can lead to a re-shuffling of alleles on the chromatids. Remember that during prophase I homologous chromosomes form tetrads during synapsis that are held together by physical bonds called chiasmata. In order for the homologous pairs to separate during anaphase I, these chiasmata must break. For a brief period of time, the arms of nonsister chromatids downstream from the chiasmata are detached from their chromosome. The chromatid segments may reattach to the original arm from where they were separated, or there may be a reciprocal exchange of pieces of chromatids resulting in recombination of alleles. See Figure 3 below for an explanation of this phenomenon. Another note of caution. This diagram of crossing over is an oversimplification of the process. Chiasmata form at several locations along the length of the chromatids. Reciprocal exchanges of chromatid segments can occur at each of these sites making things a lot more confusing.

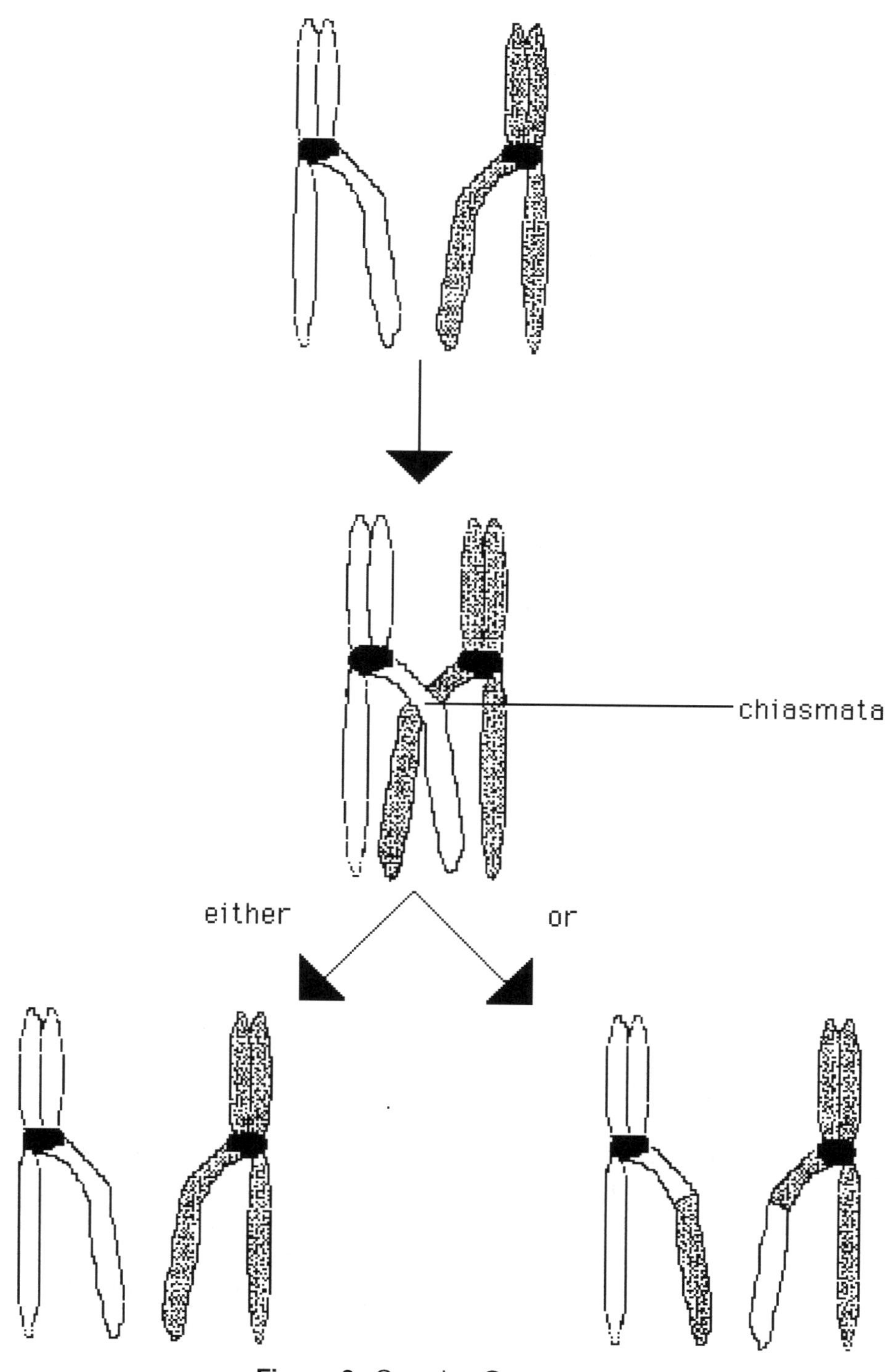

Figure 3. Crossing Over

Procedure

1. Roll out a new strip of paper towel for a second simulation of meiosis.

2. Use a single homologous pair of chromosomes for this exercise. The nine bead chromosomes from above will work well.

3. We are going to follow the fate of two genes through the meiotic process with crossing over. One gene, located on the first bead of the short arm of the chromosome, controls the production of melanin. The dominant allele (A) results in the normal production of melanin, while the recessive allele (a) results in no melanin production leading to albinism. The second gene that we are going to follow controls ear lobe attachment. The allele for free earlobes (F) is dominant to the allele for attached ear lobes (f). The gene for ear lobe attachment is found on the second to last bead on the long arm of our chromosome model.

4. Using a marking pen, label one chromosome's beads with "A" and "F" as shown below. On the second chromosome, label the corresponding beads "a" and "f." **After this exercise is completed, you will clean off your marks with 95% ethanol and tissues.** Note that at this time, the sister chromatids are identical.

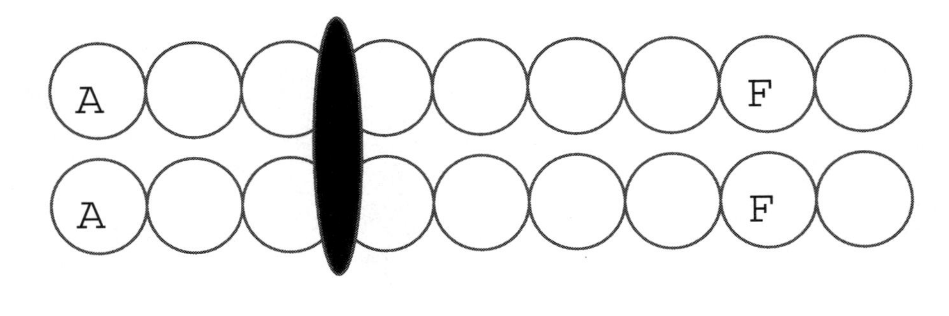

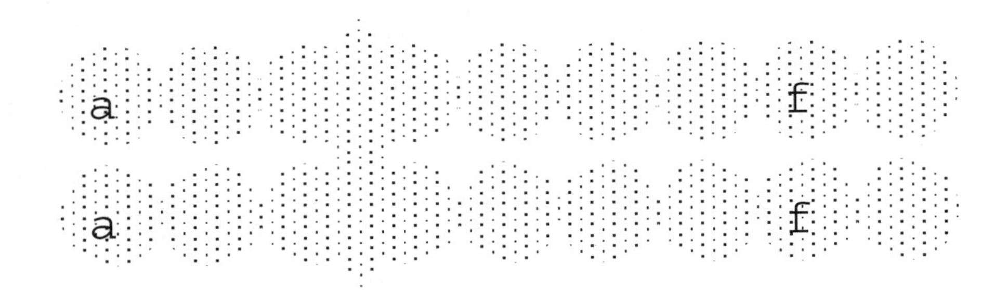

5. During prophase I's synaptic event, break three beads off the long arms of one chromatid of each chromosome and model a reciprocal exchange of pieces.

6. Manipulate your model chromosomes through meiosis I and II again, this time watching what happens to the distribution of the alleles as a consequence of the crossing over.

7. How many different types of daughter chromosomes would be present in the gametic nuclei without crossing over? How many types of daughter chromosomes are present with crossing over? A reshuffling of characteristics has occurred.

8. What would be the result of a reciprocal exchange between **sister** chromatids of homologous chromosomes?

Activity 3

Meiosis in a Mammalian Male

Obtain a slide of a testes from the front of the room. The majority of a testes is made up of a very long seminiferous tubule coiled into a tight ball resembling a ball of string. Observe the slide under 4X to see the testes' overall organizational plan. At this power, the testes almost look like a piece of Swiss cheese. The "holes" or lumen that you observe are the insides of the seminiferous tubule.

Move up to the 10X lens to study one "circular" clustering of cells. This represents a portion of the seminiferous tubule cut in cross section. The cells which are found between adjacent portions of seminiferous tubule at this power are the interstitial cells. The interstitial cells produce the hormone testosterone.

Observe the seminiferous tubule carefully under the 40X objective lens showing the process of meiosis in an animal. The outer ring of cells of the seminiferous tubule are called spermatogonia cells. These diploid cells (2N) have the same function as meristem cells in a plant. Through repeated mitotic activity, new cells destined to become sperm cells are produced throughout the life of the male. One of the two daughter cells produced by the mitotic division becomes a primary spermatocyte. The primary spermatocytes are found one cell in from the spermatogonia cells toward the lumen of the seminiferous tubule. The other daughter cell of the mitotic division remains a spermatogonium which will give rise to future primary spermatocytes by mitosis. Is there an observable difference between the spermatogonia and the primary spermatocytes?

Each diploid primary spermatocyte undergoes meiosis I to produce two secondary spermatocytes. These cells are found in the third or fourth row in from the periphery of the seminiferous tubule. Are these cells haploid or diploid cells? In what configuration are their chromosomes? (monads or dyads?) How do the nuclei of the primary and secondary spermatocytes compare (bigger, smaller, darker staining, lighter staining)? What is the significance of this difference in appearance?

The secondary spermatocytes undergo meiosis II to produce spermatids. The spermatids are found bordering the lumen of the seminiferous tubules. They have begun the morphological transformation into mature sperm cells. The spermatids appear "comma-shaped" as they have begun to elongate. Are the spermatids haploid or diploid cells? Are their chromosomes in the monad or dyad condition?

Sperm cells are visible in the middle of the lumen. These highly specialized cells have long flagella used for propulsion. The tip of the sperm cell contains hydrolytic enzymes used to "digest" a hole through the membrane of an egg cell. The midpiece of the sperm cell is loaded with mitochondria to provide the "motors" of the flagella with energy. The haploid genetic component of the male resides in the head of the sperm awaiting delivery.

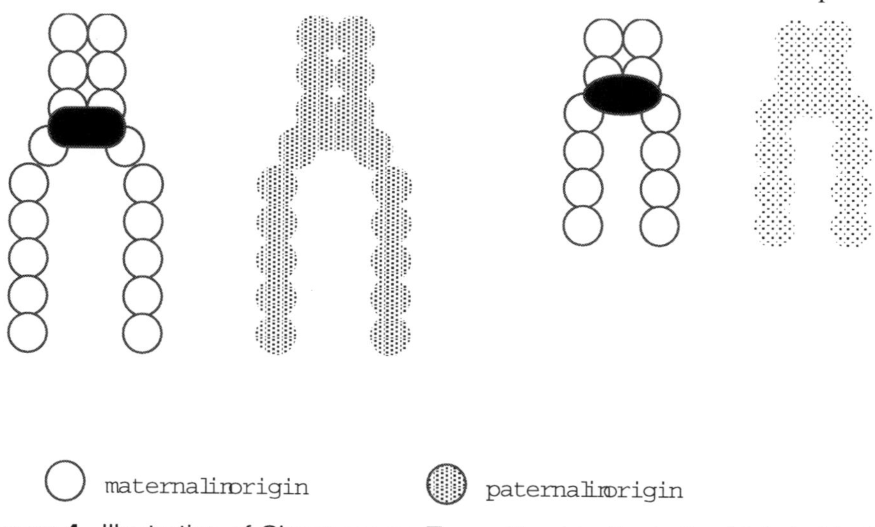

Figure 4. Illustration of Chromosome Types Used to Begin the Meiotic Model

Name ______________________________

Section ______________________________

Pre-lab Questions

1. In mitosis, the number of chromosomes in daughter cells is ____________ the number of chromosomes in the parent cell. In meiosis the number of chromosomes found in the daughter cells is __________ the number found in the parent cell.

2. What does the term "2N" refer to?

3. What types of cells does meiosis produce in animals? in plants?

4. Alternative forms of the same gene which produce a different expression of that gene are called ______________.

5. If both homologous chromosomes of a pair are found in the same nucleus, that nucleus is in the ____________ state. (haploid or diploid)

6. What is the difference between monads, dyads, and tetrads?

7. How many meiotic divisions does it take to produce functional gametes?

Name ______________________________

Section ______________________________

Post-lab Questions

1. One sister chromatid of a chromosome has the allele H. What allele will the other sister chromatid have assuming no crossing over?

2. Two alleles on one homologous chromosome are "A" and "B." The other homologous chromosome's alleles are "a" and "b."

   a. How many different genetic types of gametes would be produced without crossing over? (Refer to the left side Figure 3)

   b. What are those types?

   c. If crossing over were to occur, how many different genetic types of gametes would be produced? (Refer to the right side of Figure 3)

   d. List them.

3. Compare and contrast mitosis and meiosis.

4. From a genetic standpoint, what is the significance of fertilization?

5. How do interkinesis and interphase differ? Of what significance is the difference?

6. Be able to sketch any phase of meiosis given the diploid number of the parent cell.

7. Answer the following questions based upon the figure below.

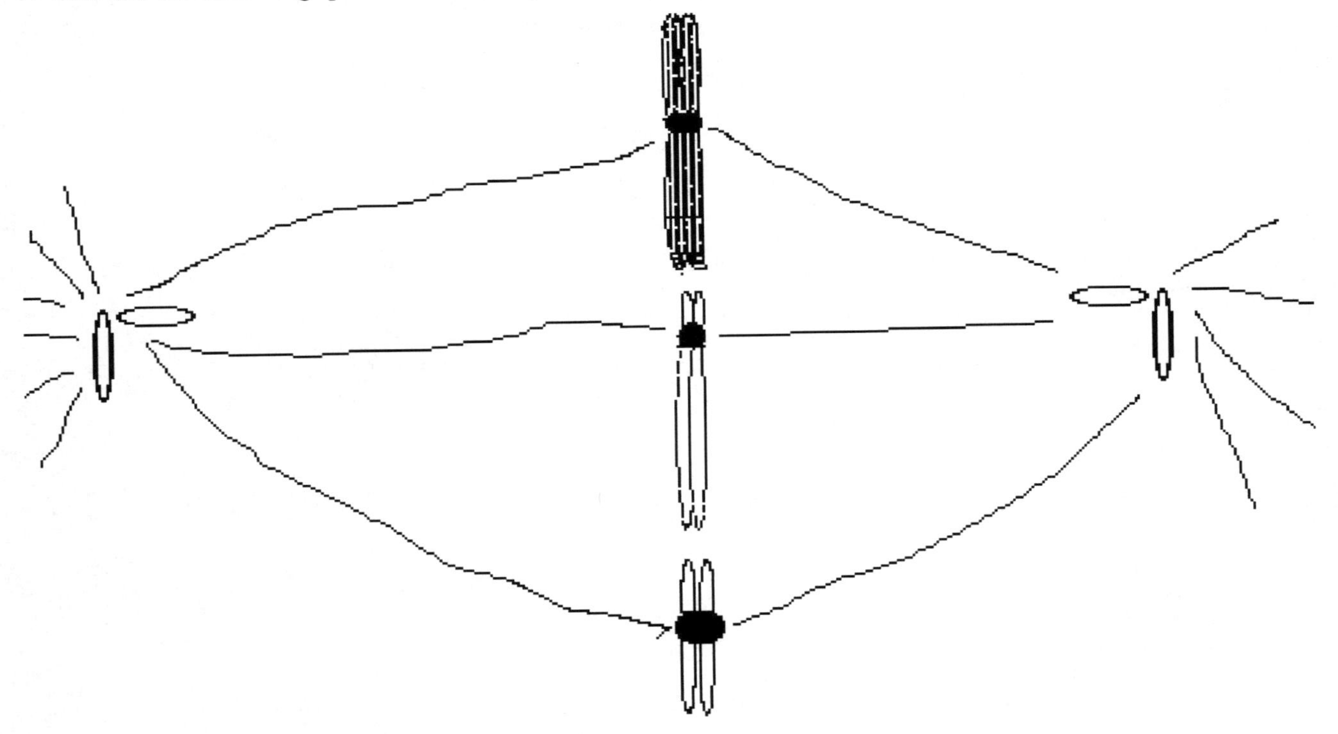

a. What stage of meiosis is represented here?

b. What is the diploid number of a cell of this species?

c. How many homologous pairs of chromosomes do you see?

d. How many chromosomes are present?

AN EXPERIMENT IN CROSSING OVER AND LINKAGE

12

Objectives

1. The student will learn how to transfer portions of fungal colonies using sterile technique.

2. The student will learn about the life cycle of *Sordaria fimicola* and how its life cycle makes it an organism of choice for genetic studies.

3. The student will learn how to make squashes of fungal asci demonstrating significant meiotic events in the fungal life cycle.

4. The student will analyze the results of a mating between two strains of *Sordaria fimicola* to determine the cross over frequency between the centromere and the gene being investigated.

5. The student will analyze the results of a cross between two strains of *Sordaria fimicola* involving two genes involved in the determination of spore color. The analysis will allow the student to make and defend a statement about linkage between the two genes.

6. The student will be able to draw figures of chromosomes to explain the various spore configurations seen in the two parts of the experiment.

Introduction

Sordaria fimicola is a fungus commonly found growing on herbivore dung. It belongs to the Ascomycota division of the fungal kingdom. The generalized life cycle for *Sordaria fimicola* is shown in Figure 1. Hyphae of different mating types cross at the top of the figure. One strain develops an antheridium, which sends nuclei to the ascogonium of the other fungal strain. The two strains have undergone **syngamy** (fusion of cytoplasm). The resulting ascogonium is now in the dikaryotic state with opposite nuclei types being paired off and separated by cell wall formation. Each dikaryotic cell will form an individual ascus where meiosis will occur following union of the two nuclei. **Karyogamy** (fusion of the paired nuclei) produces a diploid zygote within each of the individual asci. Meiosis occurs shortly after the nuclear fusion to re-establish the normal haploid condition of the fungal organism. Notice that the shape of the ascus restricts the plane of cytokinesis during the meiotic division. Daughter cells are placed in a linear order reflecting the orientation of chromosomes at the metaphase plate. The shape of the ascus also dictates that daughter cells of meiosis II lie next to each other. Following the meiotic reduction of chromosome number, a mitotic division follows producing the eight spores found within each ascus. The asci develop within a specialized ascocarp called a **perithecia** (found at approximately 7:00 on the Figure 1). When the asci mature, the ascospores are ejected with enough force to fling them up to eight centimeters away from the parent perithecium. An interesting adaptation possessed by *Sordaria fimicola* is that their perithecia are phototropic. Being able to respond to light stimuli, the perithecia aim their spores toward light. Rather than shooting spores into the dung heap, the mucilage-coated sticky ascospores are sent upwards into the air. In the wild, the ascospores will stick to blades of grass that can then be eaten by herbivores. The spores will pass through herbivore and germinate anew in this most recent dung pile. In most dung fungi, the spores will not germinate unless they have passed through an animal's digestive tract.

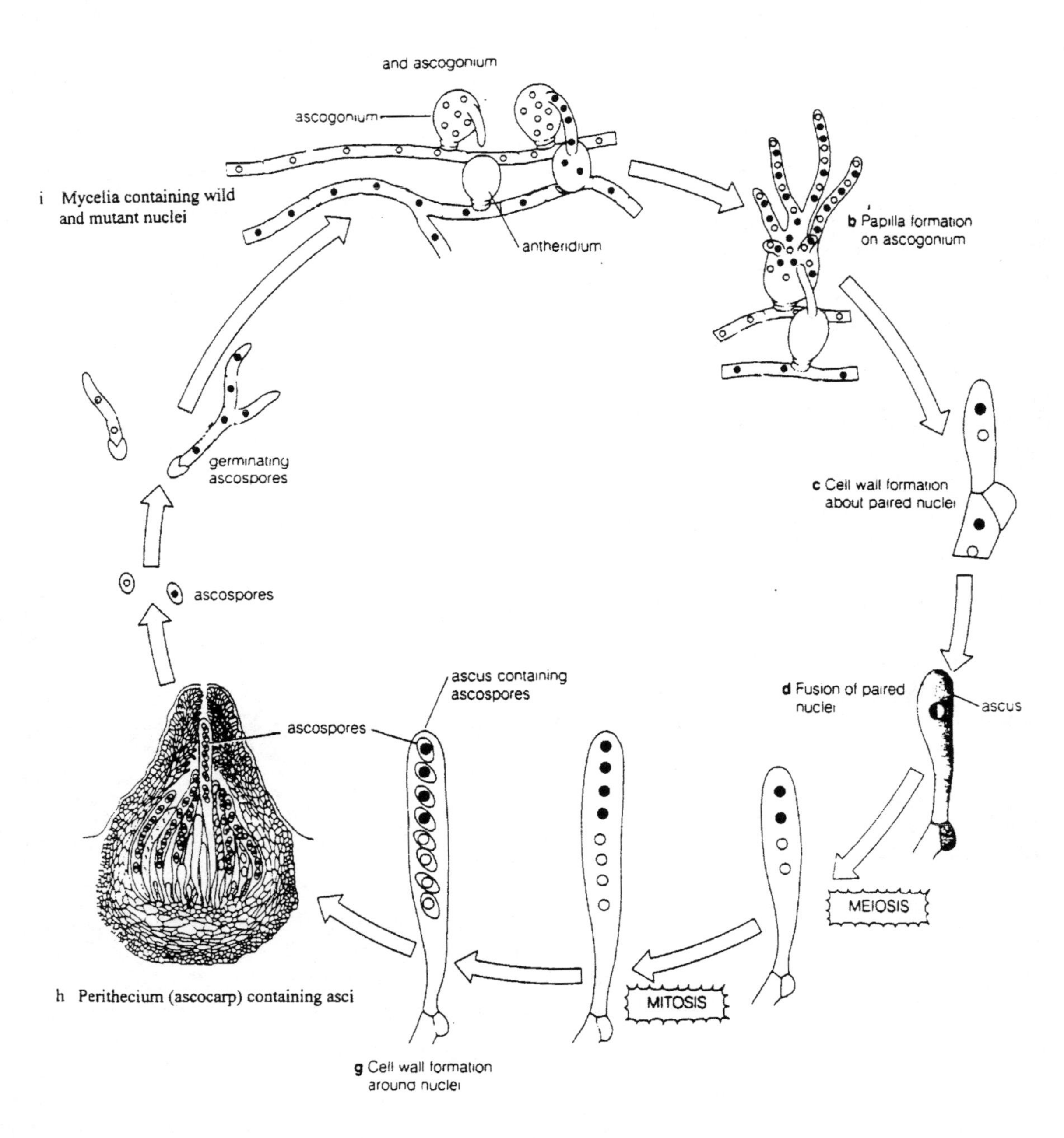

Figure 1. A generalized life cycle of *Sordaria fimicola*

As previously mentioned, *Sordaria fimicola* is a haploid organism. This quality makes Sordaria a good choice in the study of genetics. Terms like dominant and recessive don't apply to the haploid condition. With only one copy of a gene in each nucleus, there is no opportunity for one allele to express itself at the expense of a second.

In this exercise we will be investigating the inheritance of spore color. *Sordaria fimicola's* ascospores are black in the wild state. Spore color is influenced by two different genes both possessing wild type and mutant alleles. A mutation in one of the genes produces a tan mutant phenotype. The mutant allele is symbolized as "t" with the wild type allele being symbolized as "t^+." A mutation in the second gene produces a gray phenotype with the mutant allele represented by "g." The wild type allele for this gene is represented as "g^+." The four possible genotypes are shown below with their resulting phenotypes.

$g^+ t^+$	black spores
$g\ t^+$	gray spores
$g^+ t$	tan spores
$g\ t$	clear spores

We will produce two different crosses of two strains of *Sordaria fimicola* on agar plates. The first plate will be a mating of a tan strain with a wild type black strain. This plate will be used to study crossing over frequencies between the tan gene and the centromere of the chromosome on which it resides. The second plate investigated will be used to study the linkage relationship of the two genes mentioned above.

Procedure

Week One: Inoculating plates

Working in pairs, you will be provided with three cultures of *Sordaria fimicola*. One of the cultures is a wild type possessing black spores. The other two cultures are mutant strains characterized by gray and tan spores respectively. You will also be provided with two Petri plates containing a special Sordaria mating mixture.

Begin the experiment by marking the **bottoms** of the two mating plates as shown below:

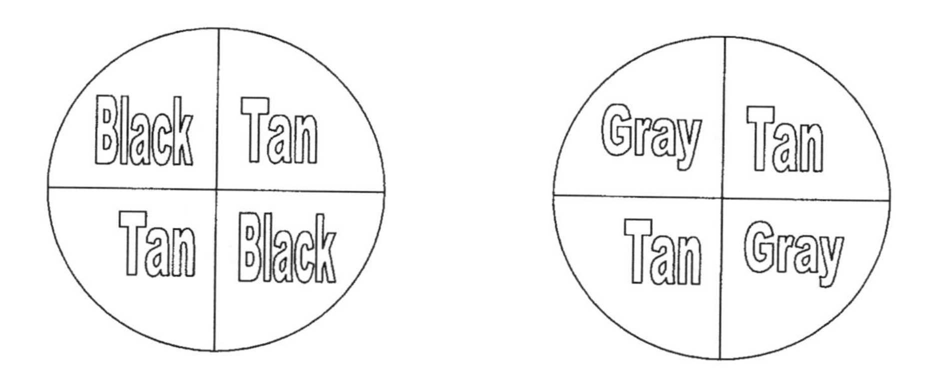

Make sure that the first plate has the mating strains "tan" and "black" on opposing quadrants and the second plate has "gray" and "tan" marked on the opposing quadrants.

Now using a flame-sterilized probe as demonstrated by your instructor, place a small block (.5 cm) of media containing the mycelia (and perithecia) of each mating type in center of the corresponding quadrants produced above. Flame sterilize the probe between each transfer to reduce the possibility of contamination. After eight to ten days of growth depending on temperature, the mating strains will meet somewhere near the lines that you have drawn. Where the two mating types meet, some of the hyphae will hybridize and produce asci containing up to four different kinds of spores. You will see a dark line of perithecia where this hybridization occurs.

Mark each plate with your name and the date inoculated. Give the plates to your instructor. The plates will be cultured at room temperature in a dark cabinet and returned to you in two weeks. The cultures will be stored in an upright position for one to two days after which they will be inverted. When the perithecia are mature, the plates will be placed in the refrigerator and stored until you are ready for them.

Week Three : Data Collection

Part I: Black x Tan Cross

This plate represents a monohybrid cross between two mating types of *Sordaria fimicola*. The tan strain is produced by the presence of a mutant allele (t). The wild type allele of the gene (t^+) produces a black spore coat on the ascospores. In the perithecia the paired (diploid) nuclei have undergone three cellular divisions. Meiosis I and meiosis II have occurred to re-establish the haploid chromosomal condition characteristic of the fungi. Following meiosis II, each cell underwent a mitotic division to produce the eight ascospores found within each ascus.

Place a drop of water on a microscope slide. Collect some perithecia for microscopic study by dragging a probe across the boundary where the two mating types have interbred. By sampling **perpendicular** to the boundary of hybridization, you increase the chances of finding hybrid perithecia. Swirl the tip of the probe in the drop of water to distribute the perithecia evenly throughout the droplet. Place a cover slip over the perithecia applying a light downward twisting pressure. This pressure should release the asci from the perithecia producing a rosette pattern of asci at the tip of the perithecia. If you apply too much downward pressure, the asci themselves will rupture destroying the desirable linear arrangement of ascospores.

Not all of the observed perithecia will contain hybrid asci. Ignore asci that contain spores of the same color. A hybrid perithecium will hold asci with ascospores of both colors. Three different patterns of ascospores are found in the hybrid perithecia. Four black ascospores will follow four white ascospores in an ascus where crossing over has not occurred. However, a crossing over event during meiosis produces spores distributed in a 2:2:2:2 or a 2:4:2 patterns. Since the likelihood of crossing over depends on the distance between the gene for spore color and the centromere, it is possible to locate the map position of the gene as a function of the cross over frequency.

1. Each student should record the phenotype (Table 1) for at least **thirty hybrid asci**. Record the phenotype of each asci going from the narrower bottom to the broader top of the sac. Record the phenotypes as BBBBTTTT or BBTTBBTT.

2. In the column for genotype, use "t+" for black and "t" for the tan mutant. As each pair of ascospores is produced by a final mitotic mutation, each pair needs to be symbolized by only one letter.

3. Check if the ascus is a recombinant (produced by a cross-over event) or non-recombinant type in the third column of the table.

4. Add up the recombinant and non-recombinant asci and add them to the class results (Table 2).

5. Using the class results, determine the crossover (recombinant) frequency between the mutant tan allele and the centromere. This is not as straightforward as you might think. When crossing over occurs during metaphase I of meiosis, we assume that only one of each sister strands crosses over. This means that in each ascus, only four of the eight ascospores found are the products of crossing over. To determine the crossover frequency, divide the # of recombinant asci by the total number of asci. This number must then be divided by ½ to get the true percent of crossing over.

   % crossover = [(recombinant asci)/ (total asci counted)]/2

   The % crossover represents the map unit distance between the gene and the centromere.

6. Using the chi-square table at the back of you lab manual, determine if your results are consistent with the reported gene-centromere value of 27 map units. In the chi-square analysis, the expected numbers of recombinant asci will be 54% of the total observed.

Part 2: Tan x Gray Cross

This plate represents the outcome of a cross between two different mutant types of *Sordaria fimicola*. The gray and tan spore phenotypes are produced by mutations at two different genes located on different areas of the chromosomes. This cross gives some interesting results. Even though not present in the parental crossings, it is possible to produce four different spore colors as a result of this cross. For example, when both wild type alleles of the two genes assort together (g^+t^+), black ascospores will be produced. Ascospores will be clear when both mutant alleles (g t) assort together. Gray will be the phenotype when the genotype is g t^+, and tan will be seen when the genotype is g^+t. **The purpose of this analysis is to determine if the two genes being investigated are traveling on the same chromosome (linked) or on separate chromosomes (unlinked).**

The **unlinked** condition is shown below. The position of the genes in these diagrams does not imply their actual locations. The chromosomes carrying these two genes are differentiated by their shapes. The open and closed centromeres are used to indicate the two different parental strains mated. The situation portrayed on the left side indicates orientation at the Metaphase I plate when parental chromosomes are aligned on the same side of the equatorial plate. The right side of the diagram portrays the chromosomes of the two strains shuffled such that nonparental chromosomes assort together. The law of independent assortment predicts that one configuration is as likely to occur as the second. The resulting colors and arrangements of ascospores produced are shown in the two asci below the diagram.

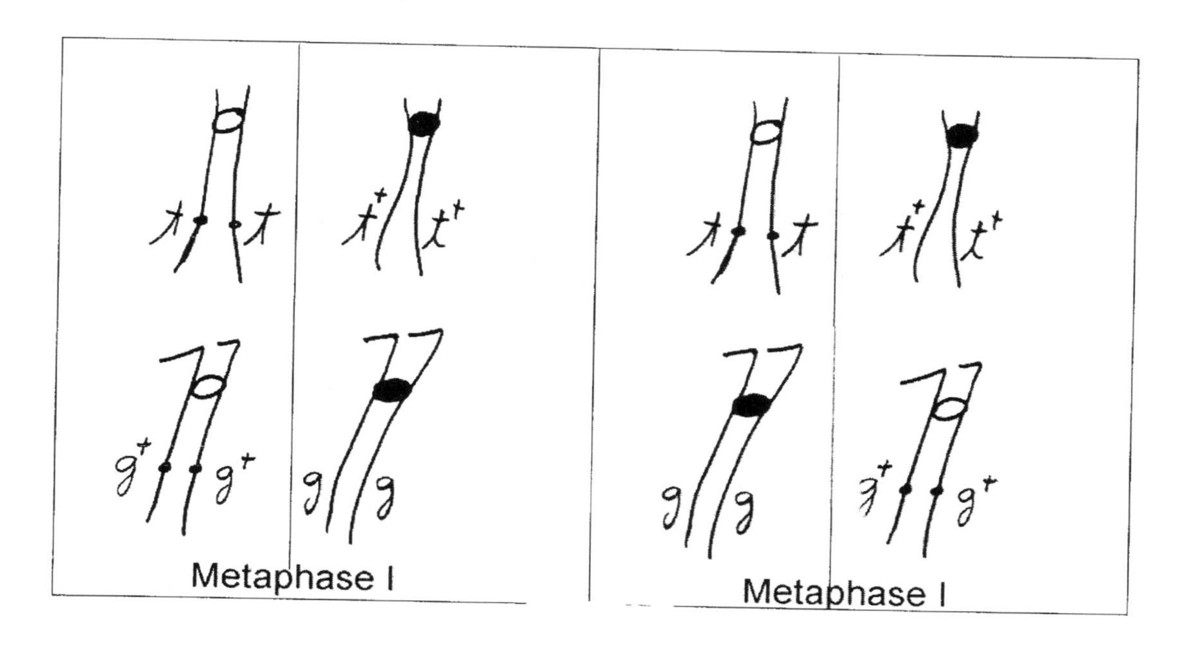

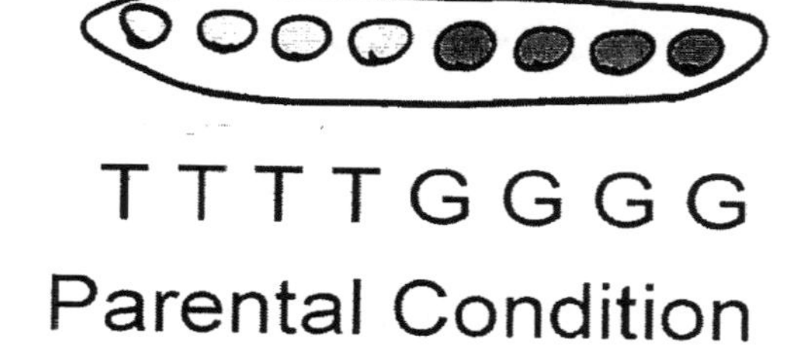

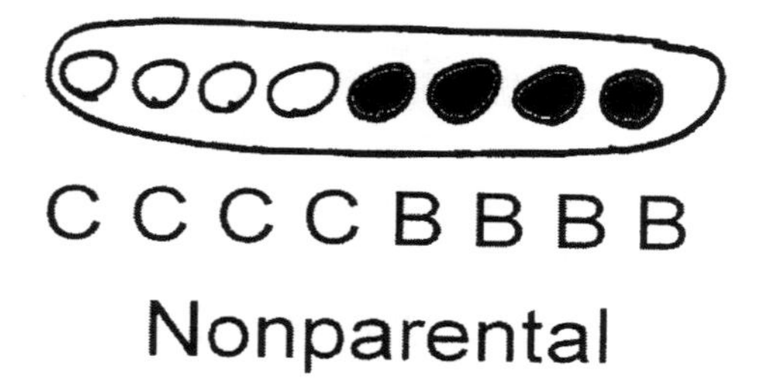

Even though the genes may be unlinked, crossing over will still occur between the individual genes and their centromeres producing many different recombination spore patterns. Crossing over between the genes and their centromeres will give rise to different tetratypes found in the asci (i.e. BBTTCCGG). This type of crossing over has no bearing on the outcome of the frequency of parental and nonparental types to be expected from the above analysis. **If the two genes are unlinked, the frequency of parental and nonparental asci should be equal as they are produced by independent assortment.**

The diagram below indicates what you might expect if the two genes are **linked** with both genes traveling together on the same chromosome. The chromosomes are lined up on the equatorial plate for metaphase I of meiosis. Notice that independent assortment is not possible as the genes are on the same chromosome. The spore patterns that would be produced in the absence of crossing over are shown in the asci below.

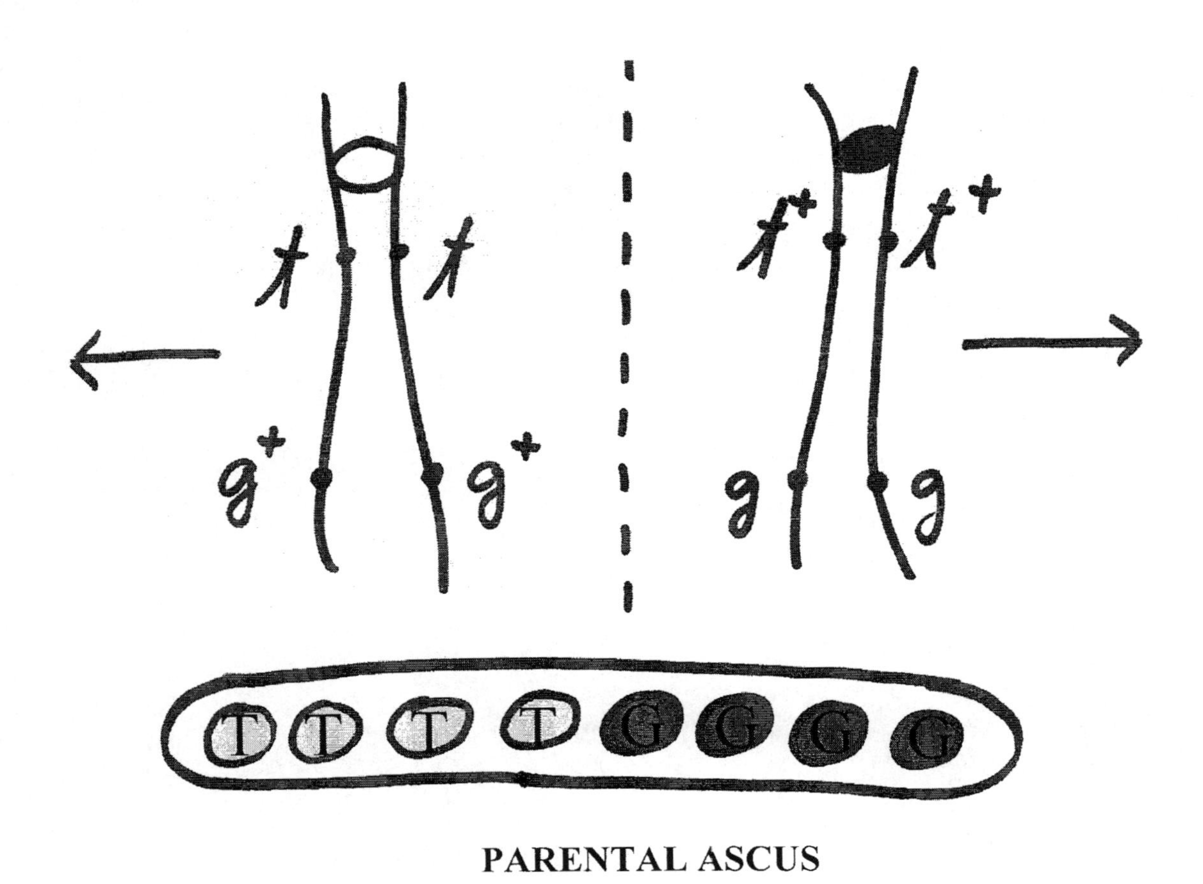

PARENTAL ASCUS

In the linked state, the nonparental spore pattern of four clear and four black ascospores will only be seen if a rare double cross over event occurs. This unusual occurrence is shown below. You would expect this type of rearrangement to occur with a very **low frequency**.

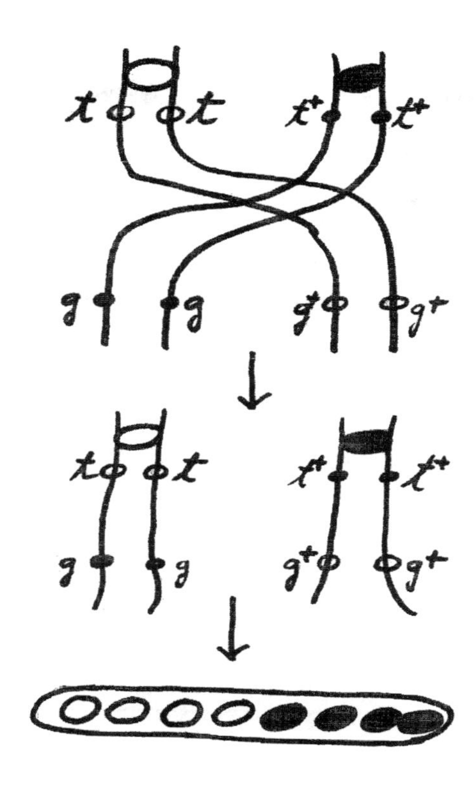

NONPARENTAL ASCUS

AIf the tan and gray spore color genes are linked, then the two genes will not assort independently, and the frequency of the parental asci types will be much greater then the nonparental types.

1. Record the phenotype of at least thirty asci that are clearly the result of hybridization between the two mutant types in Table 3. Hybrid asci will possess two or more ascospore types.

2. Complete the column for genotypes. Black will be g^+t^+, clear will be g t, gray will be g t^+, and tan will be represented by g^+t. Note that adjacent pairs of ascospores should have the same genotype as a result of the final mitotic division before the ascospores are laid down in the asci. Therefore you need only include one genotype for each pair of ascospores.

3. Check if the ascus is a parental type, nonparental type, or a tetratype produced by the many different crossover combinations. Be certain that the asci counted as parental or nonparental types only include the patterns of 4:4 ascospores. Other combinations of asci containing only gray: tan or black: clear are possible, but they are not to be considered as parental or nonparental types.

4. Add up the parental (P), nonparental (NP) and tetratypes. Add your numbers to the class total (Table 4).

5. Are the tan and gray genes on the same or on different chromosomes? Provide a supporting statement and chromosomal sketch.

6. Choose any two of the tetratype asci (cross-over combinations) and diagram the crossover events that would produce that combination of spore phenotypes. Be sure that your sketches reflect your answer to question #5. Remember that crossing over may occur between the genes and their centromeres, there may be single or double crossing over events, and if the genes are linked there may be crossing over episodes between the two linked genes.

Table 1. Individual results for the Black x Tan

Phenotype	Genotype	Nonrecombinant	Recombinant

Table 2. Class Results for the Black x Tan Cross

Nonrecombinant asci	Recombinant asci
Total	Total

% crossing over = [(recombinant ascospores/total number of ascospores)]/2

________________ % crossing over

________________ map units (gene to centromere)

Chi square results:

Phenotype	# observed	# expected	(obs – exp)2/exp
Recombinant ascospores			
Nonrecombinant ascospores			
Total number			

Do your crossing over results agree with the published map unit distance of 27 map units according to the results of the Chi square analysis?

Table 3. Individual results of the second plate Tan (mutant) X Gray (mutant)

Phenotype	Genotype	Parental type	Non-parental type	Tetratype

Table 4. Class Results of Tan (mutant) X Gray (mutant)

Group Number	Parental	NonParental	Tetratype
Totals			
			Totals

1. What is the percentage of each ascus type reported above?

2. Are the tan and the gray genes linked or unlinked? Give your supporting evidence including a chi-square analysis of the above data.

3. Cross-over demonstrations:

   Original dyads illustrating alleles in zygote before crossing over
   (Be sure that your demonstration reflects your answer to the above question #2.)

   Tetratype spore pattern in asci to be demonstrated

   Crossing over patterns needed to produce spore distributions illustrated above

   New alignments on the chromatids following the cross over events

PLANT REPRODUCTION WEEK ONE

13

Laboratory Objectives

1. To become acquainted with the concept of "alternation of generations" as it applies to plant reproduction.
2. To be able to fit the moss and fern life histories into the "alternation of generations" concept.
3. To recognize and sketch the various stages of the moss and fern life cycles.

Introduction

Plant reproduction, as animal reproduction, employs the process of meiosis to reduce chromosomal number by one-half so that the number of chromosomes remains constant from one generation to the next. In animals, meiosis directly results in the production of gametes (sperm and eggs) which come together immediately to re-establish the normal diploid condition. Plants go about the game of reproduction differently than animals with an intervening step or stage existing between meiosis and fertilization.

Plant meiosis occurs in specialized organs known as sporangia where diploid cells give rise to haploid (1N) **spores.** Spores, not being gametes, do not fertilize one another to give rise to zygotes. Spores are released from the parent plant to settle down some distance away to start a new generation of the plant's life cycle. The spore, dividing by mitotic divisions, produces a haploid mass of plant tissue referred to as the **gametophyte generation.** Someplace on the haploid gametophyte plant, the plant's gametes are then produced by **mitosis.** In the lower plants (mosses and ferns), the sperm producing structures are called **antheridia** and the egg producing structures are called **archegonia.** The haploid generation is called the gametophyte generation because it produces the plant's gametes by mitosis.

Fertilization re-establishes the diploid condition in the plant life cycle when the zygote is produced. The zygote is the first cell of a stage of the life cycle referred to as the **sporophyte generation.** When the zygote grows to maturity by mitosis, it will possess a diploid **sporangium** where meiosis will take place once again producing more haploid spores. Since the diploid generation produces the spores, it is also called the **sporophyte** generation. See Figure 1 for an examination of a typical plant life cycle.

It should be noted that female haploid tissue protects the egg, zygote, and the early stages of the sporophyte generation from environmental dangers.

In this laboratory exercise, you are going to examine the reproductive structures of two different groups of plants.

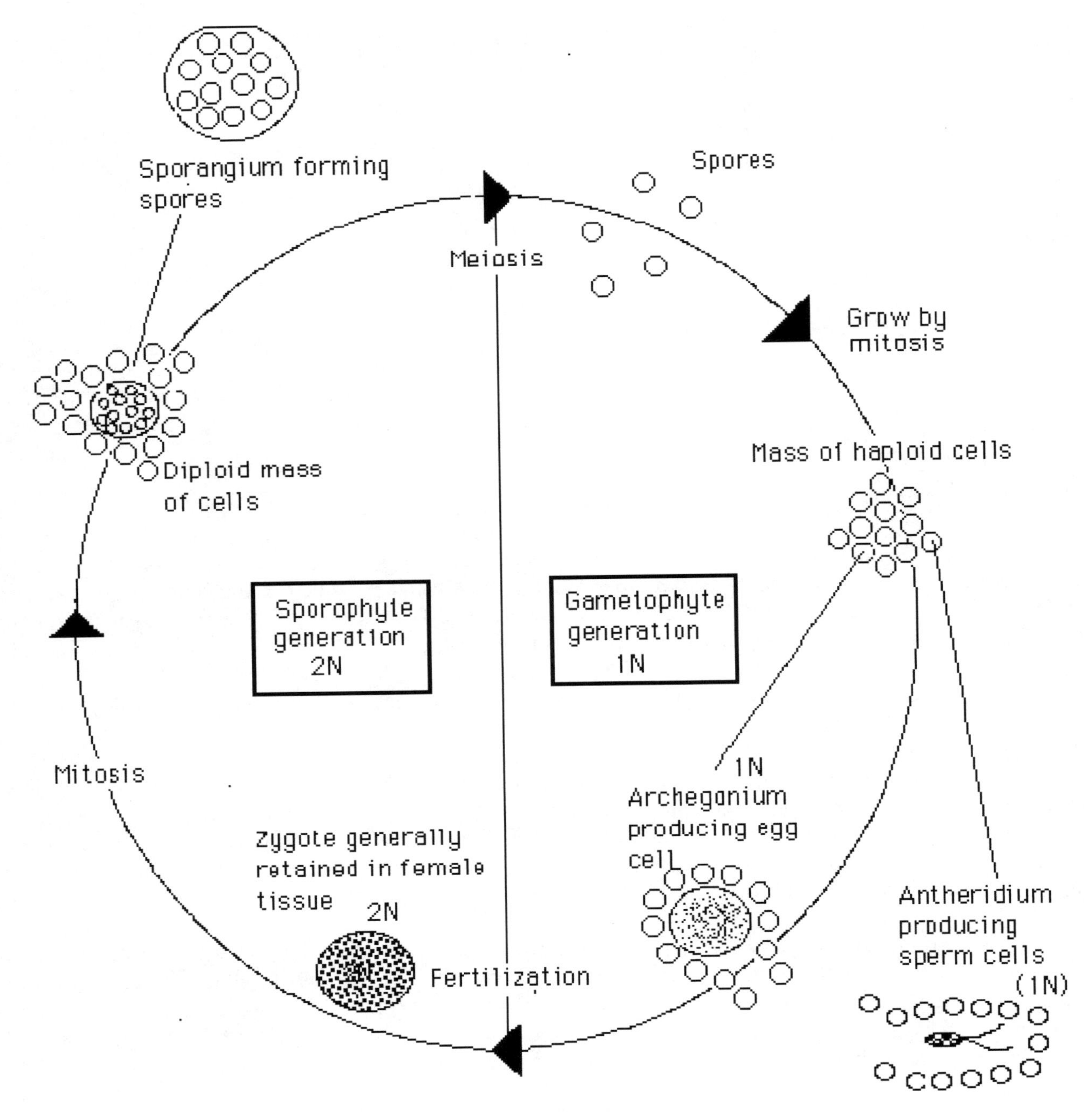

Figure 1. Alternation of Generations

Activity 1
Seedless Nonvascular Plants, the Bryophytes

The bryophytes (mosses and their relatives) are relatively small plants that live in habitats where water is plentiful, since water is necessary for their survival and reproduction. Bryophytes have been called the amphibians of the plant kingdom because amphibians (frogs, salamanders, and toads) also have a similar need for water.

Bryophytes resemble sponges in many ways. During periods of rain, they soak up and store a lot of water in the many spaces between the leaf-like portions of their plant body. Bryophytes also rapidly lose water by evaporation. Lacking specialized vascular tissue (xylem and phloem), the water that is lost by evaporation cannot be replaced from the soil. Bryophytes do not possess the plant organs (roots, stems, and leaves) studied earlier in the semester. Their relative simplicity in structure limits the height that mosses can grow. Rhizoids, root-like structures, provide holdfasts for the mosses but do not absorb water from the soil.

The gametophyte generation is the dominant phase of the Bryophyte life cycle. The sporophyte generation is never free-living, and is permanently attached to and dependent upon the gametophyte tissue for its nutrition.

General Procedure

There are four slides which you are going to use to observe the moss life cycle. Sketches of each slide (along with identifying labels) are to be placed in the appropriate sections of Figure 2 below. A low power view should be sketched to give an overall impression of the structure and then a high power view should be sketched to provide details not seen at the lower power. Be neat and careful in your sketches. Ask your instructor or a student near you for help if confused on any point. You will be observing the *Mnium* (moss) antheridial head, moss archegonial head, moss capsule (sporangia), and moss protonema slides.

A. Obtain a plastic mount of moss gametophyte tissue. The moss that you are examining is **dioecious,** meaning that there are separate male and female plants. The male gametophyte can usually be distinguished by the flattened **rosette** of leaflike structures at its tip. Embedded within the rosette are the male sex organs, **antheridia** (antheridium, sing.). Sketch a male moss stalk below as it appears in the plastic block. Examine a prepared slide of the antheridia of a moss using the 4X objective lens. Identify and sketch the antheridia (large oval-like structures containing smaller cell clusters within). Scattered among the antheridia, find the numerous sterile **paraphyses.** These filamentous strings of cells do not have a reproductive function, but they do hold water by capillary action protecting the antheridia from drying out. Using the high-dry objective lens, locate the jacket layer surrounding the sperm forming tissue that gives rise to numerous biflagellate sperm. With the 40X objective lens, you can see nuclei with chromosomes (condensed chromatin) undergoing mitotic divisions to form sperm cells. The nuclei are too small to identify stages of mitotic activity. You should be impressed with the numbers of sperm that this one little plant could have produced if left undisturbed. Why are so many sperm cells required?

B. Examine and sketch a female (archegonial) moss stalk from the same plastic embedded specimen used earlier. The female stalk lacks the rosette pattern seen in the male stalk. Perhaps it can better be described as resembling the tail of a horse as it tapers towards its end. The apex of the female stalk contains numerous **archegonia** which house the eggs of the moss plant. Obtain a prepared slide of the archegonial head of the moss plant. Start with the low power objective to gain an appreciation for the overall organization of the female stalk. Again the archegonia are surrounded by sterile **paraphysial stalks** which help keep the archegonia moist by capillarity. There are numerous, but fewer, archegonia at the tip of this stalk. You will have to examine several of the archegonia to be able to formulate an overall impression as to their shape. Since the stalk is a three dimensional structure, some of the archegonia have been cut more anteriorly and others more posteriorly.

Moving to the 10X objective lens, find and sketch an archegonium which possesses an egg cell in its swollen base. If the stalk that you are observing does not have an archegonium sectioned in such a way that reveals an egg, borrow a neighbor's slide to work with. The entire archegonium looks like a bud vase. The long neck of the archegonium leads down to the egg cell which is protected within the slightly swollen base of the vase.

The central core of the archegonial neck contains cells that break down when the egg cell is mature. These core cells release a fluid rich in sucrose and that attracts sperm that are swimming in dew or rain water. The sperm cells of moss cannot swim great distances. When a rain drop hits an antheridium, it splatters the sperm cells to surrounding areas. If, by chance, a sperm cell lands near an archegonium, it is enticed to swim down the neck

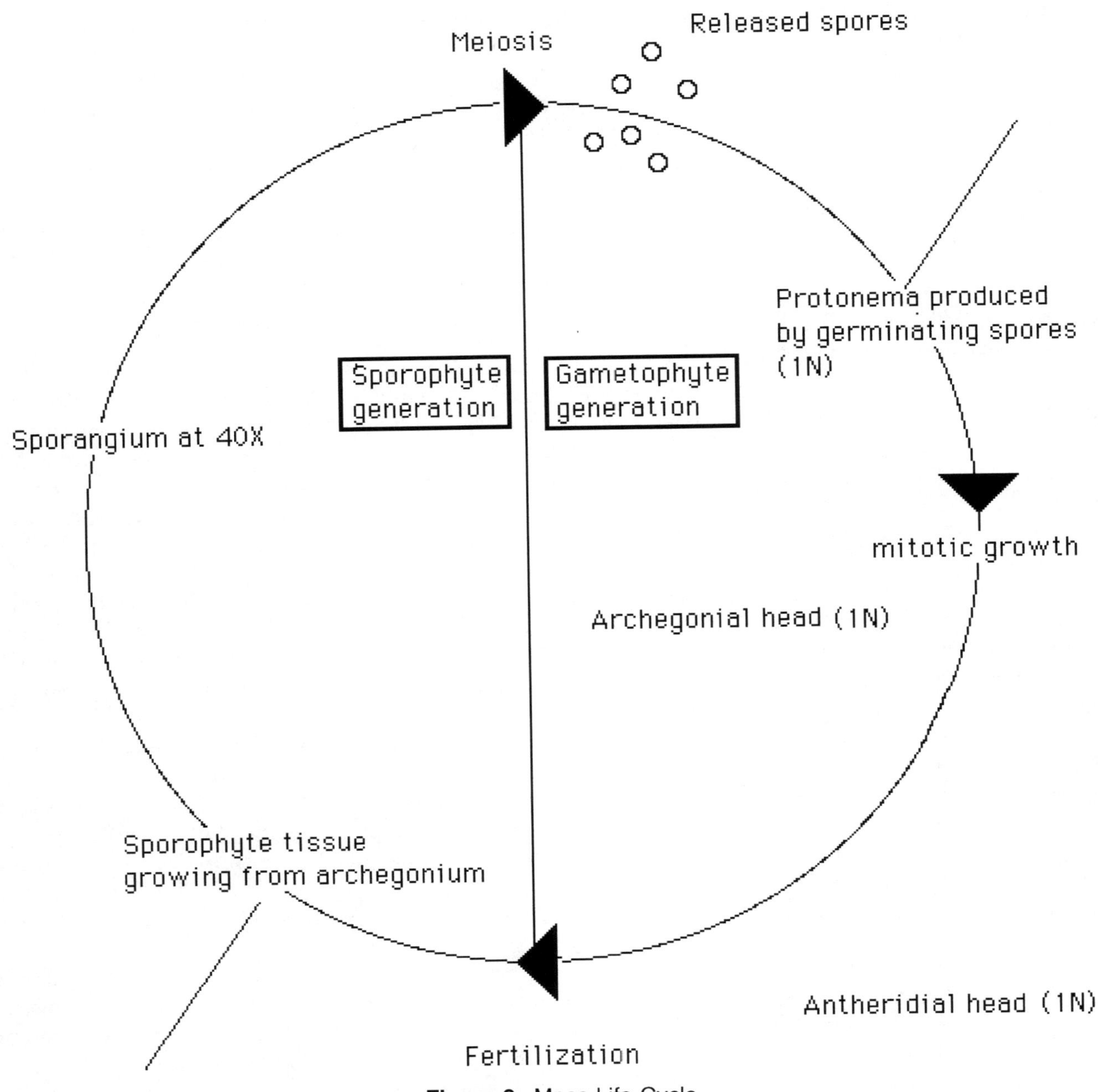

Figure 2. Moss Life Cycle

of the archegonium toward the waiting egg cell at the base. The egg cell is larger than the surrounding sterile jacket cells and sits in the middle of a cavity seen at the base of the archegonium. Fertilization, creating the **zygote,** occurs within the female archegonial tissue. This fertilization creates the first cell of the next **sporophyte** generation. With repeated mitotic divisions, an embryo is produced. This embryo grows right out of the archegonial tissue differentiating into the mature sporophyte generation of the moss plant.

C. Using the same plastic embedded specimen, identify and sketch the **sporophyte generation** emerging from the female gametophyte stalk. Is the sporophyte tissue green or brown in color? What can you conclude about its ability to produce food? The **sporangium** of the sporophyte tissue may be covered by a papery hood derived from the tissue surrounding the archegonium from which it grew. The papery covering hood-like structure is referred to as the **calyptra** (1N since its origin is archegonial).

Sporophyte tissue can be seen rising from moss mats in our area in the fall. You may have noticed these brown rather inconspicuous looking structures while on a hike. They generally rise 2-3 inches above the rest of the gametophyte moss tissue. When a drying wind blows, the sporangium's lid opens to allow spores to escape to travel on the wind away from the parent plant.

Study and sketch a prepared slide of a longitudinal section of a sporangium (moss or *Mnium* capsule). Your slide may show **peristomal teeth** that point inward from the margin of the opening just under the capsule lid. The peristomal teeth are **hygroscopic,** which means that they change shape as they absorb water. When they are moist, they swell and point downward. When they dry out, they arch upward, loosening the cap of the sporangium. What conditions favor spore release in the mosses?

Locate the tissue from which spores are produced. The sporogenous tissue of the sporangium is diploid, but the spores are haploid being the first cells of the gametophyte generation. What type of cell division took place in the sporangium to produce the spores? Observe the spores with the 40X objective lens. It may be an artifact of the slide making process, but do the spores look hydrated or dehydrated? What advantages would the spores gain from this condition?

D. Study and sketch a section of a prepared slide labelled **moss protonema** which was formed from germinating spores. This slide looks like a filamentous green algae. Some slides will possess a thick mat of protonema fibers growing in every direction. Try to find an area of the slide where the fibers are spread out enough to allow easy observation. You may be able to find the "casings" of spores which were germinated to make this slide. Find a single filament of cells and move along its length. If it is of the right maturity, you may find cell clusters that would have differentiated into rhizoids. Also along the length of the protonema will be cell clusters that will give rise to vertical male and female stalks of a new "moss mat." At the end of a filament you may also see cytokinesis occurring with the erection of a new cell wall forming between two daughter cell.

Activity 2

Seedless Vascular Plants (Fern)

Vascular plants have increasing refinements for a successful life on land. As the name implies, the specialized tissues of xylem and phloem are present allowing for the transportation of water and food. Cuticles, consisting of fatty and waxy layers, are more developed than in the bryophytes. Stems, leaves, and roots exist allowing the ferns and their allies to reach toward the skies. The spores of these plants are also more protected from drying out than are the spores of the bryophytes.

Seedless vascular plants make up only a small part of today's plant kingdom, but during the Carboniferous period, 285-360 million years ago, the seedless vascular plants dominated the landscape. Most of the seedless vascular plants present from that period are now extinct, but some large tree ferns still can be seen in tropical rain forests.

We are going to study the life cycle of the common fern. The roots of ferns originate from underground stems called rhizomes. The leaves, called fronds, also arise from the underground stems. In most species, spores are produced in rusty patches called sori found on the undersides of leaves. A leaf which possesses sori is called a **sporophyll.** Because sori give the frond a spotted appearance, many people mistake these reproductive structures for some type of disease.

General Procedure

You will be observing five slides illustrating the fern life cycle. As above, sketches of the various stages of the life cycle will be placed in the appropriate locations of Figure 3 found below.

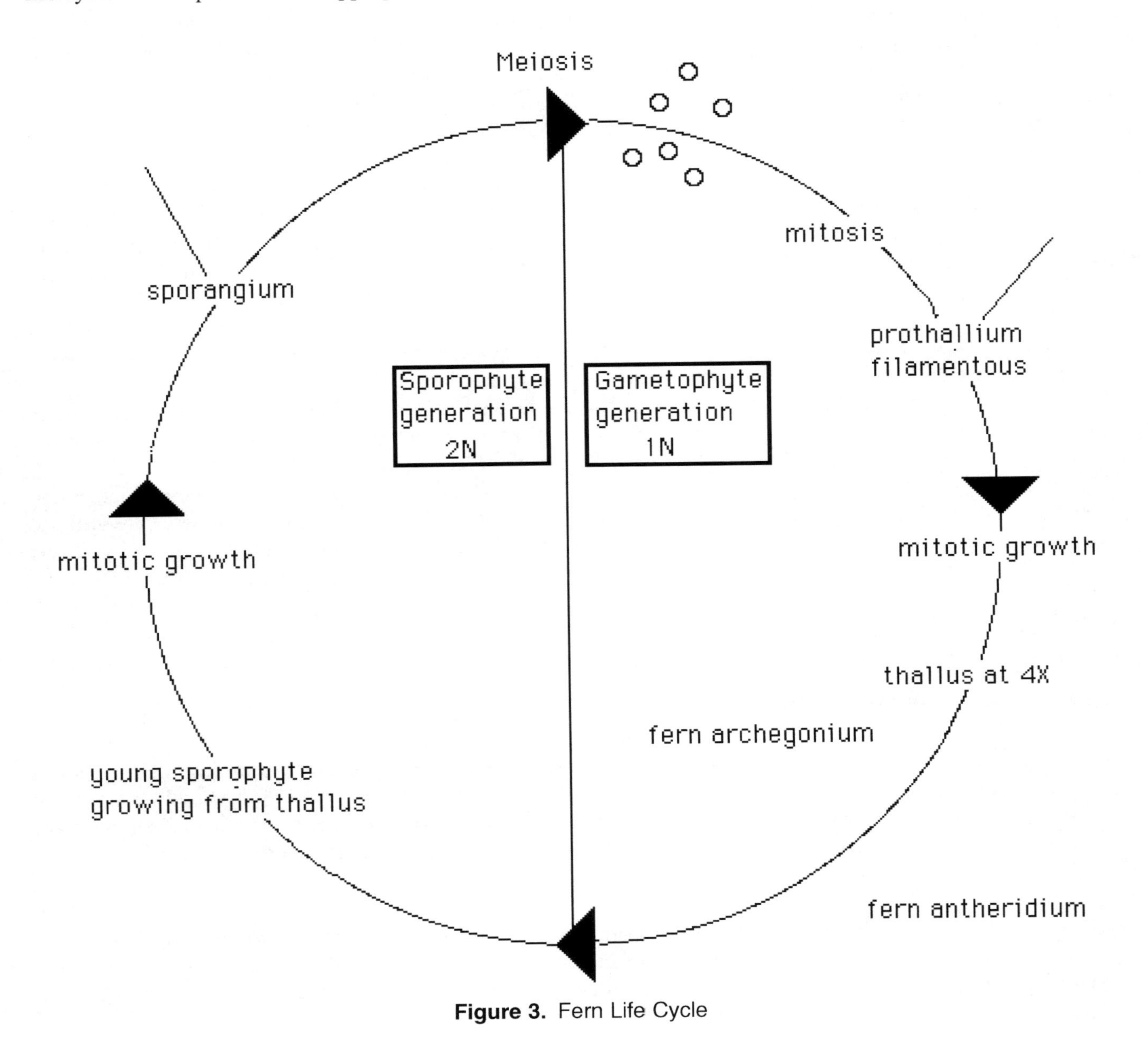

Figure 3. Fern Life Cycle

A. Obtain a specimen of a fern embedded in plastic. Find and sketch the **sori** (fruit dots) on the back of a fertile frond **(sporophyll).** One sorus (pl. sori) is made up of hundreds of fern sporangia clustered together.

B. Study and sketch a prepared slide of fern **sporangia.** On the slide you can find sporangia in different stages of development. The bluish colored sporangia look more immature with the spores inside being smaller and less well developed. The more reddish colored sporangia look "riper" and ready to release their spores. Find one mature sporangium that is somewhat isolated from its neighbors and observe it with the 40X objective lens. Note the row of thick walled cells running over the top of the sporangium, the **annulus.** It looks like a thick spring that has been stretched around the outside of the sporangium. The annulus is hygroscopic. Changes in moisture content within the cells of the annulus cause the sporangium to crack open. As the annulus opens, spores are flung away vigorously from the parent plant. When the released spores land in a suitable environment (moist and shaded), the spores germinate and grow into the haploid gametophyte stage of the life cycle.

Closely observe the spores inside of the sporangium. Sketch a few spores under high magnification. They appear to possess a thicker outer cell wall than do the spores of moss. What might be the significance of their overall shape?

C. Study and sketch a slide of the **fern prothallium filamentous.** This slide is the equivalent of the moss protonema slide. The congested mass of fibers that you observe under low power has been created by germinating spores. Is this tissue haploid or diploid?

Find an isolated **prothallium** and sketch under the 40X objective lens. The specimen looks almost like a small insect with antennae. The "antennae" will develop into a **rhizoidal** mat better seen in the next slide. What is the function of rhizoids? The rhizoids are attached to an 8-10 cell developing flattened "thallus." The term "thallus" refers to an undifferentiated leafless, rootless, and stemless plant body. In the fern, the haploid thallus will hold the fern's archegonium and antheridium.

D. Study and sketch a slide bearing the label **Fern Archegonia, w. m.** The initials "w. m." means whole mount. The thallus containing the archegonia to be viewed was simply placed on the slide and covered with a coverslip. No cuts were made to reveal structures within the thallus.

Since the thallus forms from a germinating spore, it represents the haploid gametophytic tissue of the fern life cycle. It is a free living stage possessing chlorophyll pigments. The thallus is about the size of a penny. If you were to look for the thallus on the forest floor, it would be like looking for a four-leaf clover. The haploid tissues of the fern's life cycle are much less conspicuous than the haploid moss tissues seen in the activity above.

Sketch the thallus using the 4X objective lens. The overall thallus shape resembles a "heart." At the base of the "cleavage" of the heart you can just see tighter clusters of cells that are the fern's **archegonia.** At the apex (tip) of the thallus are numerous rhizoids which would help to anchor the thallus to the substratum (soil).

Find a mature archegonium (light brown in color) to observe using the 40X objective lens. Since we are looking down upon the archegonium, it is more difficult to observe its structures than when seen in the moss. By focusing up and down with the **fine focus adjustment,** you can observe the three dimensional architecture of the archegonium. It is shaped like a volcano with a central chimney leading to the egg cell at it base.

Return to the 4X objective lens to search amongst the rhizoids for signs of antheridia which may be present on this slide (or they may not be here). You are looking for round cells about the same size as surrounding thallus cells containing 20-30 smaller spherically shaped masses in their cytoplasm. The **antheridia** that you may find here produce sperm cells that must swim to the archegonia seen above for fertilization.

Why are ferns usually restricted to moist environments?

E. Study and sketch the slide labelled **Fern Antheridia.** The fern antheridia are more easily visible in this slide than the slide above. The spherically shaped antheridia are jammed with developing sperm cells. With very careful observation, it is possible to see some of the **mitotic** stages (especially anaphase) that these cells go through to form sperm cells.

Find an antheridium that looks like it has just recently ruptured. Using the 40X objective lens, look for mature sperm cells. The flagella will form "squiggly" patterns sometimes taking the shape of a "?" mark. The "heads" of sperm cells may be visible. The sperm cells appear to be biflagellate.

The sperm swim to the archegonia to fertilize the eggs within the protective confines of the female reproductive tissue.

F. Observe and sketch the slide labelled **Fern Young Sporophyte** using the dissecting microscope. The diploid plant can be seen emerging from the thallus with a young leaf rising vertically from the forming horizontal stem (rhizome). The young root is growing in the opposite direction as it heads down into the soil. The young diploid plant is growing out of the archegonia in which it was conceived. As you can see, the haploid thallus portion of the life cycle is going to be dominated by this newly forming diploid plant. As the sporophyte continues to develop, the gametophyte thallus withers away.

Name ______________________________

Section ______________________________

Post-lab Questions

1. Explain the concept of alternation of generations as it applies to the plant world.

2. Mosses and ferns show heteromorphic generations as they cycle between the haploid and diploid world. What does the term "heteromorphic" mean?

3. What do the protonema and the thallus share in common?

4. Why must water be present for the bryophytes to complete the sexual portion of their life cycle?

5. In what structure does meiosis occur in the bryophytes?

6. Describe in your own words the difference between a sporophyte and a gametophyte.

7. List two features which distinguish the seedless vascular plants from the bryophytes.

8. The environments in which ferns grow range from standing water to very dry areas. Nonetheless, all ferns are dependent upon free water in order to complete their life cycle. Explain why this is the case.

ANGIOSPERM REPRODUCTION

14

Laboratory Objectives

1. To place angiosperm reproductive structures into the concept of alternation of generations.
2. To learn the parts of the flower and their functions in reproduction.
3. To examine and sketch the development of a pollen grain.
4. To examine and sketch the tissue of the female gametophyte tissue.

Introduction

A major breakthrough in the movement of plant life away from standing water was the evolution of the seed. Seed producing plants include the Gymnosperms (cone bearing plants) and the Angiosperms (flower producing plants). Even though ferns had evolved vascular tissue and a tougher more water resistant cuticle than present in the Bryophytes (mosses), they were still tied to a moist environment for their reproductive cycle. The evolution of the seed is one of the most important reasons why seed plants replaced (for the most part) the ferns and its allies which had dominated the earth's surface during the Carboniferous period. The following characteristics of seed plants clearly have helped them in their movement into more arid environments.

1. All seed plants produce pollen grains. One of the limiting features that keeps ferns and its allies relegated to moist environments is the need for sperm cells to swim to egg cells for fertilization. Seed plants "air-mail" their sperm cells to the female reproductive structures and are thus independent of free water for reproductive purposes. The evolution of the pollen grain allows these plants to reproduce in more arid environments.

2. Seeds usually have a supply of "stored food" which the embryo can draw upon during the germination process. The embryo can grow and become established before it needs to begin photosynthesis to produce its own food.

3. All seeds have a seed coat. The seed coat for most of its existence is a water proof "wrapper" that protects the young plant embryo and its stored food supply from potentially harsh environments. The presence of an intact seed coat also enforces a period of dormancy (almost like a hibernation) preventing premature germination of the seed during an inhospitable season. A seed which has been produced late in September in our area might be "fooled" into germinating during a late Indian Summer in November. For a few days the young germinating plant might be fine, but its survival will be short-lived. Seed dormancy prevents germination until favorable conditions prevail.

Seed plants still exhibit an alternation of generations as seen in the mosses and the ferns. The gametophyte haploid (1N) stage is present, but is much reduced in size. The dominant sporophyte generation produces spores which give rise to gametophyte tissue by meiosis as before, but the gametophyte tissue is represented by only 2-8 cells. Seed plants are also **heterosporous** producing two types of spores (micro and macrosporangia). Mosses and ferns are **homosporous** as they produce only one spore type. The existence of heterospory has given rise to a different set of terms, but the terms can still be related to the reproductive patterns studied earlier.

Gymnosperms rely for the most part upon the wind to carry the male pollen grain to the female cone. In addition, the gymnosperms produce naked seeds on the surface of the scales of pine cones. When the seeds are released, they are protected only by their seed coat. No extra layers are provided by the parent plant for the seed's protection or its dispersal.

Angiosperms have taken the game of reproduction one step further than the Gymnosperms and have evolved flowers. There are more flowering plants in the world today than any other plant group. If numbers of species means success, then angiosperms are the most successful plant group to have evolved. The evolution of the flower has been important for the success of this group for two reasons. The flower has promoted more efficient transport of pollen to the female tissues of the plant. Rather than throwing their pollen into the general air currents allowing for chance pollination, many of the angiosperms take some of the chance out of the pollination process by having animals deliver their pollen for them. This reduces pollen wastage and increases the efficiency of pollen transfer. The second major advance in reproductive strategy employed by the angiosperms is the evolution of floral parts which mature into fruits. A fruit is a container that protects the seeds, allowing them to be dispersed without coming into contact with the rigors of the external environment. Besides protecting them from hostile environments, many fruits aid in seed dispersal. Fleshy fruits are often eaten by animals. Most animals are not as fussy as humans and the entire fruit is consumed. Sometime later, the nondigestible portions of the fruit including the seed must be eliminated. By the time this elimination occurs, the seed has usually travelled some distance away from the parent plant.

Activity 1
Floral Dissection

Flowers are whorls of modified leaves attached to a common branch that has been modified during the course of evolution to increase the probability of fertilization. For example, some floral parts are colorful, attracting animals that serve to transfer pollen to the receptive female portions of another flower. The number of flower types is immense, but there is also a great deal of similarity among flowers. After studying a representative flower, you will be able to recognize the floral parts of most other types of flowers.

Procedure

Obtain a flower provided for your dissection and label its parts in Figure 2.

The flower parts are arranged in whorls on top of a swollen stem tip, the **receptacle.** The term "whorl" simply means a concentric ring. The outermost whorl of leaves is frequently green and is the **calyx.** Individual leaves of the calyx are called **sepals.** The calyx surrounds the rest of the floral parts while they are in the bud stage.

Moving inward, locate the next whorl of modified leaves of the flower, the usually colorful **corolla** made up of **petals.** It is usually the presence of the beautiful petals which draws our attention to the flower, but the petals are not "pretty" for man's benefit. The bright colors and shapes of the petals are the "advertising" which attracts the animal pollinators to the appropriate flower. In this manner, plants stand a better chance of being pollinated, producing seeds, and perpetuating their species. Some types of angiosperms still employ wind pollination. How large and brightly colored would you expect the petals of wind-pollinated flowers to be?

Both the sepals and the petals are sterile parts of the flower. They are not directly engaged in the reproductive process, but are present only as accessory structures.

The next whorl of modified leaves are the **stamens.** A whorl of stamens collectively is known as an **"andromecium."** The stamens are the male part of the plant as they produce the pollen grains of the flower. The **filament** is the stalk which holds the **anther** (microsporangia) aloft. Stamens are also called **"microsporophylls."**

Recall that the fern fronds that bear the sori are called "sporophylls" since they carry the sporangia where meiosis produces the fern spores. Angiosperms are heterosporous producing two different types of spore cells. The anthers are in actuality modified leaves which carry sporangia (microsporangia in this case) just like the sporophylls of the fern.

Next locate the female portion of the flower, the **pistil.** A pistil consists of one or more **carpels,** also called megasporophylls. A megasporophyll is a leaf bearing megasporangia. In the case of the flower, the ovules are the megasporangia. If the pistil consists of more than one carpel, the carpels are usually fused together, making it difficult to distinguish the individual components.

Many students have difficulty understanding the relationship between the term "pistil" and "carpel." If you understand that a whorl of sepals is collectively called a calyx, and that a whorl of petals is known as a corolla, extend the concept to include that a whorl of carpels forms a pistil. The only difference is that in the last case, the individual components of the whorl (carpels) are fused to produce a single structure (pistil). The figure on the next page should help to clarify this relationship.

Identify the different parts of the pistil. The **stigma** is at the top of the pistil. The stigma serves as the receptive site where the pollen of flowers of the same species will attach. The neck-like **style** leads down to the swollen **ovary.**

With a sharp razor blade make a cross section of the ovary. Examine the sections with a dissecting scope. Find the numerous small **ovules** within the ovary. The ovule is the megasporangium of the flower. Within the ovule, a **megaspore mother cell** will undergo meiosis to form the first cell of the next female haploid gametophyte generation of the angiosperm. An ovary can have many ovules within it. After fertilization, the ovules will develop into seeds and the ovary will enlarge and mature into the fruit of the angiosperm.

There are two groups of angiosperms, the monocotyledons and the dicotyledons. The number of flower parts indicates which group a plant belongs. Generally monocots have floral parts present in multiples of three. Dicot flowers have floral parts represented in multiples of four or five. Count the number of petals that you have been examining to see if you have a monocot or a dicot?

Activity 2
Pollen Germination

At this point, perform the following steps to see what happens to a pollen grain when it lands on a stigma of a female pistil. Proceed with the rest of the lab while the pollen grains germinate.

1. Using forceps, pluck a stamen from the flower provided. Holding the stamen by its filament, brush the anther face down in the well of a concavity slide. You may want to use a dissection microscope for this portion of the procedure.

2. Put a drop or two of 10% sucrose solution in the well of the slide with the pollen. You do not need to cover the slide with a coverslip.

3. Examine the slide using the scanning objective of the compound microscope. Sketch a few pollen grains at this stage of their development.

4. Check the slide in **30 minutes** to see if any germination has occurred. If the slide is drying out, add another drop of sucrose. Work on other exercises while you wait.

For the flower that you used for this activity, how far will the pollen tube have to grow in order to reach an ovule? Why doesn't pollen carry its own food supply?

Activity 3

Floral Diagram

Construct a floral diagram of the live flower which you just dissected. a floral diagram represents a cross-section of a flower as it would appear if all the parts were at the same level. It might also be thought of as an aerial view of a flower in diagrammatic form. For uniformity and convenience, the various parts are represented in diagrams by standard symbols as shown in Figure 1. In general, if the parts of the flower are connected, the diagram will show a line connecting these parts. An example is also shown in Figure 1.

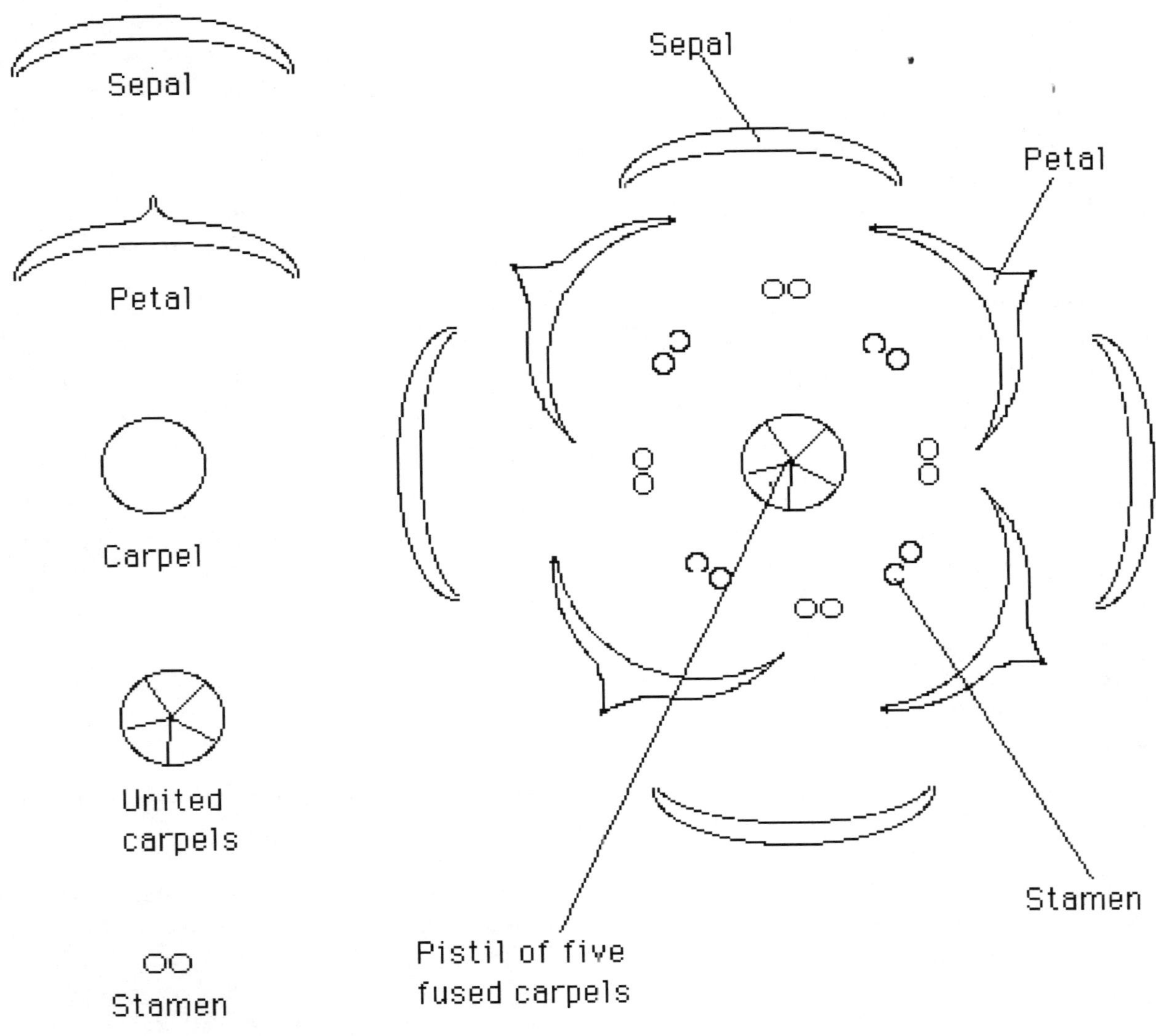

Figure 1. Floral Diagrams

Activity 4

Microscopic Examination of Pollen Development in an Anther (Microsporangium)

In the case of the moss a spore upon germination produces a protonema. In the case of the fern, the germination of a spore begins the development of the fern thallus, the heart-shaped structure bearing the antheridia and the archegonia. In the case of an angiosperm, development of a haploid microspore by mitosis produces a pollen grain which represents the male gametophyte (1N) of the angiosperm.

A. Examine a single stamen in greater detail using a dissecting microscope. Each stamen consists of a stalk-like **filament** and an **anther.** The anther consists internally of four **micorsporangia** (also called pollen sacs). The microsporangia are homologous to the sporangia clustered in sori on the back of the fern fronds, except the microsporangia produce one of the two types of spores involved in the reproductive cycle of the angiosperms. Recall that angiosperm plants are heterosporous producing spores of two different sizes. The anthers produce microspores by the process of meiosis.

B. Obtain and sketch a slide labelled "Lilium: Early Anthers" using the 4X objective lens. Place your sketch in the appropriate section of Figure 2. In the cross section of one anther, you can see the four circular **microsporangia.** If cut in the right plane, you can also see the vascular tissue of the **filament** which supports the anther. Inside the microsporangia are many **microspore mother cells (2N)** which will give rise to four microspores (1N) by the process of meiosis. How many anthers does the flower of the Lily possess?

The sketches of the following slides should be placed in the appropriate sections of Figure 3.

C. Obtain a slide labelled "Lilium: First Division." In this slide you can see microspore mother cells (2N) either in prophase I, metaphase I, or anaphase I. These cells are undergoing the reduction division phase of meiosis I where the number of chromosomes is being reduced by 1/2.

D. Obtain a slide labelled "Lilium: Second Division. In this slide you will the see the daughter cells of the first division in various stages of meiosis II. Recall that in meiosis II, sister chromatids are separating from one another (dyads separate to form monads). Did the daughter cells of the first meiotic division separate or remain together?

E. Obtain a slide labelled "Lilium: Pollen Tetrads." In this slide you can see the four haploid daughter cells still encased in the original cell mass of the microspore mother cell. Most of the pollen tetrads visible appear to be in the telophase II configuration as their chromosomes are still visible.

F. Obtain a slide labelled "Lilium: one celled microspores. In this slide you can see that the daughter cells of the meiotic division have separated, they contain a single nucleus in their cytoplasm, and their outer cell wall is beginning to develop ornamentation unique to each species of seed bearing plant. The developing wall is tough in construction to protect the more fragile nuclei that will be housed with its confines. Pollen has been recovered from early burial sites and has been induced to "germinate" after thousands of years. The pollen wall also has a unique protein configuration which allows it to recognize a stigma of its own species.

G. Obtain a slide labelled "Lilium: Mature Anthers." Before pollen is released from the anther, a single mitotic division occurs forming two nuclei housed within the same pollen grain. The larger of the two nuclei is called the **tube nucleus.** Its function is to direct the synthesis of the pollen tube which grows from the stigma, down through the style to the ovary. The second nucleus in the pollen grains is the **generative nucleus.** As the pollen tube begins it descent in search of the egg cell, the generative cell undergoes a second mitotic division forming two sperm cells.

H. Obtain a slide labelled "Lilium: Stigma and Pollen Tubes." This is a more difficult slide than those viewed above. Make sure that you look at the top of the stigma for the germinating pollen grains. Look for the "ornamental casings" of pollen grains which have been left at the top of stigma. By careful observation, you should be able to find a pollen tube which has just emerged from the pollen grain. The tube nucleus will be right behind the growing tip of the pollen tube. The generative nucleus lags a bit behind.

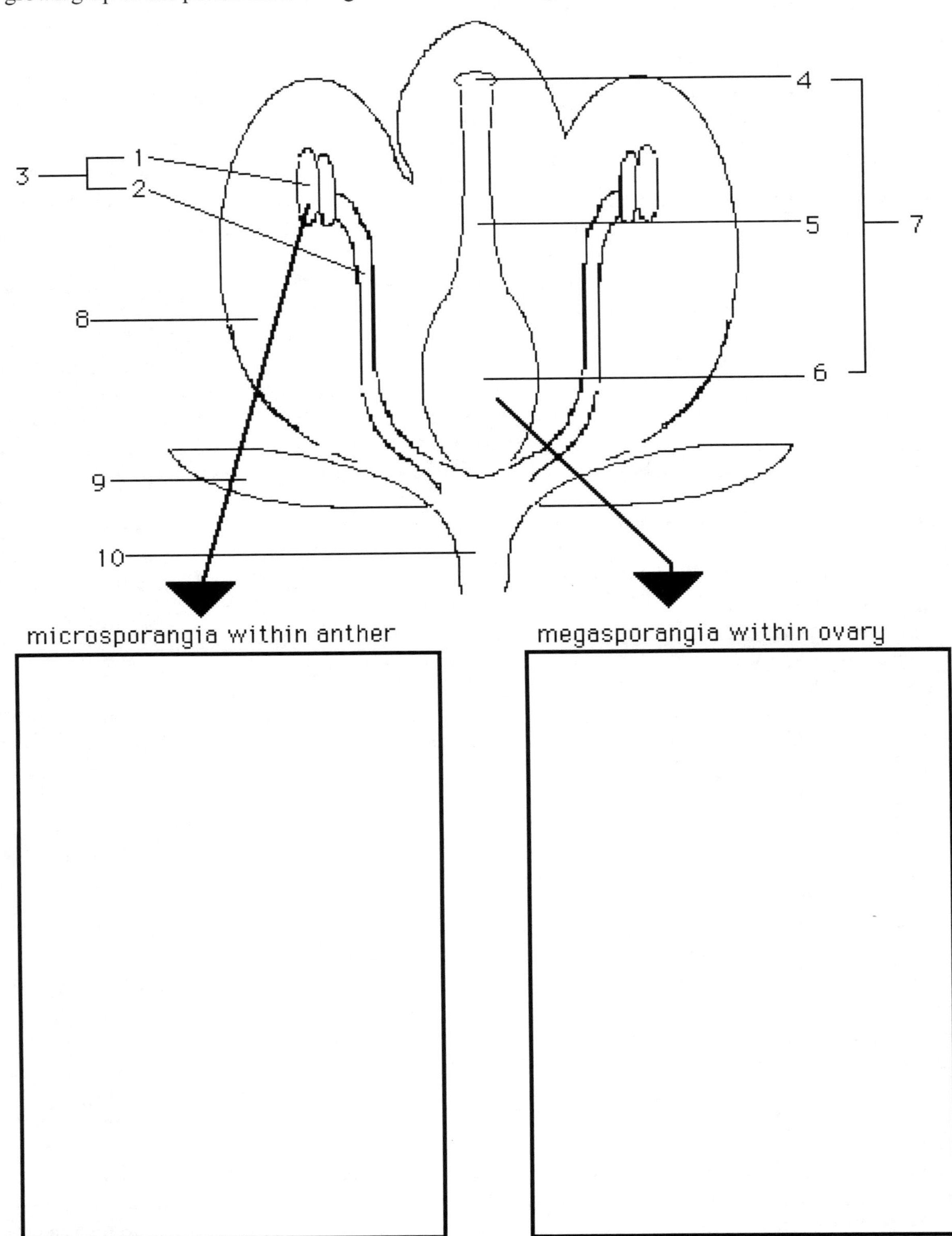

Figure 2. Sketches of Micro and Megasporangia

First Division 1	Second Division 2
Pollen tetrads 3	One-celled microspores 4
Mature anthers 5	Pistils with germinating pollen 6

Figure 3. Development of Male Gametophyte Tissue

Activity 5
Megasporangia and the Female Gametophyte

A. Obtain and sketch a prepared slide labelled "Lily ovary: embryo-sac 1 nucleus." Place your sketch in the appropriate section of Figure 2. Several **ovules** (megasporangia) will be visible. One or two ovules will probably be sectioned in a plane that will reveal a very large and obvious diploid (2N) **megaspore mother cell.** The megaspore mother cell is housed inside the megasporangium (ovule). The ovule is enveloped by two flaps of tissue called the **integuments.** By curling your thumb and index finger so that their tips almost touch, you can produce a model of an ovule. The megaspore mother cell would reside in the space between your index finger and thumb. The space separating the tips of your thumb and index finger is representative of the ovule's **micropyle.** The micropyle is the small opening through which the pollen tube grows to gain access to the egg that will be housed in the ovule. The integuments of the ovule later develop into the seed coat which protects the embryo and its food source. The ovule is attached to the ovary wall by a region of tissue known as the **placenta.**

B. Study and sketch the demonstration slide of the seven-celled, eight nucleate embryo sac (the plant's female gametophyte tissue). Place your sketch in Figure 4.

Figure 4. Female Gametophyte Tissue (Embryo sac)

The formation of the female gametophyte tissue which houses the egg cell is very complex and we will not follow all of its events. The megaspore mother cell (2N) undergoes meiosis I and II to form four megaspores (1N). Only one of the four megaspores goes on to develop further by mitosis. The remaining three megaspores degenerate. Through a complex series of mitotic divisions, the surviving megaspore produces an **embryo sac** within the ovule. At one end of the embryo sac nearest to the micropyle are three cells: the **egg cell** itself and two flanking cells called **synergids.** The function of the synergid cells is not well known, but they are believed to somehow help in the fertilization process. At the other end of the embryo sac are three **antipodal cells.** These cells apparently wither after fertilization occurs. There are two other nuclei, called **polar nuclei,** which share the cytoplasm of the large central cell of the embryo sac. These two nuclei will be fertilized by one of the two sperm cells delivered in the pollen tube to form the seed's triploid **endosperm.** The endosperm develops into the stored food of the seed.

As the pollen tube passes through the micropyle, it discharges two sperm cells which have been produced by the mitotic division of the generative nucleus. **At this time return to the slide of the germinating pollen grains to observe their progress.** One of the sperm cells fuses with the egg cell to produce the new zygote (2N) which represents the first cell of the next sporophyte generation. The other sperm cell enters the central cell and fuses with the two polar nuclei to form the primary endosperm nucleus. The primary endosperm nucleus is therefore a (haploid, diploid, or triploid ??) nucleus.

The process just described is often referred to as a **double fertilization.** The endosperm nucleus, along with the central cell, will develop into a stored food reserve for the embryo

Activity 6

Fruits and Seeds

A fruit develops from the ovary wall and its function is two fold: protection of the seed within and dispersal of the seed away from its parent plant.

A seed is a mature ovule with its outer limits being the seed coat. The seed coat develops from the ovule's integuments. Within the seed coat will be an embryo which grew from the zygote and a stored supply of energy which formed from the developing endosperm.

A. Obtain a bean pod and carefully split it open along one seam. The pod is a matured ovary. The pod is therefore more appropriately called a __________.

1. Find the sepals at one of the pod and the shriveled *style* at the other end of the pod. The beans inside the pod are more appropriately called ______________.

2. Note the point of attachment of the bean to the pod. What is the point of attachment of the ovule to the ovary called? _____________

B. Remove a bean or obtain a bean that has been soaked overnight to soften it.

1. Find the scar where the seed was attached to the fruit wall. Near the scar, look for the tiny opening left in the seed coat. What is this tiny opening called? The pollen tube grew through it.

2. Remove the seed coat to expose the two large cotyledons. Split (open) the two cotyledons to expose the plant embryo. The two large halves of the bean seed are the **cotyledons.** The cotyledons have absorbed the carbohydrates produced by the endosperm. Thus the bean cotyledons are very fleshy, storing carbohydrates which will be used during seed germination. In other plants (monocots), the endosperm is not absorbed by the cotyledons and remains a separate part of the seed (i.e. wheat germ is the endosperm of the wheat seed).

3. Observe the embryo. Locate the attachment point between the embryo and the cotyledons. The epicotyl is that part of the embryo which is above the attachment and the hypocotyl is the lower portion. Most of the shoot develops from the epicotyl while the root grows from the hypocotyl. Note the miniature leaves (plumule) attached to the tip of the epicotyl and the root bud (radicle) at the end of the hypocotyl.

ANIMAL TISSUES TWO WEEK ACTIVITY

15

Laboratory Objectives

1. To introduce the student to the concept of multicellularity and its advantages over the cellular level of organization.

2. To become familiar with the characteristics of the four categories of animal tissues.

3. To produce detailed sketches of the different animal tissue types described below.

Introduction

We have discussed in lecture that the cell is the basic unit of life. It is the simplest unit of organization which possesses all of the properties associated with life.

The Protistan kingdom, as previously studied, is composed solely of unicellular organisms. With a very successful organizational plan, protistans have evolved to fill many different niches in the environment. Some are producers, others are heterotrophs, while still other protistans have taken up a parasitic existence. All protistans have, however, one major limitation, that of *size*. A single cell cannot continue to grow in size and remain effective at performing the basic activities central to maintaining life. A large undivided mass of protoplasm becomes both physiologically and structurally ineffective. A large cell lacks mechanical strength if flattened or elongated and also lacks the surface area needed for exchange of materials with its environment. Recall that the nucleus is the "headquarters" of cellular activity. By producing different pieces of mRNA, the nucleus directs the ribosomes of the cell to produce different classes of proteins. The mRNA heads out to the protein factories through the relatively slow and inefficient process of diffusion. Large cells cannot effectively communicate with ribosomes found at great distances from the nucleus.

Even saddled with all of the problems mentioned above, bigger is sometimes better. A larger animal can obtain food more easily by using its larger mass to capture smaller organisms. A plant which possesses a larger leaf surface area possesses more chlorophyll allowing it to manufacture more sugar. How can growth occur if cell size is limited by the inherent problems described above?

Evolution of multicellularity allows growth in size while minimizing the problems mentioned above. A mass of smaller cells also allows for specializations to evolve. Cells may become modified to become masters of movement, coordination, protection, or support. In the vertebrates (animals with backbones) four different tissue types can be identified that have evolved to perform very specific functions. Essentially there has been a division of labor as these tissue types have evolved. **Epithelial** tissues cover, **muscle** tissues move, **nervous** tissues coordinate, and **connective** tissues support the bodies of animals made up of many cells.

In this laboratory exercise, you will study each of the four basic types of vertebrate tissues mentioned above. You will be introduced to each tissue's basic characteristics, and then you will find and sketch different specialized examples of each tissue type.

Instructor Expectations

1. You are expected to produce large and detailed sketches of the following animal tissues on separate pieces of paper. Do your sketches in pencil. You do not need to be good artists to produce accurate sketches, but you do need to take time and care in their production. You may use your textbook as a resource, but do not copy their illustrations. Provide labels for the specific structures **(bold faced type)** described in the following exercises. Your sketches will serve as your evaluation of this material, therefore your grade is in your own hands. Ask your lab instructor or fellow students for help when you are confused. If you do not have enough time to complete your work, come into lab at other times to finish your assignments. Be neat and organize your sketches following the order of presentation in the lab exercises.

2. At the end of each work session, the slides will be put back in the proper order listed on the top of the slide box. Microscopes will be returned to their proper shelf in the storage cabinet.

Activity 1
Epithelial Tissues

Epithelial tissues cover surfaces of multicellular animals. The surfaces may be external as well as internal surfaces. The structure of the epithelium reflects the location of where it is found. External surfaces come into contact with harsh environmental conditions. The multilayered structure of external epithelial surfaces reflects this situation. Internal surfaces of an organism (ie the lining of the intestine or the lungs) are protected from harsh outside forces. Their protected location allows for the exchange of materials between the animal and its environment. A single layer of epithelial cells cover these exchange surfaces.

Even though there are many different types, epithelial tissues share some common features.

1. Epithelial tissue is usually found on a surface of an animal. Remember that animals possess both external as well as internal surfaces. Your stomach, lungs, bladder, and blood vessels (to name a few) possess internal surfaces.

2. Since they cover body surfaces and regulate (to some extent) what may enter and leave the body, epithelial cells are packed closely together with little intercellular material between them. Material entering or leaving an animal's body generally has to pass through an epithelial cell rather than between two cells.

3. Being found on a surface, epithelial tissues possess a free border or free surface that is generally bathed by either air or fluid. Using the United States as an example, the Atlantic and Pacific coasts would be examples of free borders. You don't have to go through Customs to get into the Pacific Ocean.

4. Epithelial tissue is characterized by the presence of uniform sheets of similar cells.

5. Epithelial tissues do not possess blood vessels. Exchange of needed supplies occurs by diffusion from blood vessels found in underlying connective tissues.

6. A basement membrane tends to anchor the epithelial tissues to the underlying connective tissue. If you were to apply tape to the top of a table, the basement membrane would be analogous to the adhesive found on one side of the tape.

Classification of Epithelial Surfaces

Epithelial tissues can be structurally classified based upon the number of cell layers found between the free surface and the basement membrane of the tissue, as well as by the shape of the outermost cell layer.

Epithelial tissues may be either **simple, stratified,** or **pseudostratified** in structure. A simple epithelium is only one cell layer thick. A stratified epithelium is composed of multiple layers of cells, while a pseudostratified epithelium (pseudo = falsely) looks as if it is composed of many cell layers, but in reality there is only one.

The **outermost** cells of the epithelial covering may possess three distinct shapes. Some epithelial surfaces possess **squamous** cells (flattened). A frying pan filled with cooking eggs would resemble a field of squamous cells. The yolks would be the nuclei with the egg whites representing the cell cytoplasm. Other epithelial surfaces possess **cuboidal** cells (box-like), while still others may possess **columnar** cells (much taller than they are wide). As a generalization, the taller an epithelial cell is, the more work the cell does in moving materials into or out of the body.

By combining these two sets of terms, almost all of the epithelial types that we will observe today can be named. What should a simple cuboidal epithelium look like? Sketch your answer below.

Procedure

Find and sketch the epithelial tissues named below. Label the **free surface,** the **basement membrane,** and **magnification** used for each sketch. Remember as a general rule that epithelium is found on surfaces. Also beware of first impressions. You are going to be observing animal organs in this exercise which are made up of **different** tissue types. Therefore, what you are sketching may not be what you are supposed to be sketching. Check with neighboring students and your instructor to see if you are on the right track before investing a lot of time sketching the wrong material.

1. **Stratified squamous epithelium,** as its name implies, is many cell layers thick possessing flattened cells on its outmost free border. This type of epithelium covers surfaces of the body that are exposed to frictional forces eroding their surface (surfaces that need to be protected from excess wear). You can find stratified squamous epithelium covering the surface of your skin, or lining the esophagus (food tube connecting the stomach to the mouth), vagina, or anal canal.

   The slide that you are observing is probably from an esophagus. Under low power, observe its general structure. There is abundant muscle tissue needed to propel food toward the stomach. Find the internal convoluted surface of this organ which is covered by the tightly packed stratified squamous epithelial cells. Move to high power and carefully sketch a patch of these cells. Near the basement membrane the cells are cuboidal in shape, but they flatten into squamous cells as you progress toward the free surface. In some preparations the cell membranes are difficult to see, but the cells' nuclei are clearly visible.

2. **Simple columnar epithelium,** as its name implies, is composed of a single layer of very tall and slender cells. It is found on surfaces of the body where active absorption and secretion occur. Simple columnar epithelium can be found lining the walls of the intestine or kidney tubules. Even though the locations are different, simple columnar epithelium looks the same in both locations. Make sure that you see examples of simple columnar epithelium from both locations.

   If you have an intestinal slide, you will see a tube cut in cross section. The inner surface of the tube will have finger-like projections pointing toward the lumen of the intestine. The villi (finger-like projections) increase the available surface area for the absorption of the products of digestion. Focus on the fringe of a villus to find the simple columnar epithelium. If your slide is very good, you might notice that the apical surface of the simple columnar epithelium (toward the lumen) appears "fuzzy." Sometimes this surface is referred to as the **brush border** (named for its appearance). This "fuzziness" is produced by convolutions of the cell membrane into many smaller fingers (microvilli) which further increase the surface area of this tissue.

   Your slide illustrating this tissue type might be from the kidney. The kidney is composed of millions of smaller functional units called "nephrons." A nephron is a long tube that possesses different modifications along its length necessary for the production of urine. Urine is composed of a filtrate of blood that must be concentrated and modified as it passes along the nephron. Since these processes involve transport across an epithelial surface, the entire length of a kidney nephron is composed of simple epithelium. Unlike other slides where you will looking for an obvious free border, the tubules' free border is the **inside** of the tube. The portion of the nephron that you are looking for is deep inside of the kidney. The tubule will be cut longitudinally running across the field of view. Look for rows of very tall columnar cells. In a functioning kidney, there would be forming urine within the tubule.

3. **Simple cuboidal epithelium** is also found within the kidney. Read the paragraph above (if you haven't already done so) describing the organization of a kidney. On the slide illustrating simple cuboidal epithelium, search (using the low power objective lens) the slide looking for "ring-like" cross sections of the kidney tubules. Focus on a "ring-like" cluster of cells to see an example of simple cuboidal epithelium. Imagine a box of donuts that are tightly packed together. The walls of the donuts are composed of simple cuboidal cells with the hole of the donut representing the free border of this epithelial tissue.

4. **Transitional epithelium** lines the bladder which collects urine after it has been produced in the kidney. By now you are probably wondering what is the fascination with urine? The name "transitional epithelium" seems to violate the system of nomenclature that we have been using so far. The name does reflect the fact that the appearance of this tissue changes with the state of the organ which it lines, the bladder. The bladder is a very distensible organ that can be very full at times and relatively empty at other times. A moderately full bladder can hold 500 ml of urine. In this state the walls of the bladder would be tightly stretched. The extent of filling influences the appearance of the epithelium lining the bladder. The slide that you are observing is of an empty, relaxed bladder.

   Observe the wall of the bladder. Most of it is composed of smooth muscle tissue. Locate the inside convoluted surface made up of transitional epithelium. It is a stratified tissue with its surface cells cuboidal to columnar in shape. In its relaxed state, it would be called "stratified cuboidal epithelium" if it did not change in appearance as the bladder fills.

5. **Pseudostratified ciliated columnar epithelium with goblet cells** is a very specialized epithelial tissue that is found lining the trachea and respiratory passages leading into the lungs. As its name implies, the epithelium looks like it is multilayered because the nuclei present are at different levels of the epithelium. All of the cells do come in contact with the basement membrane, however, so it is in reality a simple epithelium. The majority of the cells in the lining are tall and columnar in shape possessing cilia at their apical end. The cilia look like hairs extending toward the lumen of the trachea. Intermingled between the columnar cells are

large "goblet-shaped" cells filled with a lighter staining cytoplasm. These Goblet cells produce the mucous which lines the respiratory passageway.

The entire epithelium works very much like the "fly-strip" that you might have seen at fruit stands. The mucous produced by the Goblet cells is very sticky. Inhaled dust particles and other debris can become stuck (like the flies) in this mucous trap. It is the job of the numerous cilia to sweep the mucous out of the respiratory passages toward the oral cavity. When the dust laden mucous reaches the mouth it can be eliminated by swallowing.

6. **Simple squamous (frog skin).** This slide of simple squamous epithelium is a different type of preparation than the others you have seen. You would expect this epithelium to be very thin composed of a single layer of flattened cells. Indeed, it is constructed this way, but the slide that you are observing has been prepared by laying the frog skin directly onto the slide without cutting it. You are looking down on the epithelium from its free border, rather than looking at it in cross section. It is analogous to looking down onto a floor which is covered by tiles. Through experience you know that the tiles are thin and that there is only one layer of tile between you and the sub-floor. However, you cannot see this arrangement. The same is true of the frog skin. The label says that you are looking at simple squamous epithelium, but from this preparation there is no way to know that by just looking.

   From this surface view, you can appreciate how the cells of epithelium fit together like the pieces of a jigsaw puzzle. On the surface of the frog skin, you might find some small holes or pores which drain mucous glands located below the frog skin. These secretions help to keep the frog skin moist as it serves as an important respiratory surface. Frog lungs are not that well developed, and a substantial amount of gas exchange occurs across its skin.

Activity 2
Nervous Tissue

Nervous tissue is highly specialized to carry information from one part of an animal's body to another. This communication allows different cells of the multicellular animal to coordinate their activities to accomplish some common task. Nerve tissue is said to be "irritable." In other words, nerve tissue is able to generate an electrochemical signal in response to some environmental stimulus.

Nerve tissue is made up of two different classes of cells. The **neuron** is the "workhorse" of the nerve tissue. A hypothetical generalized neuron possesses a nerve cell body which houses the cell's nucleus and much of its metabolic machinery. **Dendrites,** relatively short and branched extensions of the nerve cell body, carry nervous impulses toward the neuron. The **axon** is a longer and less highly branched extension of the neuron that carries a nervous impulse away from the cell body. Some of the longer axons may reach up to a meter in length. By stringing nerve cells in series (one after another), nervous signals can be transmitted from one part of the body to another.

The other type of cell found in nerve tissue is the **glia** cell. This cell type plays a supporting role to the neuron in the functioning of nerve tissue. Glial cells help to organize, insulate, and take care of some of the metabolic needs of the highly specialized nerve cells. The terminal portion of an axon may lie one meter away from its own nerve cell body.

Procedure

Observe the slide labelled "Spinal Cord-Smear" under low power. Notice the many fiber-like nerve cell extensions that carry impulses through this tissue. Also notice the numerous nuclei of the glial cells. In a spinal cord, there are 50 times more glial cells than there are actual neurons.

Scanning across the slide look for a "typical" neuron. To sketch a neuron, you may have to make a composite drawing after observing several less than desirable candidates. This large neuron should possess a box-like **cell body** with cytoplasmic extension leading away from it. Ideally, one or two unbranched extensions should be seen representing the **axon**(s) of the neuron. The shorter, more highly branched, extensions would be the **dendrites** of the neuron. A **nucleus** should be visible in the cell body of the neuron. Include in your sketch the numerous **glial cells** found in the vicinity of the neuron.

Activity 3
The Connective Tissues

Of the four types of animal tissue, the connective tissues appear to be the most heterogeneous in structure. For example, both bone and blood belong to the connective tissue group. Even with this lack of apparent similarity, connective tissues do share some common features.

In general, the connective tissues are characterized by abundant, intercellular material usually referred to as **ground substance** or **matrix.** Coursing through the ground substance are **fibers** which provide structural support to the connective tissues. Scattered among the fibers and ground substance are **cells** of different types and functions. To make a model of a general connective tissue, you would place peas and spaghetti into a bowl of Jello. The pasta would represent the fibers of the connective tissue, the cells would be portrayed by the peas, and the Jello would be representative of the ground substance or matrix.

Connective tissues make up the framework for the organs of the body. Connective tissue is found throughout the body and serves primarily to support the body and to bind its parts together. It also provides a framework for movement, and acts to store minerals.

Differences in consistency of the ground substance accounts for much of the wide variation seen in the connective tissues. Ground substance may be liquid as seen in blood, gel-like as it appears in cartilage, or solid as in bone. The basic properties of the connective tissues are influenced by the types of and relative proportions of fibers found in the ground substance. **Collagenous** fibers are very strong, and do not stretch very much. Under the microscope, collagen fibers will be seen to be relatively straight and dense in construction. As form and function go hand in hand, collagen fibers are best seen in tendons which join muscle to bone. **Elastic** fibers have much less strength than do collagen fibers, but they are much more flexible. More elastic fibers are found in ligaments (connecting bone to bone) than are collagen fibers. It is important that joints between bones exhibit a certain degree of flexibility. Elastic fibers are also present in large numbers in the walls of large arteries. These vessels expand in diameter greatly when the heart contracts forcing blood through them. Microscopically elastic fibers are long, thin, wire-like threads which may appear to look like a coiled thread. **Reticular fibers** are the third type of fiber common in connective tissues. These relatively thin, short branching threads form the framework of many of the body's organs. If all the cells of the body were removed leaving only the reticular fibers, an outline of each organ would still be present. Special staining techniques are required see reticular fibers, so we won't see them in today's laboratory.

The third component of connective tissues are the widely scattered cells. Though quite numerous and varied, these connective tissue cells are basically of three predominant types: the **fibroblasts,** the **macrophages,** and the **mast cells.** As their name implies, fibroblasts produce the connective tissue fibers as well as the ground substance. Macrophages have small, uneven dark nuclei and their cytoplasm often contains ingested matter. The primary function of this cell type is to engulf, by phagocytosis, foreign matter and cellular debris found in the connective tissue proper. Mast cells are the third major type of cell found in connective tissue. This cell type produces histamine in response to a localized injury or infection. Histamine causes changes in the circulatory supply to the injured area limiting or preventing the spread of the infecting agent. These cells are circular and show numerous cytoplasmic granules.

Procedure

For each of the following slides, draw a sketch and label the indicated structures appearing in bold print.

1. **Areolar connective tissue** is also called loose irregular connective tissue. It most closely resembles the "model" of connective tissue mentioned above. Its second name relates to the organization of fibers coursing through its ground tissue. The fibers are not tightly packed together (loose) and they run in all directions (irregular). Areolar connective tissue is found throughout the body, but is most easily seen attaching skin to the underlying muscle tissue. As you skin an animal, you have to tear away at the areolar connective tissue connecting the hide of the animal to its underlying muscle.

   Focus on a patch of areolar connective tissue using the 40X objective lens. **Cells** will be seen to be relatively widely scattered. It is very difficult to distinguish the three main categories of cells in this preparation. Two distinct fiber types can be seen. The narrower and darker **elastic fibers** can be seen to form coils along some of their length. The **collagen fibers** are less distinct, wider, and more lightly staining than the elastic fibers. The spaces between the fibers and the cells contains the **ground substance** of the areolar tissue.

2. **White Fibrous Connective Tissue** was obtained from a tendon. Tendons connect a muscle to a bone. The high concentration of **collagen fibers** give the tendon its great strength. The fibers may have been teased apart to illustrate their lack of branching. This connective tissue can be classified as a dense regular connective tissue. Its fibers are tightly packed together and they run in the same direction. The nuclei of the fibroblasts can be seen closely associated with the collagen fibers.

3. **Compact Bone** is one of the most visually distinctive tissue types that we will look at. The first impression under low power is that bone is full of holes. The larger black holes are called **Haversian Canals.** These are the centers of subunits of bone called **Haversian Units.** Scan your bone slide to find a Haversian Unit that looks like it has complete concentric rings of cells around the central Haversian Canal. Move to high power and sketch with labels a complete Haversian unit.

   The **Haversian Canal** is at the center of the Haversian Unit. It contains nerves and blood vessels which nourish this living tissue. Bone, being a living tissue, is capable of growth and repair. Concentric rings of hard calcified ground substance encircle the Haversian Canal. These rings are called **lamellae.** The fibroblasts of the bone **(osteoblasts)** are located within the lamellae enclosed in small cavities called **lacunae.** If an osteoblast were plucked out of the solid ground tissue of the bone, a cavity (lacunae) would remain. (If water were removed from a lake, a depression would be left in the ground.) Adjacent lacunae are connected by small tiny cracks called **canaliculi.** The canaliculi allow nutrients to pass between the entrapped bone cells.

4. **Hyaline Cartilage** can be found covering the ends of long bones where they meet other bones at joints. The hyaline cartilage reduces the friction of movement at the body's joints. Cartilage is a nonvascular tissue. No blood vessels flow through nor directly supply cartilage with oxygen or nutrients. This is one reason why damaged cartilage (in the rim of the ear for example) is slow to heal after being traumatized. The **chondroblasts** (fibroblasts) of cartilage are located in **lacunae** scattered throughout the glassy **ground substance.** Notice that the chondroblasts seem to occur in pairs. These pairs represent the daughter cells of a mitotic division. As the fibroblasts become active secreting fibers and ground substance, the daughter cells will grow apart. Also notice the apparent lack of fibers. Fibers are actually present, but are translucent being hidden in the ground substance. (It is difficult to see a glass rod placed in water for the same reason.)

5. **Blood** is an interesting example of a specialized type of connective tissue in that the **matrix or ground substance** is liquid. The **fibers** of blood are in solution with the liquid plasma (ground substance) until the blood clots. When blood clots, the previously soluble fibers form a net which stems the flow of blood from the dam-

aged vessel. The cells of the blood can be divided into three classes: **red cells, white cells,** and **platelets.** The red blood cells are the most numerous. They are shaped like biconcave discs which accounts for the lighter central portion of the cell. They also lack nuclei making more room for the oxygen carrying hemoglobin molecule. The much less numerous white blood cells are larger than the RBC's, and are stained a darker purple color. Some of the white blood cells can be seen to have multi-segmented nuclei while other types of white blood cells do not. The third formed cell element of the blood are platelets. Platelets are less obvious than the first two cellular components as they are much smaller in size. The platelets, appearing like sand grains scattered throughout the plasma, are very important in initiating the process of blood clotting.

6. **Adipose tissue** can be found in many locations of the body. The slide that you are looking at is from the subcutaneous region, but you can also find fat around the heart, kidneys, and other internal organs. Adipose tissue looks a bit like the honey comb of a bee. Unlike other connective tissues, the cells of adipose tissue are relatively tightly packed together. The cytoplasm and nucleus of each cell is pushed to the periphery as the "stored fat" occupies most of the cell's volume. In the process of fixing and staining this tissue, the fat itself was dissolved away leaving the void in the center of each cell.

   Fat molecules, gram for gram, contain twice as many calories as the other food groups. This makes adipose tissue a very efficient storehouse of reserve energy supplies. Can you think of other functions adipose tissue may serve.

Activity 4

Muscle Tissues

Muscle tissue is the contractile tissue of the body. Although the cells are generally elongated, they are capable of shortening, and this shortening gives the tissue its ability to do mechanical work. It should be noted here that muscle cells cannot lengthen on their own. They must be pulled back to their original pre-contraction length by opposing forces which are often generated by antagonistic muscle groupings. There are three types of muscle tissue found in an animal's body each with its own sets of characteristics.

Procedure

For each of the following slides, draw a sketch and label the indicated structures appearing in bold print.

1. **Striated muscle cells** are attached to and move the bones of the skeleton. For this reason, striated muscle is sometimes called **skeletal** muscle. Movement for the most part is initiated by voluntary nervous signals emanating from the central nervous system. The action of these muscles is controllable through nervous impulses. For this reason, striated, skeletal muscle can also be called **voluntary** muscle.

   Make sure that you are observing a longitudinal section of the voluntary striated muscle. The fibers should be seen running across the slide. If you have a cross section (not desirable), the slide would look like a piece of "meat." Striated muscle is easily recognized by its distinct **striations** (bars) running across the muscle cells. If your muscle slide does not clearly show the striations, check the slide of a neighboring student. Often striated muscle cells are referred to as muscle fibers as they are generally giant, cylindrical, multinucleated cells which range in size between 3 mm and 7.5 cm in length. The long muscle fibers are arranged in a parallel fashion. At higher power, the alternating light and dark bands characteristic of this tissue can be observed. Label the peripherally located **nuclei** in the muscle cell.

2. **Smooth muscle** is not as distinctive looking as is skeletal muscle. It surrounds hollow body organs allowing the organs to move and to maintain their shape. For this reason, smooth muscle is also referred to as **visceral**

muscle. Visceral muscle cannot be controlled consciously. Our stomachs move about to mix the stomach's contents with digestive enzymes. We cannot consciously control when these muscles are going to contract or with how much vigor. For this reason, visceral muscles are also called **involuntary muscles.** Involuntary muscles are stimulated to contract by signals from the involuntary nervous system, hormones, or from being stretched by outside forces. The smooth muscle making up the walls of the bladder become stretched as this organ fills with urine. When the stretching reaches a certain threshold, the muscle contracts spontaneously.

Observe the slide of smooth muscle with your naked eye. The slide will most likely show a hollow organ such as an intestine. The majority of the wall of the organ will be composed of visceral muscle. Place the slide on the microscope and examine it under low power. There will be two layers of visceral muscle surrounding the organ; an inner circular layer and an outer longitudinal layer running the length of the organ. What do you think is the functional importance of this dual muscle layer.

Focus on the circular muscle layer under high power. These muscle cells are not as long as voluntary muscle fibers being only 40-100 microns in length. The smooth muscle cells lack the striations or bars that were characteristic of the striated muscle cells. These cells also possess a single central **nucleus** with the cytoplasm tapering toward both ends of the cell. These cells are often referred to as being spindle shaped.

3. **Cardiac muscle** is found in the heart. It seems to be a hybrid between the two muscle types already looked at. Cardiac muscle possesses the striations seen in striated muscle, but the smaller cells also possess one nucleus like the smooth visceral muscle cells. Cardiac muscle contracts involuntarily, but its contractions are self-initiated.

   Find a **darkly stained** slide of cardiac muscle. Focus on an area where the fibers are running longitudinally. Each cardiac muscle cell branches at either end to intermesh with adjacent cardiac muscle cells. Their ends interlock as the fingers of your two hands can be made to intermesh. These areas of overlap show up as the **intercalated discs** of this tissue. Cardiac muscle cells possess a single centrally located nucleus. You will observe a lot of blood vessels in this tissue as it is very vascular.

Name ______________________________

Section ______________________________

Post-lab Questions

1. What advantages do multicellular organisms have over single celled organisms?

2. What are the four basic types of animal tissues?

3. What are four characteristics of epithelial tissues.

4. How does a "simple" epithelial tissue differ from a "stratified" epithelial tissue in structure? In function?

5. In what ways do axons differ from dendrites?

6. What are the major components of a typical connective tissue?

7. What are the three fiber types found in connective tissue and how do they differ?

8. What similarities do striated muscle fibers and smooth muscle cells share? How do they differ?

9. What are three functions of adipose tissue?

10. What properties do tendons and blood share in common? How do they differ?

11. Why must a striated muscle cell be a multinnucleated cell?

12. What is the significance of the anatomical arrangement of the circular and longitudinal smooth muscle layers found around hollow body organs?

AN INTRODUCTION TO PORIFERA AND COELENTERATA

16

Objectives

1. To observe and sketch the Poriferan body plan.
2. To observe and sketch the Colenterate body plan.
3. To investigate how an animal's form is related to its way of life.
4. To compare and contrast the life histories of these two phyla.

Introduction

Evolutionary schemes trying to illustrate ancestral relationships are based upon a number of different criteria. Some of the more traditional taxonomic schemes are based upon common embryological stages, anatomical similarities, fossil records, vestigial structures and the like. With advances in molecular biology and molecular genetics, new tools are available for constructing evolutionary trees. Sometimes the newer methodologies corroborate the traditional views of evolutionary relationships, and at other times newer molecular evidence challenges some of the long held traditional ideas of who is related to whom. For the sake of the following several laboratory exercises, the traditional portrayal of the animal's evolutionary tree as seen in the figure below will be followed. The tree is based upon the presence or absence of traits/characteristics listed in Table 1. These adaptations are considered to be major milestones which opened up new possibilities in the phylogeny of animals. Today's laboratory concerns the first few phyla which are found on the "main trunk" of the tree. It is not implied that these phyla are more primitive than other existing animal phyla nor that they are in any way less successful. However these simpler phyla more closely resemble animal's early ancestors than some of the more complex phyla to be studied later in the semester.

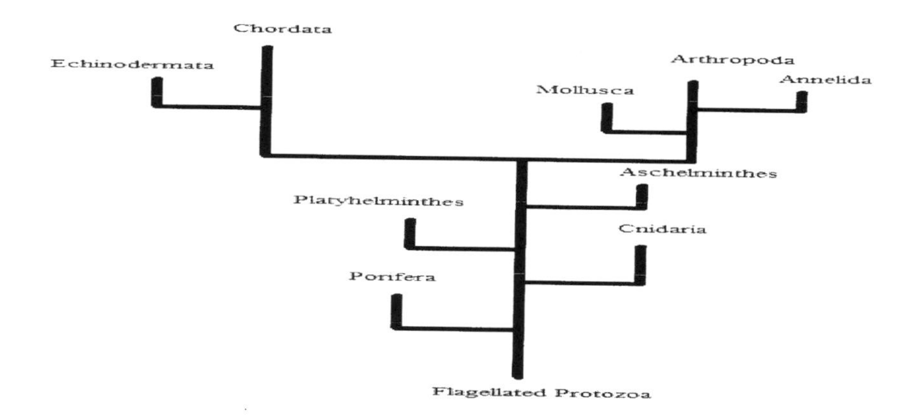

Table 1. Primitive versus More Advanced Traits

Figure 1. Typical Portrayal of the Evolutionary Tree of Animals.

Table 1. Primitive versus More Advanced Traits

	Most primitive	Primitive	Advanced	Most Advanced
Body plan	None	Sac plan	Tube within a tube	Specialized regions of tube within a tube
Symmetry	Asymmetric	Radial	Bilateral	Bilateral with cephalization
Germ layers	None	Diploblastic	Triploblastic	Triploblastic
Level of organization	Cellular	Tissues	Organs	Organ systems
Coelom	Acoelomate	Acoelomate	Pseudocoelom	True coelom
Segmentation	Nonsegmented	Nonsegmented	Segmented	Segmented with specialization of parts

The above evolutionary tree is based upon the criteria listed in Table 1. As we observe the different specimen in the following exercises, we will see how each of the animals fits into the above scheme.

Body plan is often related to the mechanism that the animals use to obtain food. A "sac" body plan involves having a central cavity with one opening that serves as both an entrance and an exit. The "tube within a tube" plan refers to a complete digestive tract with both a mouth and an anus. A complete digestive tract allows different regions of the tract to become specialized to take on different functions.

As mentioned before, animals that exhibit radial symmetry resemble a soup can. Any longitudinal cut passing through the center of the animal will produce mirror images. An animal with radial symmetry cannot be surprised by someone sneaking up from its behind as the animal has no "back". The radially symmetrical animal is able to deal with environmental stimuli reaching it from any direction. An animal with bilateral symmetry not only possesses a dorsal (top) and ventral (bottom) surface, but also a right and a left hand side with an anterior head and posterior tail. The bilaterally symmetrical animal can only be divided into mirror images with a single cut passing down its midline. Bilaterally symmetrical animals tend to have their sense organs concentrated at their anterior end (cephalization) which allows them to move forward in their lives increasing their mobility.

Germ layers appear during an animal's embryological development. As cells move to specific regions of the embryo, their developmental fate becomes fixed. Most animals are triploblastic meaning that they possess three germ layers. The ectoderm cell layer is found covering the outside of the embryo and will develop into the animal's skin and nervous system. The animal's endoderm lines its primitive gut and will develop into the lining of the animal's digestive tract. Its mesoderm, or cells found between the previous two germ layers, develops into the animal's various internal organs.

Some of the simpler animals are diplobastic meaning that they possess only two germ layers, an ectoderm and an endoderm. These animals do not develop organs and instead have only a tissue level of organization. Some of the simplest animals in the world even lack the tissue level of organization and only possess specialized cells in their structure.

A **coelom** is a body cavity that houses an animal's internal organs. This arrangement provides a number of benefits including freedom of movement for an animal's internal organs. The coelom also offers the possibility of a hydrostatic skeleton if the animal lacks an internal or external skeleton. An **acoelomate** animal lacks a body cavity and is one solid mass of tissue. An animal which possess a **pseudocoelom** possesses a body cavity that is not completely lined by mesoderm. An animal with a true coelom has its body cavity completely lined by mesoderm.

When an animal is segmented, it has repeating body parts as you might see on an earthworm. The possession of these segments allows for specialization of body parts resulting in an increasing complexity of body structure.

Activity 1

Phylum Porifera: Sponges

Early in the study of life, sponges did not seem to fit clearly into the plant or the animal world. Sponges are very sessile and some forms possess symbiotic algae within their body that give them a "greenish" tint. Therefore sponges were considered to be plant-like. However, most sponges lack symbiotic algae and possess a greater sensitivity to external stimulation than typically exhibited by plants. Therefore sponges appear to be more animal-like. Porifera are now clearly recognized as being animals at their simplest level of organization.

Sponges evolved from protistan-like ancestors that took up a colonial existence. They diverged early from the main line of animal evolution as shown in the above animal family tree. All forms of sponges are aquatic with most of the sponges found in the marine environment. As adults, sponges are sedentary or sessile, but they may disperse from their point of anchorage by employing swimming larval forms. Sponges do not show a true level of tissue specialization. They do have cell types that are specialized for specific functions, but the cell layers do not form true tissues. Sponges therefore show a cellular-specialization level of organization.

Some sponges show a crude form of radial symmetry, but for the most part sponges are asymmetric.

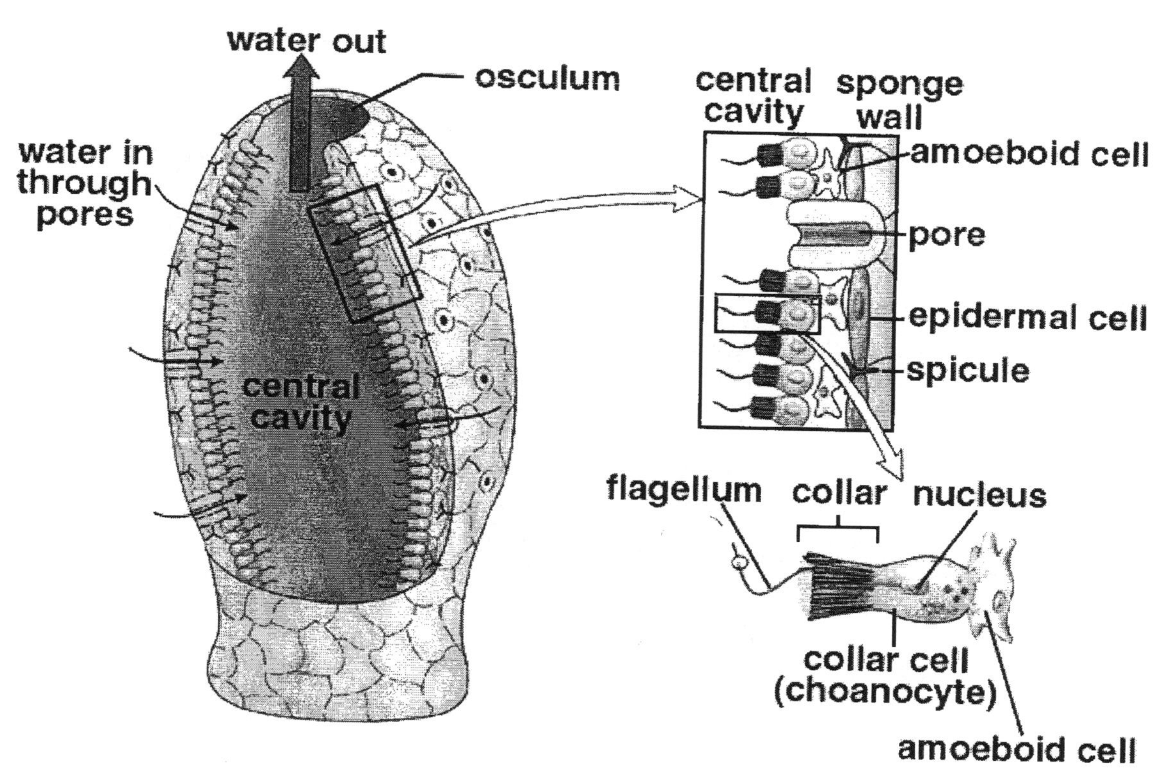

At its simplest level, a sponge can be thought of as being a vase-shaped organism with a body wall composed of two cell layers. Please see the figure below. The outer epidermal layer is composed of rather unspecialized flattened cells with the more complicated **choanocytes** (collar cells) forming the inner cell layer. The choanocytes possess long flagella which produce a gentle flow of water through the wall of the sponge. The wall of the sponge is perforated by small openings called **ostia** (pores). The beating of the inwardly directed flagella of the choanocytes produces a current of water that flows into the sponge's **spongocoel** through the ostia. The larger **osculum** of the sponge serves as the excurrent opening. The choanocytes are able to filter suspended organic materials from this stream of water that serve as the sponge's food source. The area between the two cell layers is occupied by wandering **amoebocytes** which can do a number of jobs. The amoebocytes may help to distribute the products of digestion accomplished by the choanocytes as well as become modified to produce egg and sperm cells for reproduction of the sponge.

Also notice in the figure of the sponge the presence of **spicules** that help to support the body wall of the sponge. The crude endoskeleton of the sponges can be made out of spicules composed of calcium carbonate crystals, silica crystals, or protein fibers called **spongin**. The spicules take on a variety of shapes dependent upon where in the sponge they are located. The spicules' mineral composition serves as the basis for separating the sponges into their various classes.

The asconoid body plan seen in the diagram above is an example of the simplest level of sponge organization. The more complex body plans are the syncoid and the leuconoid levels of body organization. A syncoid body plan is similar to the asconoid pattern except the body wall is everted to form incurrent canals between (and radial canals within) the resulting pockets. In a leucondoid sponge, the incurrent pores open into choanocyte lined sub-chambers that communicate with the spongocoel via excurrent canals. See the diagram below depicting these various body types. The amount of water passing through the sponge increases with the more complex sponge body types.

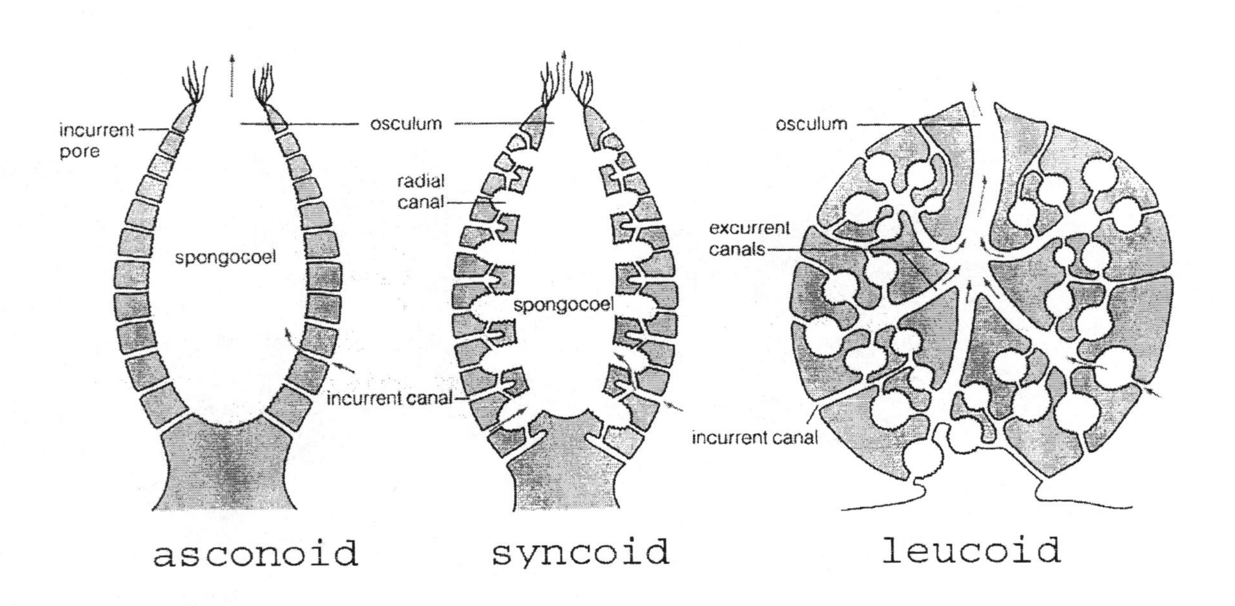

Procedure

We are going to investigate the structure of a sponge belonging to the genus *Scypha*. Members of this genus contain spicules composed of calcium carbonate. Members of this genus all have the syncoid body plan.

1. **External observation.** Remove an individual *Scypha* and examine the specimen under a dissecting microscope. Locate the **osculum** of the specimen. Recall that this is the excurrent opening from the animal's **spongocoel**. Notice the ring of spicules surrounding the osculum. You will also notice individual spicules sticking out of the sponge's body wall. Locate the numerous smaller **ostia** (pores) covering the outer wall of the sponge. Identify the **hold fast** which anchors the sponge to its substratum.

2. **Spicule demonstration.** You are going to make a slide of the spicules making up the endoskeleton of the *Scypha* specimen by digesting away the fleshy components of the sponge. Place a small piece of the *Scypha* specimen onto a microscope slide. Add a couple drops of bleach to cover the specimen. Stir the bleach and the specimen together allowing time for the bleach to digest the "meat" of the sponge. Place the microscope slide of the digested mass of *Scypha* under the 4X lens of the microscope. Focus on the edge of the mass in an area where the spicules are less dense. Make a sketch of what you observe in the space provided below.

3. **Microscopic view of the sponge wall.** Obtain a microscope slide labeled *Scypha*, c.s. (cross section) and l.s. (longitudinal section). Begin the microscopic observation using the cross section of the specimen. Since the *Scypha* specimen possesses the syncoid body arrangement, it will take some work to elucidate its structure. You will need to visualize finger-like outgrowths extending outward from the **spongocoel** of the sponge. The inside of each finger-like growth would be lined with **choanocytes** normally found on the inside of the spongocoel of a simpler asconoid sponge. These finger-like extensions of the spongocoel are referred to as **radial canals.** The passageways between adjacent fingers which communicate with the fluid found outside of the sponge are called the **incurrent canals.** As water enters an incurrent canal through an external pore, it can crossover into a radial canal by passing through a smaller pore-like opening. As the fluid passes into the spongocoel from the radial canal, organic detritus enters the choanocytes by phagocytosis for intracellular digestion. Water will flow out of the spongocoel via the **osculum**.

   Use the figure below to identify the following parts of the wall of the sponge. The radial canals are lined with choanocytes that appear as numerous small, dark bodies. Each cell bears a flagellum which projects into the lumen of the radial canal giving the border along the lumen of the radial canal a fuzzy appearance. Your microscope cannot resolve the separate individual flagella of the choanocytes. The incurrent canals are lined with squamous-like flattened cells which produce a much sharper border.

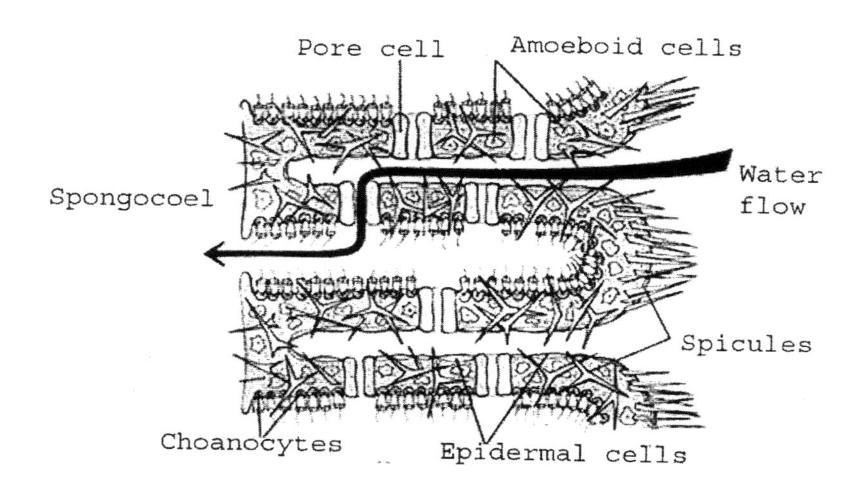

The architecture of the choanocytes (collar cells) cannot be appreciated under the light microscope (even using the oil immersion lens). Their structure is illustrated in the figure below. It is the beating of the cilia which creates the current of water that flows through the sponge. As the current flows past the collar of the choanocyte cells, the microvilli of the collar region acts as a strainer to collect organic detritus from the water flowing by. After the organic material has been filtered from the water, the choanocytes and perhaps other cells engulf the filtered material by phagocytosis and digest the material by intracellular digestion. Intracellular digestion is a trait or characteristic of less highly specialized animals. We will see in the course of the next few weeks that there is an evolutionary movement toward extracellular digestion allowing animals to consume larger prey items.

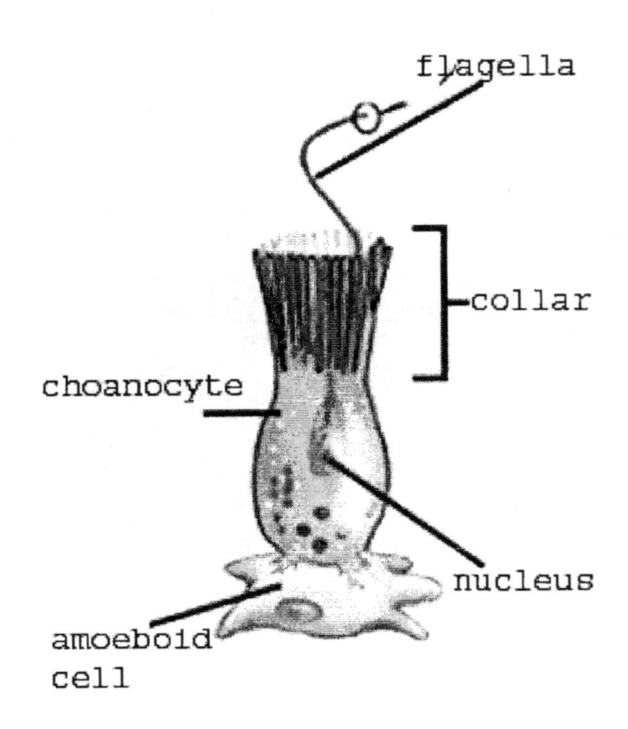

Flat protective cells cover the outer surface of this sponge and its incurrent canals forming its epidermis. The flat protective cells are also seen covering the spongocoel. Between the epidermis and the layer of choanocytes in the radial canals is a gelatinous layer sometimes called the **mesohyal**. The mesohyal is home to the spicules of the sponge's skeletal system as well as its **amoeboid cells.** The amoeboid cells have a number of functions. As the name indicates, these cells are able to move actively throughout the mesohyal. They may participate in the intracellular digestion of the filtered detritus as described above. The amoeboid cells may also become transformed to form the sponge's gametes. The amoeboid cells also carry products of digestion to other regions of the sponge.

Sponges possess sexual reproduction. Most species of sponge are **hermaphroditic**, but the male and female reproductive organs develop at different times in the life cycle of the sponge. This differential maturation of the reproductive structures minimizes the possibility of self-fertilization. The eggs and sperm are produced from amoeboid cells within the mesohyal. In some species it is also possible that sperm cells are produced from choanocytes. You will not see gametes on your slide, but it may be possible to observe some embryonic stages protruding into some of the radial canals. These young embryonic stages appear as relatively large and dark collections or masses of cells. When the larvae break loose, they find their way into the radial canals, move into the spongocoel, and then are released through the osculum.

Asexual reproduction also occurs in the sponges by the process of budding. A cluster of epidermal cells near the base of the sponge differentiate and begin to grow a "baby" sponge by mitosis. The growths would look a bit like a bud on the side of a tree stem. At some point the bud becomes large enough to break off the parent plant to survive on its own. Sometimes the buds remain attached to the base of the parent plant to form a cluster of related sponges. Budding will not be seen on the section of *Scypha* , but you may see some evidence of budding on the preserved specimen of *Scypha* or the whole mount slides of another genus of sponge (*Leucosolenia*).

Some species of sponge are able to reproduce by budding internally to form structures called gemmules. Obtain a whole mount slide of **gemmules**. Note that the slide is extremely thick and therefore very easy to crack under the high power objective lens. Be careful. Focus first under the low power and then move to the high dry lens. Around the periphery of the gemmule, observe the spicules sticking outward with their uniquely shaped cap. During unfavorable conditions, such as winter, the sponge disintegrates and releases the previously encased gemmules. Under more favorable conditions, cells emerge from the protective case of the gemmule to develop into a new individual. What might be the function of the gemmules outer cap of spicules? Consider its shape.

Questions

1. Would sponges be more successful if they were mobile? Why or why not?
2. What is the role of spicules in the sponge body plan? Do you think that the possession of mineral-based spicules is advantageous over the body plans where there is a lack of mineral based spicules?
3. Can sponges feed on large sized detritus floating in the water surrounding them? Why or why not?
4. Referring to Table 1, characterize the biology of the sponge following the criteria in Table 1.

Activity 2

Phylum Cnidaria (Colenterata)

Cnidarians represent the next level of complexity of animal structure as you move up the evolutionary tree of animals. A number of improvements over the sponge body plan are seen in this phylum. Increased cellular specializations have led to a tissue level of complexity. Cnidarians possess a nerve-net that allows the animal to respond to external stimuli in a coordinated pattern. Some of cnidarian's epithelial cells are able to contract forming a primitive type of "muscle" tissue. Cnidarians also possess a gastrovascular cavity and stinging cells which allow them to handle larger prey items than the sponge.

The totally aquatic and mostly marine Cnidarians are radially symmetrical. There are two types of body forms that exist in this phylum. The more sessile form is the **polyp** stage. The opening of the **gastrovascular cavity** is surrounded by the animal's **feeding tentacles** which are directed upward. The other end of the polyp is attached to the substratum. The other body type is called the **medusa** stage. The common jellyfish represents the medusa body form. The opening of the gastrovascular cavity with its surrounding tentacles is directed downward. The medusa form is mobile as its body wall is capable producing a bellows-like pulsatile contraction moving the animal from one point to another. Either or both forms may be found in any individual Cnidarian life cycle. Also common to Cnidarians is the presence of a swimming larval form known as the **planula** larva. What might be the advantage of having a larval stage that is mobile?

Cnidarians are efficient predators even though it seems strange that the sessile polyp stage would be capable of capturing prey items. Both life forms possess nematocysts (stinging cells) that enable them to capture prey items. Observe the diagram of a nematocyst below.

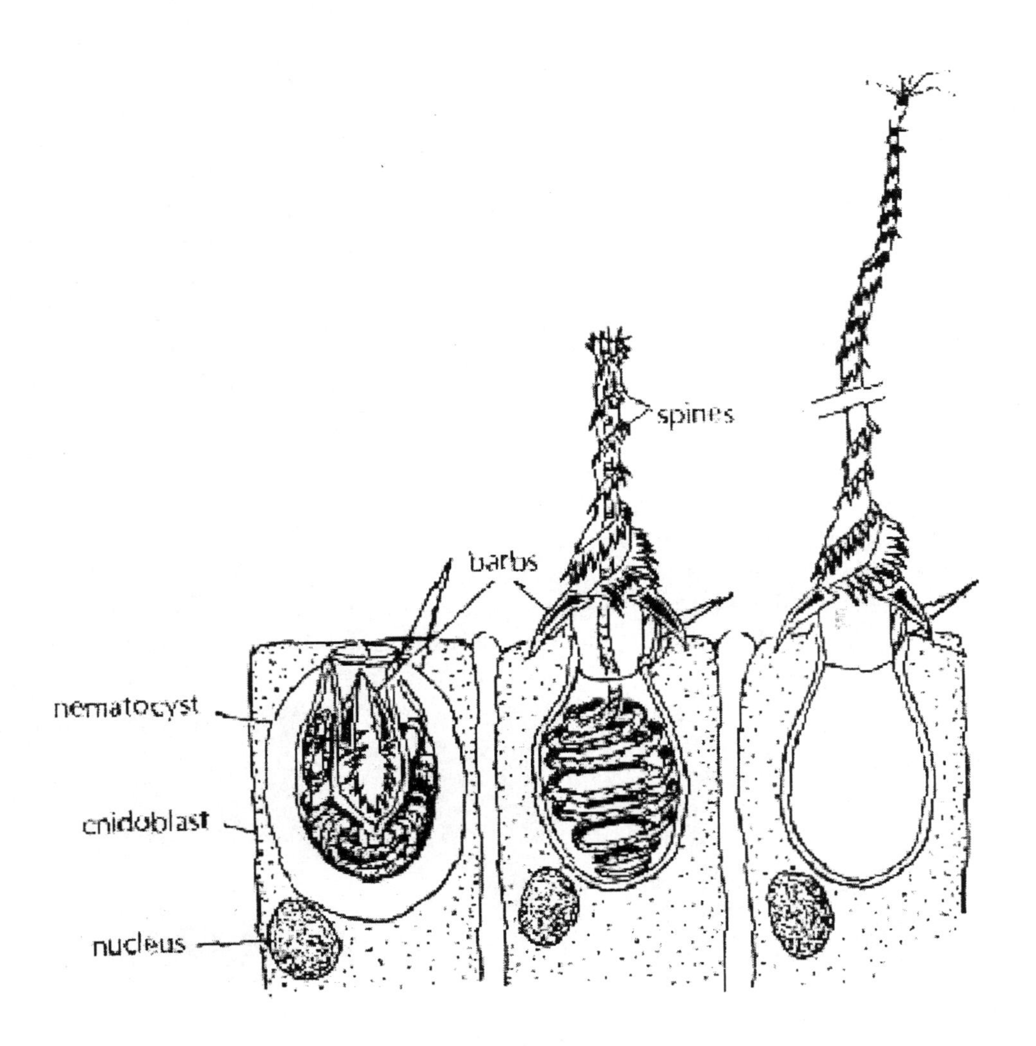

Nematocysts are found within cells called **cnidoblasts**. Cnidoblasts are found along the margins of the tentacles and also the general body surface. There are pressure sensors which detect prey items touching the surface of the tentacle. When the pressure sensor fires, the nematocyst shoots out of the cnidoblast acting much like a harpoon. Notice the presence of the spines and barbs found on the shaft of the nematocyst's "spear". The tip of the nematocyst becomes buried in the flesh of the prey item while the other end of the nematocyst remains anchored to the wall of the cnidoblast. In some species the nematocyst contains a toxin which immobilizes the prey item. In other species the nematocyst may simply entangle the prey item in its filament. In either case the prey item is pulled toward the mouth of the cnidarian to be deposited into the animal's gastrovascular cavity. Here the prey item undergoes both extracellular and intracellular digestion. Digestive enzymes are released into the gastrovascular cavity to begin the digestion of the prey item. Pieces of the prey are then engulfed by the cells lining the gastrovascular cavity where the digestive process is completed.

There are three different classes in the phylum Cnidaria. The Cnidarian class Scyphozoa includes the larger jellyfish. The Anthozoa (which means flower animals) includes sea anemones and the corals. We are going to investigate the Cnidarian body plan and biology by working with two representatives of the third class known as the Hydrozoa.

Activity 3

Class Hydrozoa

This is mostly a marine group, except for the hydras. Both the polyp and medusa stage may be found in the life cycles of hydrozoans. Marine polyp life forms tend to be colonial.

1. *Hydra*

   We are going to first investigate the structure of a freshwater hydrozoan that only exists in the polyp stage. *Hydra* lacks the medusae stage in its life cycle. Most hydrozoans are marine and the polyp stage usually exists in colonies.

a. Use a dropper to get a hydra from the container marked "hungry". Place the hydra into a shallow bowl containing fresh stream water. You may have to flush the hydra out of the dropper by sucking water into the dropper and then rapidly squirting it out. Observe the living hydra using a dissecting microscope and sketch the hydra. Get a brine shrimp or Daphnia and put it in the dish with the hydra. Describe the hydra's feeding response. When you have finished with the hydra, put it into the container marked "fed" whether it consumed the prey item or not.

   The digestive system of the hydra only has one opening. The meal enters through the same opening that the undigested remains leave the gastrovascular cavity. Why might this be an inefficient type of digestive system?

b. Observe a whole mount slide of a preserved hydra under the compound microscope. Compare the specimen that you are looking at with the figure seen below. Note its polyp body form and the other details shown in the diagram.

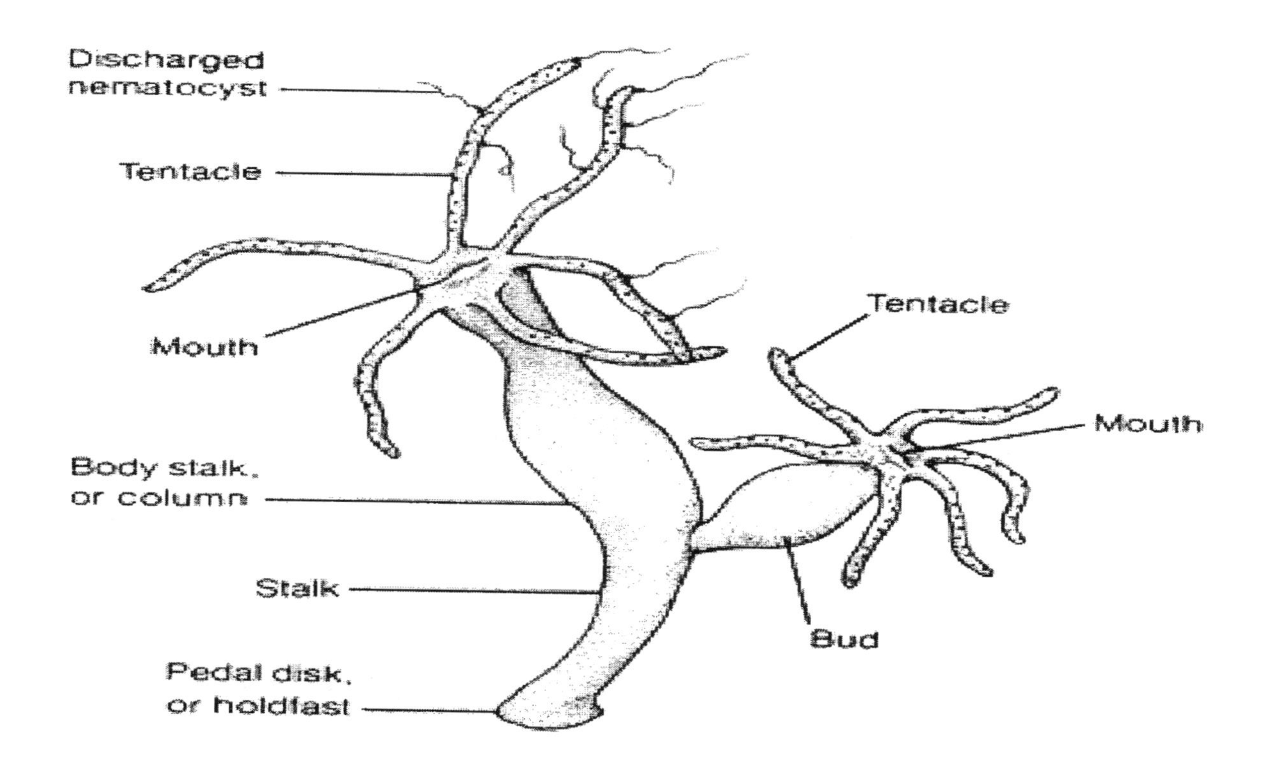

c. Find the tentacles at one end of the hydra. Note the swollen area where the tentacles attach to the body. This swollen area represents where the mouth is found that leads into the gastrovascular cavity. You may notice small swollen areas on the tentacles. These represent cells which contain the nematocysts described earlier. The area in the center of the stalk represents the gastrovascular cavity. As it name implies, this structure has two distinct functions.

d. Examine the specimen for the cells associated with reproduction. Hydra reproduce both asexually as well as sexually. An **asexual bud** forms from an evagination of the wall of the parent hydra. A mouth forms distally on the bud with tentacles forming around the newly formed mouth. When mature, the bud will drop off of the parent stalk. The bud looks like a miniature of the adult form. Hydra animals are also monoecious. The term **monoecious** means that the same individual possesses both male and female reproductive structures. This condition is also referred to as hermaphroditic. Testes are found as small rounded protuberances attached to the stalk of the Hydra near the tentacles and the opening to the gastrovascular cavity. The ovary looks like a larger swollen area further down the body of the Hydra that encircles the entire stalk. Occasionally you may see an individual stalk with an attached embryo that will be released from the body wall of the adult form. This embryo appears darker and more condensed than an asexual bud, and it also lacks the general adult body plan of its parent.

e. Get a slide of a hydra presented in both a cross and a longitudinal section. Observe the slide under the 4X objective lens to find the cross section of the hydra's body. At this low power two traits are very obvious. The central gastrovascular cavity appears as a central lumen. Surrounding this **gastrovascular cavity** are two distinct layers of cells representing the **epidermis** and the **gastrodermis**. Three types of cells might be visible in the epidermis. These include the **cnidoblasts** with their nematocysts, the **epitheliomuscular cells** which reach from the outer border of the hydra to the boundary between the two layers of the hydra's body wall, and the smaller interstitial cells that congregate together intermingling with the two other cell types.

   The larger scattered cells in the epidermis which appear to have a relatively clear cytoplasm are the cnidoblast stinging cells that are more commonly found on the tentacles of the hydra. The second most common cell type seen is smaller and darker cells with purple stained nuclei. I would think that these cells represent the epitheliomuscular cells. Interspersed among these more numerous cells are cells containing a "pinkish" stained nucleus representing the interstitial cells. Interstitial cells can become modified to produce sperm cells, eggs, and any other types of cells found in the animal.

   Examine the **gastrodermis**. There are two cell types in this layer of the hydra's body. The gland cells (larger and clearer cytoplasm) release digestive enzymes into the gastrovascular cavity to initiate the digestion of ingested prey items. The more numerous internal epitheliomuscular cells are sometimes also called nutritive muscular cells. Might the second name imply a secondary function of these cells?

   The boundary between the epidermis and the gastrodermis is referred to as the **mesoglea**. In the hydra, the thin gelatinous membrane representing the mesoglea lacks cells.

   The contractile elements in the hydra's epidermis are arranged in a longitudinal fashion. The contractile elements found in the gastrodermis run in the opposite direction in a circular fashion around the body of the hydra. These two layers are antagonistic to each other. What might be the functional significance of this arrangement? How might these two layers of contractile elements function to maintain and also change the shape of the Cnidarian body plan?

f. Find the longitudinal section of the hydra on the same slide. Examine the specimen under low power. The central cavity represents the gastrovascular cavity of the hydra. Move up to the 40X objective lens focusing on the epidermis of the hydra. Find scattered in this layer the cells that are very round with apparently thicker cell membranes. This represents a better view of the cnidoblast "stinging" cells discussed previously.

Find a cnidoblast in which you can faintly see the barbs of the nematocyst inside the cytoplasm. Move up to the oil immersion lens to get a better view. Even under this power the cnidoblasts remain very small with it being difficult to see a lot of detail.

2. Obelia –a colonial hydrozoan

In the hydras above, buds may form on the stalk as a simple evagination of the body wall of the parent hydra. A mouth develops at the distal end of the evagination and a ring of tentacles then forms around the mouth. When mature the bud drops off the "parent" hydra to take up an independent existence. In the case of colonial hydrozoans, the newly forming buds do not drop off, but remain attached to the original parent organism. In this way all of the polyps remain attached to one another forming a colony. The three layers of the colony (epidermis, gastrodermis, and mesoglea) along with the various gastrovascular cavities are all continuous with one another. Where one individual leaves off and the next individual begins is impossible to tell.

Most hydroid colonies are only a few inches in size. Individual polyps are approximately the size of the oral ends of fresh water hydras. Most of the hydrozoan colonies are surrounded by nonliving chitinous material forming a protective housing material referred to as the **perisarc**. The living material that the perisarc surrounds is known as the **coenosarc**. In some hydroid colonies like the Obelia that you are going to observe, the perisarc extends to protect the individual polyps of the colony providing a refuge that they can withdraw into when disturbed. This extension of the perisarc is called the **hydrotheca**.

a. Obtain a whole mount of an Obelia colony. Notice on the Obelia colony that there are two distinct types of "buds". The feeding buds are called **gastrozoids**. Their name reflects their function. The feeding polyps look like a hydra with tentacles surrounding the upwardly directed mouth of the gastrovascular cavity. The tentacles again are armed with **nematocysts**. The feeding polyps capture and ingest prey items. They are also the site of an initial extracellular digestion. The products of this partial extracellular digestion then enter the common gastrovascular cavity of the colony to be distributed to other parts of the colony. Intracellular digestion then takes place in other cells of the colony. Circulation is probably aided by rhythmic contractions of the colony and also by the movement of the gastrozoids into their hydrotheca when disturbed.

   The reproductive polyps of the colony are called **gonangia**. The gonangia of Obelia are bigger than the gastrozoids. They are also apparently equally numerous. Examine the inside of a gonangia. You will see around 15 developing asexual buds that will develop into the mobile medusa stage of the life cycle. Through growth each bud will develop into a complete medusa which swims away as a tiny jellyfish.

b. Obtain the slide labeled "Obelia-young medusae strew". This slide shows individual medusae that have been released from the gonangia. Hydroid medusae are usually small and umbrella-shaped. The gastrovascular cavity of the medusa form is more elaborate than is the polyp form. The mouth opens at the end of a tube-like extension of the gastrovascular cavity called the **manubrium**. The manubrium hangs down from the center of the subumbrellar surface (in the same location where the handle of the umbrella would attach to the ribs which support the cloth of the umbrella). The mouth opens into a central stomach which then extends toward to the periphery through four radial canals. The four radial canals are easily visible in the medusa slide if you are looking from above or below the animal. The canals are not visible if looking from the side. The radial canals join a ring canal that extends around the margin of the umbrella. The manubrium, stomach, and four radial canals are all lined with gastrodermis.

   The **mesoglea** of the medusa stage is more developed than the mesoglea of the polyp life form. In the hydromedusa it is still devoid of cells, but it does contain fibers that are probably secreted from the epidermis as well the gastrodermis. These medusae of the Obelia are also carnivorous with stinging nematocysts on their tentacles. The medusa form possesses a muscular system best developed around the bell margin. The contraction of the muscular system produces a rhythmic pulsation of the bell which produces a crude swimming of this animal. The "swimming" is limited to a vertical climb through the water column. Horizontal movement for the most part is produced by the ocean currents.

The nervous system of the medusa is highly developed for the coelenterate body plan. There are two nerve nets which run around the periphery of the bell. Associated with the nerve rings are two types of sense organs. The medusa has both **ocelli** that are light sensitive photoreceptors and **statocysts** which are organs of equilibrium. Both of these sensory organs are located near the base of the tentacles where they attach to the periphery of the bell. Notice that there is a swollen area at the point of attachment of the tentacles where both the statocysts and ocelli are located.

All medusa reproduce sexually. The eggs and sperm arise from epidermal or gastrodermal interstitial cells that have migrated to the epidermis. In some species fertilization may be external while in others fertilization may be internal. In animals with internal fertilization, early development occurs in the gonad of the parent medusa. After fertilization and related embryological development, a free-swimming *planula* larva is produced that may exist for several hours to several days. The planula larva lacks a mouth and a gastrovascular cavity. It is essentially a solid mass of ciliated cells whose function is dispersal away from the parent organism.

AN INTRODUCTION TO THE PHYLUM PLATYHELMINTHES

17

Objectives

1. To become familiar with the typical platyhelminthes body plan.
2. To relate body form to the life history of the animal.
3. To become familiar with the three different classes of platyhelminthes.
4. To realize that sometimes evolutionary modifications lead to loss of unnecessary structures.

Introduction

The flatworms are the simplest animals that exhibit bilateral symmetry. The name of the category obviously comes from the physical flattening of the animal in the dorsal/ventral planes. There are three classes of worms in this phylum. The **Turbellaria** are the most primitive members of the phylum being the free-living ancestors of the other two parasitic classes. The **Trematoda** class is made up of the parasitic flukes. The class **Cestoda** are the most highly specialized Platyhelminthes. All of the cestodes are endoparasites covered with a nonliving cuticle protecting them from the host's digestive tract that they inhabit. Unlike the other two classes of Platyhelminthes, the cestodes lack a digestive tract.

Organisms with bilateral symmetry can be cut only along one plane, parallel to the main axis of the animal's body, producing mirror images of the animal. Going hand in hand with bilateral symmetry is the concentration of nervous and sensory structures at the anterior end of the animal. This phenomenon is referred to as **cephalization**. A bilaterally symmetrical animal is more mobile than a radially symmetrical animal. As it moves forward through its environment, it only makes evolutionary sense to place a higher concentration of nerve tissue at the anterior end of the organism. Here the sensory organs are more likely to encounter stimuli alerting the animal to the presence of such things as prey items or dangerous predators.

The flatworms are also the simplest animals to show a true organ system level of development. The flatworms possess a nervous system with a rudimentary brain. A primitive excretory system is found with its **flame cells**. There are multiple organs in the flatworm's digestive system even though it still only has one opening to its digestive tract. Its gut is referred to as being an **incomplete gut** as it has only one opening to the outside. Material enters the system through the same orifice that it exits. It is difficult therefore for any region of the tract to become specialized for any one purpose.

The flatworms are also **triploblastic** meaning they have an ectoderm, endoderm, as well as a mesoderm layer sandwich in between the two. Flatworms may have moved ahead evolutionarily from the coelenterates on some fronts, but the flatworms still are **acoelomates** not possessing a body cavity.

There are three classes in this phylum. The Turbellaria are the most primitive of the three classes and have most likely given rise to the other two parasitic classes. The Turbellaria are free-living. The Trematoda comprise the flukes and the tape worms are representatives of the class Cestoda.

Activity 1
Class Turbellaria: Planarians

Free-living flatworms are found in both marine and fresh-water environments on the undersides of rocks, leaves and sticks submerged in the water. Fresh water planarians are the representatives of this class most easily studied in a general biology course. They are relatively small flatworms being two centimeters or less in length. The name of the genus usually studied is *Dugesia*.

1. Examine a whole mount of a planaria. Sketch the specimen that you see. Identify the head end of the planaria. Note the two darkened **eyespots**. These structures are not true eyes that can form images of objects seen in the environment, but represent **ocelli** that respond to the presence or absence of light. The **auricles** are the flared portions of the body wall lateral to the eyespots that are sensitive to chemicals present in the water as well as to the sensation of touch. Approximately midway down from the head end of the flatworm is its muscular and protractile **pharynx** with its mouth at its terminus. The pharynx protrudes from the ventral surface of the animal's body and essentially acts like a vacuum cleaner sucking up insect larvae, small crustaceans, and other living and dead animals that comprise its food. Between feedings the pharynx is retracted into a narrow chamber in the body. The lumen of the pharynx is continuous with the **intestine** of the flatworm. There are three branches of the intestine. One of the main branches heads off in an anterior direction with two parallel branches heading in a posterior direction. Note that along the length of the intestine's main branches that there are many side branches permeating the body tissue of the flatworm.

   In what way is the digestive apparatus of the flatworm similar to the digestive apparatus of the hydra?

   In what ways is it different?

   Would you expect an animal with the flatworm's design to have a circulatory system? Why or why not?

   What advantage or adaptation does the highly branching intestine provide the flatworm?

   You may also observe the elimination of solid wastes from the planarian's gastrovascular cavity. If you observe the process of elimination, describe the process. If not, describe the route by which this process must occur. This might not leave a pleasant taste in your mouth.

   The flatworm also possesses other organ systems that are not visible on this slide. The flatworm is capable of controlling and modifying the contents of its tissue fluids using branching tubules which permeate the tissues of the animal. The tubes are lined with ciliated flame cells that drive excess water from the flatworm's body. The branching tubes are called **protonephridia**. The flatworms also have a rudimentary nervous system with a concentration of neurons at the anterior end of the animal (cephalization) with two lateral nerve cords running caudally. The flatworms have a reproductive system with both sex organs being found in the same individual (monoecious state).

2. Obtain a slide of a flatworm that has its digestive cavity specially stained with carbon particles. This staining will accentuate the structures mentioned two paragraphs above.

3.a. Place a piece of fresh liver in a clear Petri dish containing clean and fresh stream water. Place several live planaria obtained from the culture labeled hungry planaria in the same dish to observe their feeding behavior with the dissection microscope. Describe what you see below. After the trial, whether the planaria fed or not, place the living planaria in the container labeled "fed planaria".

b. Using a moderately bright overhead light source and aluminum foil investigate whether planaria exhibit positive or negative phototaxis. Obtain a Petri dish and draw a diagonal line with a wax pencil across the bottom of the dish separating the dish into two halves. Label one side as "light" and the other side as "dark". Place a planaria in the dish covering the dark side with aluminum foil. Make sure and cover the side of the dish as well as the top and bottom. Place the dish under a moderate light source for ten minutes. After this time interval, remove the planaria with an eye dropper. Sprinkle a small amount of carmine powder onto the surface of the water allowing time for the powder to settle onto the mucous trail left behind by the flatworm. In one smooth motion, pour off the water and the remaining carmine powder. You should be left with a history of where the planaria has been for the past ten minutes. Does the flatworm exhibit positive or negative phototaxis? Why does this response make sense considering the way the flatworm makes his living?

c. If the above did not work, use a penlight to investigate phototaxis. In otherwise dark conditions, shine the pen light from above at the anterior end of the living flatworm. What is its response? Now shine the light toward the head of the animal from its side instead of directly overhead. What is the animal's response now? Is the planaria positively or negatively phototactic?

4. Obtain a **cross section** of a planaria showing three different sections through the flatworm.

Sketch the **middle section** first. The most obvious structure under the ten power lens is the muscular pharynx. It looks as if the pharynx is suspended within a "coelom" or body cavity, but remember that the platyhelminthes lack a coelom. The cavity that you see is the chamber into which the pharynx may be retracted when not in use. The pharynx is a most amazing structure. It looks as if the outside of the pharynx is covered with a ciliated epithelium. Just under the epithelium you can find a longitudinal band of muscle. Just to the inside of the longitudinal band of muscle is an antagonistic circular band of muscle. As you move toward the lumen of the pharynx passing through a relatively thick layer of looser tissue that looks a bit like connective tissue, you will find a second layer of longitudinal and circular muscle around the lumen of the flatworm's pharynx. What is the significance of this musculature?

Lateral to the pharynx, find branches of the intestine. The cells lining the lumen of the branched intestine fold in a convoluted pattern. What might be the significance of this folded pattern?

Inferior and lateral to the pharynx (but inferior and medial to the branches of the intestine) are two small more clear and less cellular looking zones representing the two longitudinal nerve cords of the flatworm.

Locate the cilia along the ventral surface of the animal. What is the function of these cilia?

Examine both the dorsal and ventral surface just below the epidermis for the presence of melanin. Melanin is a dark to black pigmented molecule that is deposited in the epidermis of animals to provide color. Where is the melanin more highly concentrated, on the dorsal or the ventral surface? What selective advantage might the flatworm obtain from this distribution of melanin?

Now move to the sections immediately above and below the section which reveals the flatworm's pharynx? Which one of the two is anterior to the pharynx? How can you tell? Look for some of the same structures that you observed above?

Activity 2

Class Trematoda (Flukes)

Flukes are typically internal parasitic leaf-shaped flatworms modified to live in a very different environment than the planaria observed above. One of these adaptations is the possession of a noncellular cuticle covering the surface of their body instead of a cellular epidermal layer. Why do the flukes need a cuticle as an adaptation?

Obtain a whole mount of a sheep liver fluke, *Fasciola hepatica*. Observe the same slide using both the scanning lens of the compound microscope and the higher powers of the zoom-lensed dissecting microscope. Make a sketch of the liver fluke noting the following structures presented in bold-faced print.

Note that its general morphology is similar to the planaria in being very flat. Observe the **oral sucker** which directly encircles the mouth of the fluke. Posterior to the mouth is a relatively enlarged **pharynx** that leads to a short **esophagus**. These structures probably possess some darkened food material to enhance their appearance. The esophagus leads into two arms of the **intestine** which run posterio-laterally along the fluke. There are small blindly ending **ceca** that can be seen along the length of the intestine. Shortly after branching from the esophagus, much of the detail of the intestine is obscured by reproductive structures of the liver fluke.

The penis and the opening to the uterus are just posterior to the bifurcation of the esophagus leading into the two arms of the intestine and just in front of the larger **ventral sucker** of the flatworm. Directly posterior to and in the midline of the flatworm find the obvious **uterus** filled with **mature eggs**. The eggs appear to be large cells encapsulated in a blackened shell. Posterior to the mature shelled eggs find the eggs that have not as yet been encapsulated by the shell gland. They appear to be the same size as the eggs, but have a very clear cytoplasm. Why might it be important to encapsulate your eggs if you were a parasitic liver fluke? In what environment might your eggs have to pass through on their way to freedom?

The remainder of the *Fasciola hepatica* body is difficult to discern. There appear to be two different duct systems in the body of the fluke. One duct system has a smaller lumen with a more distinct border. This duct system seems to emanate from the periphery of the fluke's body in the vicinity of the **yolk glands**. These **yolk ducts** collect yolk to be deposited into the yolk reserve (not discernible on this slide) found just posterior to the shell gland. The ducts of the other system appear to be larger in diameter and their contents are much more granular in appearance. In the more anterior part of the fluke, these ducts probably represent the ovary of the flatworm. The less granular larger ducts in the posterior regions of the flatworm represent the posterior testes. There seems to be a premium placed on reproductive structures with less emphasis on digestive organs in the trematode group.

Why might the digestive system be less complexly organized than the planaria and the reproductive structures so much more developed?

The reproductive scheme of *Fasciola hepatica* is not straightforward. A fertilized egg that has developed into a miracidium (larval stage) passes from the sheep gut in feces to develop and hatch in fresh water. In the fresh water the miracidium stage enters an aquatic snail to form a sporocyst with developing rediae. The redia stay in the snail with developing cercariae larval stage. Free-swimming cercaria encyst on a water plant as a metacercaria. This form is then eaten by a sheep to complete the cycle. And you thought reproduction in the vertebrate world was confusing!!

Activity 3
Class Cestodac

Tapeworms are flattened internal parasites that inhabit the small intestines of vertebrates. Their general appearance is a bit like a segmented egg noodle. They completely **lack a digestive system** as they are bathed with a pre-digested nutrient rich broth coming down their host's digestive tract. They absorb nutrients from this soup across their general body surface. Their body is covered with a thick cuticle much like the body of the trematodes seen above. The tapeworms, considering their parochial environment, have had to develop into reproductive machines ensuring the perpetuation of their species. Reproductive organs occupy the vast majority of each mature body segment.

1. Obtain a composite whole mount of *Taenia pisiformis*, the tapeworm that inhabits the small intestine of dogs. The tapeworm's body is divided into three basic segments: the **scolex, neck, and body**.

   Examine the **scolex** under the compound microscope. It is an amazing structure that is used to anchor the organism to the host's intestinal wall. Diagram the scolex using the 4X objective scanning lens. The most anterior part of the scolex is the **rostellum**. The rostellum is ringed with **hooks** that are used to attach the rostellum to the intestinal wall of the host. Notice under 10X that the hooks seem to be of two types. The superior set of hooks appears to be straight and lance-like. The more inferior row of hooks is recurved like fishing hooks. Inferior to the hooks, find the four suckers that encircle the body of the tapeworm.

   The next section on the slide shows the budding zone of the tapeworm where new proglottids are formed. There is not a lot of detail that can be seen in the interior of these immature segments. The early beginnings of reproductive structures are visible, but their exact identity is difficult to identify. The immature vagina and ductus deferens appear as a dark thin line running down the middle of the proglottid. For whatever reason, it appears that the developing reproductive tracts seem to alternate sides on successive proglottids as you pass down the section.

   Tapeworms are monooecious. They are one of the few animals that may self-fertilize. This is probably the result of their lonely existence in their host's digestive tract. Few animals could be host to multiple parasites and be strong enough to be a stable host. For their own survival, tapeworms lead a solitary existence.

2. Examine the next set of proglottids on the slide. They come from the middle region of the tapeworm. They are mature enough to see reproductive structures, but not too mature to have the details clouded by the presence of a large number of eggs. The genital pore is a common opening for both the vas deferens and the vagina. Notice that the genital pores seem to alternate sides of the tapeworm. What might be an advantage of this relationship? The uterus can be seen in this section as a fairly fine line running north to south through the proglottid. The uterus appears to have many smaller diverticula branching laterally from it. The ovaries are found below the intersection of the ductus deferens and the uterus in the midline of the proglottid. The ovary will appear as a relatively large and roundish area of darker more granular appearing material. The testes are superior to the ovaries and are represented by a paler and more diffuse mass of tissue. The testes are hard to identify. The vas deferens leads from the testes to the genital pore. On either side of the reproductive organs running longitudinally from north to south on the periphery of the proglottid are the tapeworm's excretory canals.

3. Examine the last set of proglottids on the slide. Here you will see zygotes (fertilized eggs) in the expanded uterus. Mature proglottids containing embryos are voided with the feces of the dog. As some of you know, many dogs are coprophagous (eat feces) and this behavior would allow the cycle to repeat itself. In more discriminating dogs that don't eat feces, the tapeworm can be picked up by eating a rabbit. The rabbit picks up the tapeworm embryo by coming into contact with the feces of the infected dog. The rabbit's ingested tape-

worm embryo bores through the intestine of the rabbits gut to encyst as a larval stage in rabbit muscle tissue. When an infected rabbit is eaten by a dog, the tapeworm's larval stage undergoes metamorphosis in the new dog's gut and attaches here by its own scolex.

Thought Question

Evolutionary processes within a taxon of organisms leads to more and more complex adaptations increasing the success of the individual organisms.

Support or refute the above statement based upon your experience with the three classes of platyhelminthes observed.

WORM DISSECTION

18

Laboratory Objectives

1. To carefully observe some of the external landmarks of the earthworm.
2. To perform a gross dissection of the worm to see the "structure-function relationship" between an animal's organ systems and its way of life.
3. To examine a microscopic cross-section of the worm to more closely observe the structures discovered in its gross dissection.

Introduction

The earthworm belongs to the phylum Annelida. This group contains about 15,000 segmented worms. Its members exhibit several advances that simpler phyla do not possess. The most impressive quality of the earthworm is its mobility. Its fluid filled coelomic body cavity produces an efficient hydrostatic skeleton. The annelids also possess a closed circulatory system, efficient excretory organs, and a fairly complex brain and nervous system.

There are three classes within the phylum: Polychaeta are the marine worms which possess many (poly-) well developed bristles, the Oligochaeta which typically have scanty bristles (earthworm), and the Hirundinea (leeches).

Procedure

Activity 1

External Anatomy

Get a preserved worm from the front of the room along with a dissecting microscope. Place the worm in the dissection tray. The most conspicuous external landmark of the worm is its **clitellum.** This is a swollen area on the worm's dorsal (upper) surface located approximately one-fourth of the way behind the head. This structure which secretes mucous plays a role in the worm's reproductive biology to be described below.

The worm is rounder in cross section at its anterior end and becomes dorsal ventrally flattened posteriorly. Its dorsal (upper) surface is darker (more pigmented) than is its lower ventral surface. The dorsal pigment protects the worm from the damaging UV rays of the sun. It may also play a camouflaging role to allow the worm to blend in better with its surroundings.

Run your finger along the worm's ventral surface. The roughness which you are feeling is produced by the worm's **setae.** These are small bristles extending from the worm's body. In which direction do they point? The setae are important in the worm's locomotion. As we will see in the microscopic examination of the worm, two layers of muscle make up the majority of the worm's body wall. There is an inner circular layer of muscle and an outer longitudinal layer. When an earthworm moves in the forward direction, its posterior setae are extended from its body anchoring the worm to the surrounding soil particles. The earthworm's circular muscle layer then contracts exerting pressure upon its incompressible coelomic fluid, causing the earthworm to lengthen. With the posterior setae extended gaining purchase against the soil particles, the earthworm is propelled forward. After this elongation phase is completed, the anterior setae are extended from the body while the posterior setae are retracted. The longitudinal muscles then "pull" the posterior end of the worm toward the new location assumed by the anterior end.

Locate the **prostomium** which is the first lobe at the worm's anterior end. The mouth is located directly behind the prostomium on the worm's ventral surface. The segments of the worm can be used as landmarks for finding various structures during our dissection. Segment 1 is its most anterior segment. Count how many segments does your earthworm have?

The earthworm is hermaphroditic-that is, each worm possesses both male and female reproductive structures. Some of the reproductive openings are visible on the external surface of the worm. On the ventral side of segment 14, the openings of the **oviducts** (female pores) can be seen. The two openings (male pores) of the **vas deferens** found on the ventral surface of segment 15 have swollen lips. When two worms copulate the **clitellum** (the swollen region mentioned above) secretes a slimy band around the two worms. This band aids in the reciprocal exchange of sperm cells. A worm (much to its own chagrin) cannot fertilize its own eggs. Sperm cells, released through the male pores, travel along the seminal grooves, and enter the second worm's body through small openings located in the grooves between the ninth, tenth, and eleventh segments. These openings lead to the **seminal receptacles.** The seminal receptacles are a part of the female reproductive tract where sperm are stored until the fertilization process occurs. These openings will only be seen with the aid of the dissecting microscope.

Activity 2
Internal Anatomy

Place your worm along one edge of the dissecting tray ventral side down, and put a pin through the prostomium and anus. If the worm is too long for the tray, you may cut off its posterior end. Gently make an incision with a scalpel from the **anus to the head** being careful not to cut too deeply. Keep a light touch on the scalpel. If you notice a brown "ooze" escaping from the incision, you have cut too deeply and have gone into the worm's gastrointestinal tract. As you make this incision stay to one side of the dark **dorsal blood vessel** with runs on top its its gastrointestinal tract. Once this incision is complete, gently pull the body walls apart and notice the **coelom** which is divided into a series of compartments by **septa.** Carefully pick away the septa with dissecting needles as you continue to pull the walls of the worm's body further apart to expose the internal organs. Pin both sides of the body at 10-segment intervals to the bottom of the dissecting tray. Be sure to place the pins at a 45° angle so that they do not hamper further dissection. The pins, placed at 10 segment intervals, can serve as landmarks in the location of the following structures. Cover the bottom of the dissecting tray with 1–2 cms of water to float the worm's organs and to keep its tissues moist.

Notice the concentration of organs in the anterior segments of the worm. Notice the prominent light colored male **seminal vesicles.** These can be found in segments 9–12. The seminal vesicles store sperm produced by two pairs of tiny **testes** located on the posterior walls of the septa separating segments 9 and 10 and segments 10 and 11. The testes are too small to be seen in preserved specimen. The sperm, after maturation and storage in the seminal vesicles, are released via the **vas deferens** from the male pores located on the ventral surface of segment 15.

Find the two pairs of female reproductive structures, the white and spherical **seminal receptacles,** just anterior to and slightly inferior to the seminal vesicles. The seminal receptacles store sperm from another worm, and sperm are held here until fertilization. A pair of **ovaries,** if extreme care is used, can be located on the septa between segments 12 and 13. Eggs released from the ovaries enter the coelom before passing into the oviducts, which carry the eggs to the female pores located on the ventral surface of segment 14.

Fertilization of the eggs occurs externally some time after copulation. The eggs are released through the female pore into a collar of mucous produced by the clitellum. As the collar of mucous containing the eggs travels forward over the surface of the worm, sperm are released from the seminal receptacles into the mucous collar allowing fertilization to take place. The collar slips off the anterior end of the worm to form a cocoon in which the embryos develop. In this species of worm, the first embryo to complete development uses the other fertilized eggs as a food source. Development is completed over several weeks. The parent worm continues to form cocoons as long as its stored sperm from the previous copulation lasts.

Locate the **digestive tract** running from mouth to anus. Behind the mouth is the muscular **pharynx** which sucks dirt into the digestive tract. It appears to have a fuzzy appearance due to the many **dilator muscles** traveling from its surface to the body wall. Contraction of these muscles expands the pharynx sucking particles of soil and detritus (decomposing organic matter) into the mouth. The ingested material is moved by peristaltic waves through the **esophagus** into the thin walled sac-like **crop.** Immediately behind the crop is the **gizzard.** Poke each of these two structures with a dissecting needle comparing their "firmness." Which of the two is more muscular?

Cut open both of these structures examining the contents of each under the dissecting scope. Compare the consistency of the "goop" in each. Look for the presence of small pieces of "stone" in the worm's gizzard.

What do you think is the function of the crop? What is the function of the more muscular gizzard?

Worms obviously do not have teeth. If they did, fishermen would not be so ready to use them for bait. Before food can be chemically digested, it must be mechanically broken down into smaller particles. This maceration increases the exposed surface area of the food which can then be acted upon by digestive enzymes.

Locate the worm's **intestine** which runs from its gizzard to anus. Why is it so long? What does its length infer about its function?

The worm possesses a closed **circulatory system.** Blood returning from capillaries in each body segment enters the **dorsal blood vessel** which runs along the top of the digestive tract. Blood travels anteriorly through this vessel to the five pairs of **pseudo-hearts** surrounding the esophagus. Find these five pairs of hearts. They may be partially concealed behind the septa which you broke through as you spread the walls of the body during the pinning process. The hearts will appear to be much darker than the surrounding tissues as they were filled with blood when the worm was preserved. These hearts help to pump blood into the **ventral vessel** which carries blood posteriorly. In each segment, capillaries branch from the ventral blood vessel passing into the subneural vessels under the ventral nerve cord. Blood is not only propelled through the system by the contractions of the five pairs of hearts. Both the dorsal and ventral blood vessels also move blood along by peristaltic contractions.

Each segment, except for the first few and last few, possess a pair of **nephridia.** Each nephridium is a collection of tubes which filter the coelomic fluid removing and excreting waste products through small excretory pores found in the ventrolateral side of each segment.

The last system to be examined is the worm's **nervous system.** Remove the digestive tract from the posterior third of the worm. Look for the **ventral nerve cord** with its chain of ganglia. A **ganglion** is a concentration of nerve cells bodies found outside of the animal's brain.

Trace the path of the ventral nerve anteriorly. In segments 1–5, find the s**uprapharyngeal ganglia,** the **subpharyngeal ganglia,** and the connecting **circumpharyngeal connectives.** Obviously the worm's **brain** essentially surrounds the worm's pharynx. The nerve tissue will appear to be much "whiter" than the surrounding tissues.

Activity 3

Cross-sectional Analysis of the Earthworm

Make labelled sketches of the cross section of the earthworm. Locate the structures listed below. Not all of the structures are to be found in the same section of the worm. For example, recall the location of the **seminal vesicles** from your internal dissection. Were these structures found in the anterior or posterior region of the worm? A cross section from one of these two locations would reveal the presence of the seminal vesicles, while a section from the other region would not.

Begin by observing the middle of the three cross sections on the slide under the 4X objective lens. Make sure that the slide is placed with the **dorsal (upper) side facing up.** Remember that the dorsal surface darker and the ventral surface of the worm tends to be flatter.

The noncellular **cuticle** is the outermost layer of the body wall of the earthworm. The cuticle is produced as a secretion of the underlying **epidermis.** Moving through the 10X and 40X lenses, carefully observe the outermost cells of the "purple staining" epidermis. Their cytoplasm is packed with small round vesicles loaded with the mucoid material characteristically found on the outside of earthworms.

Return to the 4X objective lens. Below the epidermis is a fairly thin band of pink **circular muscle** wrapping around the worm. Internal to the circular muscle layer is the larger and more prominent **longitudinal muscle layer.** The longitudinal muscle has a "feathery" appearance. If the worm has been cut in an appropriate section, you may seen "horn-like" **setae** projecting through the muscle layers to the outside of the worm. Recall that the setae help to anchor the worm during locomotion.

Using the 4X lens, the most obvious structure in the **coelom** (body cavity) is the large circular **intestine.** Notice that the intestine is not just a simple tube. Suspended from the dorsal wall of the intestine is the solid tubular **typhlosole.** This structure increases the available surface area for absorption of digested organic matter.

The **dorsal and ventral blood vessels** can be seen above and below the intestine. Each is cut in cross section. The lumen of these two vessels may contain some blood cells. Recall that when you made your initial incision through the dorsal body wall of the earth you were directed to cut to one side of the dark line running along the dorsal surface of the intestine. By making this cut, you were avoiding the dorsal blood vessel.

Inferior (or possibly lateral) to the ventral blood vessel, locate the **ventral nerve cord.** Recall that in your gross dissection the ventral nerve cord was a white fiber tract located below the removed portion of the worm's small intestine. The ventral nerve cord should be a solid mass of tissue.

The **coelom** is the earthworm's body cavity in which the intestine is suspended. In your slide, the coelom is the open space located between the intestinal wall and the longitudinal muscle layer of the body wall.

Located in the coelomic cavity will be numerous thin-walled tubes which may be cut either longitudinally and/or in cross section. These tubules make up the pair of **nephridia** located in each segment of the worm.

Remove the slide from the microscope and examine the three sections with the naked eye. Find the section which possesses the large dark purple-staining organ. Put this section under the objective lens to identify this large purple mass. Examine it carefully under high power to help you identify this organ. Find a second organ in this same cross section which is located centrally within the worm. It is located approximately where the intestine was on the first section which you observed. Identify these two organs?

AN INTRODUCTION TO THE FRUIT FLY AND ITS GENETICS

19

Laboratory Objectives

1. To become familiar with the fruit fly's life cycle.
2. To observe the giant salivary chromosomes of the fruit fly.
3. To establish a dihybrid fruit fly cross to determine the pattern of inheritance of two genes.

Introduction

The fruit fly (*Drosophila* sp.) is to animal genetics as the pea plant is to plant genetics. The fruit fly is an excellent experimental model with which to investigate Mendelian genetics. The fly possesses a diploid number of eight chromosomes (four homologous pairs). This allows studies of linkage to be done fairly easily as there are smaller numbers of linkage groups than in more complex organisms.

Fruit flies are also excellent experimental animals as they have a short generation cycle. An experimenter does not have to wait a long time to analyze the results of her experiments. The fruit fly hatches, mates, lays eggs, and has progeny hatch in ten to fourteen days depending on their incubation temperature. Mating occurs four to eight hours after emergence from the pupal stage. Other traits which made the fruit fly the experimental animal of choice for early geneticists was its small size, minimal growth requirements, and the possession of distinct and easily recognized morphological traits. Finally, the salivary glands of the third instar larval stage possess giant chromosomes that are easy to see under the light microscope. These chromosomes possess distinct banding patterns that can be correlated with the presence or absence of specific genes.

Fruit Fly Life Cycle

The male and female flies have a complex courtship behavior that seems involved for a lowly fruit fly. The male initiates the mating ritual with any fly which happens to come along. The purpose of the mating dance is to discriminate the sex of the other fly. The male fly circles the second fly vibrating the wing closest to the object of his attentions. The vibrating wing of the male fly is directed toward the olfactory organ of the second fly indicating that there may be a pheromone (chemical messenger) involved. The male fly also attempts to stimulate the second fly by tapping it with his foreleg. If the second fly is a nonreceptive female or an obvious unwilling male, it ignores the amorous young male and leaves the area.

The male fly has sex combs on his forelimbs and claspers on the undersides of his abdomen which aid in the transfer of approximately 1000 sperm to the female during the reproductive act. These sperm cells are stored in the female's reproductive tract and are used to fertilize eggs as they are released. The female will release a few hundred eggs beginning approximately on the second day after she emerges from the pupal stage. The eggs are laid onto the growth medium and are fairly small in size (.5mm).

Eggs hatch at varying times depending upon the temperature that they are held at. Typically in a laboratory incubator set at 25° C, the eggs will hatch in about 24 hours. A maggot emerges from the egg to begin feeding. Between the earliest maggot and the next adult generation, the organism will proceed through three different growth stages. Being an invertebrate and possessing an exoskeleton, a fly maggot can grow only as far as its present exoskeleton will allow. The animal must molt its undersized exoskeleton to allow further growth. The maggot sheds its skeleton twice before it enters the pupal stage. Each growth cycle is referred to as an instar. The pupal stage is analogous to the cocoon that a caterpillar produces before it emerges as an adult butterfly. The maggot emerges from the pupal stage as an adult fly ready to copulate in four to eight hours. The adult fruit fly will live on average to a ripe old age of 37 days.

Activity 1

Observations of Fruit Fly Life Cycle and Various Phenotypes

The first type of adult fly to observe will be referred to as the wild type fly. Its phenotype is that of a normal fly that might be flying around the wild environment of your kitchen. The adult fly can be observed best after it has been over-anesthetized. Its wings will be held rigidly at a 45° angle from its body. Examine several adult wild type flies under the dissecting microscope. Carefully observe the antennae, mouthparts, and especially the color of its compound eye. Observe the wing to see the extent of wing size and venation. After you are familiar with the normal wild type fly, anesthetize some flies from the culture labelled "vvss." These flies are **homozygous recessive** for two traits that will be followed later in a dihybrid cross. The "v" allele produces **vestigial** wings which are shriveled in size. The "s" allele produces an eye color which is different from the brick red wild type color seen above. This shade is referred to as **"sepia."** Make a point to write a description of both of these phenotypes as they will be an important part of your lab report to be described below. Finally see if there are halteres (reduced second pair of wings) on any of the flies. It is thought that a four-winged fly was the ancestral form to the modern two-winged variety. Check out the legs looking for other variations.

After making these preliminary observations, sort the flies by sex. Males and females look very different. The male is smaller than the female fly possessing five instead of seven abdominal segments. The male abdomen is also more blunt and darker than the tapered and lighter female abdomen. Remember also that the male possesses sex combs on his forelimbs and claspers on the ventral portion of his abdomen.

Now check out the immature stages of the fruit fly's life cycle. Remove from the media some eggs and larvae to be viewed under the dissecting scope. You should be able to see mouthparts and the outline of a complete digestive tract in the larval stages. The maggot is an eating machine.

The pupal stage will be found on the sides of the growth chamber. The pupa resembles a brown rice-shaped kernel stuck to the side of the growth chamber. Remember that this stage is analogous to the cocoon of a butterfly. If you place the pupa on a slide with a drop of water, you may be able to see the transforming maggot using high light intensity and a low power objective lens.

Activity 2

Observing the Giant Salivary Gland Chromosome

The chromosomes of the third instar stage are easily seen under the light microscope. The chromosomes are enlarged and easily seen for a number of reasons. Unlike most chromosomes which uncoil to the chromatin configuration after mitosis is over, the salivary chromosomes remain partially coiled and condensed in the third instar stage. To increase the amount of mRNA which can be transcribed along the DNA template, the DNA has also replicated itself to produce 1000-5000 identical strands. These multiple sister chromatids serve as templates for transcription (RNA production). The sister chromatids also remain in register amplifying the distinct morphological regions along their length. Finally, the homologous chromosome pairs have taken a chapter out of the story of meiosis I. For some unknown reason, the homologous chromosomes lie side by side as if they were in synapsis. Some of these terms should sound familiar from the unit you studied in meiosis.

Procedure

Find a third instar larva crawling up the side of the growth chamber. Do not use the pupal stage which is the darker nonmobile kernel attached to the side of the growth chamber. Place a third instar larva in a drop of "dye" solution placed on a clean microscope slide. This dye is caustic and the student should be wearing eye protection and gloves. This procedure must be done under a **dissecting microscope.** You will attempt to separate the maggot's mouth parts from the rest of its body by "teasing" the two apart with the dissecting needles. Place the tip of one needle through the head just posterior to the mouth parts. The head is easily differentiated from the posterior of the animal as the larvae moves typically in a forward direction with its mouthparts working the majority of the time. The second needle is placed approximately half way back through the middle of the body. The two needles are then pulled apart smoothly from each other. In the process the mouth parts with the associated salivary glands will be pulled out of the maggot's body. The salivary glands will look like clusters of grapes still attached to the mouth parts. Don't confuse them with the very shiny fat bodies which may be close by or with any stray pieces of intestine. Make sure that the area of dissection stays moist by adding a drop of dye to the slide when it appears to be drying out.

Clean off any debris from the salivary glands as best you can with the dissecting needles. Move the salivary glands to a clean part of the slide. Again make sure that stain has been covering the salivary glands for at least five to seven minutes before proceeding. This stain is specific for DNA and should dye the chromosomes a purplish color for easier viewing underneath the compound microscope.

To be able to see the chromosomes, it is necessary to "squash" the salivary gland itself freeing the chromosomes from their nuclear envelope. Place a coverslip over the salivary glands and place a few Kim Wipes over the cover slip. Press directly downward on the coverslip with the end of a pencil. Make sure that you are pushing directly down so that you don't smear the chromosomes across the slide.

Use the 10X objective lens to find the squashed salivary gland. Locate a relatively clean area and then proceed to the 40X objective lens to observe the giant chromosomes (if present). Can you see the banding patterns along the length of the chromosomes which experts can link to specific genes of the fruit fly? If you have had no luck, try making a second, third, or fourth slide. This is a tricky technique that usually does not work on the first try. If you are successful, be sure and show your results to your neighbors. They may not be as lucky as you were.

Activity 3

Analysis of Dihybrid Cross

You are going to investigate the inheritance of two genes of the fruit fly. One of the genes influences wing development. The dominant allele (V) produces the normal wing phenotype. The recessive allele (v), when present in the homozygous condition, produces a fly with vestigial and withered wings. The second gene that you will be following affects eye color. Normal wild type eye color is a bright reddish orange color. Its allele is signified by the letter "S." The recessive allele (s) when present in the homozygous condition produces a purple/brown eye color (sepia). The purpose of your investigation is to establish whether these two genes are inherited independently or if the two genes are linked.

We have purchased the offspring of a cross initiated in a commercial lab between parents with the genotypes VVSS X vvss. One of the parental types was homozygous dominant for both normal eye color and normal wing development. The second parental type was homozygous recessive for these two traits. You will be presented with the F-1 generation offspring for further investigation of the inheritance of these two genes. What is the genotype of the F-1 flies that you are working with? _______________

Procedure #1: Setting Up the F-1 Cross

Your first job is to set up the growth media for the mating that you are about to produce. Add approximately one inch of powdered media to the fly vials over a sink. Add an equal volume of water and let the mixture sit. Wash any spilled media down the drain to avoid later contamination. After the growth media has solidified, **add 3–4 grains** of yeast to the culture tube. **Warning:** too much yeast will kill the eggs and larvae that you wish to raise to produce the F-2 generation.

With the growth media prepared, transfer 10–12 adult (equal numbers of male and female) flies from the stock F-1 cultures to the new fly vials just produced. To transfer the flies, the stock culture must be anesthetized for two to three minutes using the "Fly-Nap" found at the front of the room. Do not contaminate the foam stoppers with the "Fly-Nap" to avoid any residual effects to the remaining flies. **Do not overdose the flies** as dead flies will not lay eggs. During the anesthnetization, **leave the fly vial lying on its side** so that the flies do not become mired in the growth media. If flies begin to come around, you can re-anesthetize the flies in an empty fly vial for a shorter period of time. There is no growth media to absorb the "Fly-Nap" in this situation. Transfer the flies from the first culture tube to a clean sheet of paper. This paper can then be used like a funnel to place the flies into the new fly vial. Place the flies onto the side of the new culture tube. This prevents the flies from getting stuck in the media at the bottom of the new vial. **Do not place the fly vials upright until the flies have fully awakened from their sleep.**

Clear the parents from the fly chambers at the next lab meeting. At this time you should probably notice some larvae crawling around the growth media.

Procedure #2: Counting the F-2 Offspring (two weeks following procedure #1)

Approximately two weeks after the F-1 generation flies are placed into their new growth chambers, F-2 flies should emerge from the pupal stage. For this exercise to be of value, many F-2 flies have to be sorted and counted based on their eye color and wing type. Do not count your progeny until a decent number have passed through the pupal stage. To count and categorize the progeny, you should:

1. Anesthetize the F-2 offspring. You do not need to be as careful here with overdoing the anesthesia as the F-2 offspring are headed to the fly morgue after you have counted them.
2. Under a dissecting microscope on a clean piece of paper, push the flies with wild type eye color upward and move the flies with brown sepia colored eyes downward.
3. Within each subgroup, sort the flies based on the condition of the wings. Push the flies with normal wings upward and the flies with vestigial wings downward. At the end of this procedure, you will have four groups of flies with unique combinations of characters. You should have produced the following four groups of flies with the following phenotypes:

S_V_	Wild type for eye color and wing design
S_vgvg	Wild type for eye color with vestigial wings
ssW_	Sepia eye color with normal wings
ssvv	Sepia eye color with vestigial wings

Enter the number of flies in the data table provided and on the board at the front of the room. You can let the F-2 fly vials recuperate for a few days and count the number of new young which emerged since the first count.

Table 1. Results of the F-2 Cross

	Your Data	Class Data	Expected Numbers
S_V_			
ssV_			
S_vv			
ssvv			

Analyze the above cross for the monohybrid results also. Add together the numbers of flies with wild type eye color and brown eyes disregarding the condition of the wings. Secondly, combine the numbers of normal winged individuals and vestigial winged individuals disregarding eye color. Place these numbers in the table below. What phenotypic ratios emerged for these two genes when analyzed in the monohybrid fashion?

Table 2. Results Analyzed as Two Monohybrid Crosses

	Your Data	Class Data	Expected Results
Wild eye			
Sepia eye			
Normal wing			
Vestigial			

Activity 4

Lab Report of Fruit Fly Crosses

Be sure and follow the lab report format of Appendix A.

In your **Introduction** describe how the Mendelian laws of inheritance should work in the inheritance of eye color and wing design in the fruit fly. Describe why fruit flies were used as the experimental animal. Describe the mutations which are being followed. State the questions you seek to answer. What is your major purpose in performing a dihybrid cross? Which Mendelian law(s) are you investigating?

In your **Results** section include your data tables for both the monohybrid and the dihybrid situations. Include a Punnett square analysis for the above monohybrid and dihybrid crosses. Use the Punnett square analyses to predict the expected monohybrid and dihybrid phenotypic ratios of the F-2 offspring. Both data tables should have a predicted and actual "ratio" column. Calculate chi-square values for the monohybrid and dihybrid crosses and compare the calculated values to the tabular values as shown in the example in Appendix C.

Make sure that you address (at a minimum) the following points in your "**Discussion** section":

1. What patterns are shown by your data for the monohybrid and dihybrid situations as compared to the expected ratios.
2. Does your data agree with the predicted ratios of the Punnett squares or does your data deviate from the expected results? Why or why not? Use the chi-square analysis to validate your position.
3. Are these two genes linked or do they sort independently during meiosis? Justify your decision using the experimental data obtained in this exercise.
4. If your dihybrid numbers do not follow the expected pattern, analyze the monohybrid patterns to see which of the two traits caused your dihybrid ratios to be off?
5. In your analysis, discuss any circumstances which may have produced discrepancies in the actual data which differ significantly from what you might have expected.

AN INTRODUCTION TO GENETICS

20

Objectives

1. To be able to use the vocabulary of genetics.
2. To use human genetic traits to document the diversity of genotypes in the human population.
3. To count the phenotypic ratio of the kernels of corn to determine the genotypes of the parents and to construct supported inferences concerning the way in which the trait was inherited.
4. To be able to solve typical genetics problems.
5. To be able to interpret pedigrees.

Introduction

Gregor Mendel, an Austrian monk, authored a scientific paper entitled "Experiments in Plant-Hybridization" in 1866. This work, later to become the basis for the study of genetics and inheritance, went largely unnoticed until it was re-discovered independently by several European scientists in 1900. Mendel's main contribution to the study of genetics replaced the blending theory of inheritance, which stated that parental traits mixed with each other in the offspring. His particulate theory of inheritance states that characteristics are controlled by two unit factors (now called alleles) and that an individual receives one allele from each parent. These alleles (unit factors) remain separate and distinct in the offspring.

The unique feature of the particulate theory of inheritance is that the genetic factors (alleles) remain independent from one another in the offspring. When gametes are formed, the unit factors (alleles) separate during meiosis so that only one of each pair is contained in a particular sex cell. Each gamete, due to the alignment of homologous pairs of chromosomes at metaphase I of meiosis, has an equal probability of possessing either one of the two unit factors present in the original diploid cell. The particulate theory of inheritance helps to explain Mendel's First Law, or the **Law of Segregation.**

Some Basic Terminology

Before beginning the following exercises, a review of some of the basic terms of genetics is useful. A **gene** is a unit factor controlling an individual trait of an organism. As the semester progresses this definition will be refined, but now it might be useful to consider a gene to be a location (locus) on a chromosome where genetic information is stored. The specific information which is stored at a gene represents the **alleles** of that gene. In other words, alleles are the alternate forms of expression which the gene may produce. The gene producing eye color possesses two basic alleles. One allele results in deposition of the pigment melanin in the iris resulting in the production pigmented eyes. The alternate allele results in no deposition of pigment producing blue eyes. Usually in simple cases, one allele is **dominant** (represented by a capital letter, ie P) over (overshadows) the other allele. Alleles whose effects are masked by dominant alleles are said to be **recessive** and are usually represented by the lower case letter (p). Since individuals are diploid and have two alleles for every trait, their **genotype** includes one allele from each of their parents. A person's genotype includes all of the alleles which are present in the nucleus whether they appear in the individual's phenotype or not. The **phenotype** is the physical appearance of the individual (ie tall or short, bald or hairy). An individual can have a **homozygous** genotype if the two alleles present are identical. If the alleles are different from one another, the individual is said to be **heterozygous** for that trait. A heterozygous individual will show the dominant trait in its phenotype. An individual exhibiting the recessive phenotype must be homozygous to express the recessive trait.

Activity 1

Some Simple Mendelian Human Traits

With the exception of identical twins, it is very unlikely that any two humans would have the same combination of genetically determined traits. Remember how important variability is to the working of natural selection in Darwinian evolution theory? Working with a partner, determine whether or not you possess the following characteristics. Indicate your phenotype and possible genotypes for each trait. Remember **phenotype** is a description of the physical appearance of a trait. The **genotype** is expressed using letters to signify what alleles are present in the nucleus of the cell. If not completely sure about the genotype of a dominant phenotype, indicate the second allele as being (-).

1. **Ability to taste PTC.** Place a small piece of filter paper impregnated with PTC (phenylthiocarbamide) on the tip of your tongue. If you do not have any sensation, chew the paper. The chemical does not pose a health risk. If you only taste "paper" then you are a nontaster (tt). If you experience a definitive taste sensation, you are "taster." You will definitely know the distinction between the two conditions. The ability to taste PTC (T-) is dominant over the nontasting condition (tt).

   Your phenotype:_______________ Partner's phenotype:_______________

   Your possible genotypes________ Partner's possible genotypes________

   If you and your partner were of different sexes and were to have children, list all the possible genotypes that your children might have:

   If your first child were a non-taster, then your genotype would have to be __________ and your partner's genotype would have to be__________.

2. **Pigmented Iris.** The pigment of the eye is located on the front part of the iris. A person with blue eyes has no pigment in front of the eyes (ee). The eyes become darker with increasing amounts of pigment, passing through hazel, green, to gray, to brown and finally to dark brown. Several pairs of genes are probably involved in the production of eye color, but for convenience, let us consider having any pigment (E-) in the iris as dominant to the blue eyed condition (ee).

   Your phenotype:_______________ Partner's phenotype:_______________

   Your possible genotypes________ Partner's possible genotypes________

   If you and your partner were siblings, what possible combinations of genotypes might your parents have? List **all possible** combinations.

If your mother possessed blue eyes, then your genotype would have to be___________. Your partner's genotype would have to be_________.

3. **Mid-digital hair.** The complete absence of hair in the middle segment of the fingers is a recessive trait (mm). The presence of hair on one or more of the fingers is controlled by the dominant allele (M).

   Your phenotype:_______________ Partner's phenotype:_______________

   Your possible genotypes_______ Partner's possible genotypes_______

4. **Ear lobes.** Attached earlobes (ll) are recessive to free earlobes (L-). Free ear lobes do not directly connect to the cheek area of the face.

   Your phenotype:_______________ Partner's phenotype:_______________

   Your possible genotypes_______ Partner's possible genotypes_______

5. **Widow's Peak.** If your hairline shows a downward pointing extension in the middle of your forehead, you exhibit the dominant phenotype for hairline (W-). If your hairline is straight or curved along your forehead, then you exhibit the recessive condition (ww).

   Your phenotype:_______________ Partner's phenotype:_______________

   Your possible genotypes_______ Partner's possible genotypes_______

6. **Tongue Rolling.** With tongue extended, some people can roll their tongues into a U-shaped trough (R-) while others cannot. Tongue rolling is dominant to the inability to roll one's tongue (rr).

   Your phenotype:_______________ Partner's phenotype:_______________

   Your possible genotypes_______ Partner's possible genotypes_______

7. **Hand clasp.** Clasp your hands together so that your fingers interlock. If the left thumb folds over the right thumb, you exhibit the dominant phenotype (H-). If your right thumb falls naturally over your left thumb, you exhibit the recessive phenotype (hh).

   Your phenotype:_______________ Partner's phenotype:_______________

   Your possible genotypes_______ Partner's possible genotypes_______

Activity 2

Inheritance in Maize (Indian Corn)

The genetic makeup of an individual can never be seen directly but must be inferred from its appearance (phenotype). In other words, you must look at the characteristics of an individual (or a group of individuals) with the same parents to figure out what their genotypes might be. Corn is a good model to work with because a large family of siblings reside beside each other on each ear. When the ear of corn is developing a strand of silk is attached to each young kernel. Taken together the silk and the young kernels are the pistils of the corn flower. Remember from plant reproduction that the female pistil is composed of a stigma (pollen deposition), style (long neck of the pistil), and ovary (home of the pistil's ovule). The young kernel is the ovary which contains a single egg cell while the silk is the style of the pistil. When corn is pollinated, the pollen grains land on the end of the silk and a pollen tube grows down the silk carrying sperm to fertilize the egg located in each ovule. Each kernel of corn is fertilized individually and represents a unique genetic individual. When you examine an ear of corn, you are looking at several hundred offspring of one genetic cross.

Your task in this activity is to determine the probable genotype of the parents for each ear of corn and the probable genotype for each phenotype represented.

To appreciate the genetics, an understanding of the biology of the corn kernel is necessary. A kernel of corn possesses three distinct layers. The outer skin layer is called the **pericarp.** A middle layer just under the pericarp containing protein is called the **aleurone.** The starch core of the kernels is called the **endosperm.** In the following activities the pericarp is clear, so the colors you see will be coming from the aleurone and the endosperm beneath it.

If the aleurone contains anthocyanin pigments, the corn will be **purple,** with the purple color masking the underlying color of the endosperm. If the aleurone contains no pigment, the endosperm will be visible beneath it and the kernel will be yellow. The first trait which you will follow is the inheritance of color of the kernel.

The endosperm may have a high or a low starch content. Remember from plant reproduction that the endosperm of a seed is triploid tissue representing a food resource for the developing embryo. Field corn is high in starch (low sugar). The presence of the starch causes the kernel to absorb water making the kernel appear plump. Sweet corn is low in starch and high in sugar content. Due to the low starch content, the kernel loses water during the "milk" stage of its development making the kernel appear wrinkled when it is dry. The second trait which you will be following is the starch/sweet characteristic.

Remember that the phenotypic ratios you discuss in class are based on an infinite number of individuals. Even though you will count several hundred corn kernels, your ratios will not be exactly the same as those predicted by a Punnett square.

A. Monohybrid Crosses

In the monohybrid crosses, you will be following the inheritance of one trait at a time. Two of the pieces of corn that you will be examining differ in the color of the kernels. The kernels are either a purple or a yellow color. The second characteristic to be investigated is the property of whether the carbohydrate of the kernel is stored as starch or glucose.

Trait	Phenotype	Allele
1	Purple:Yellow	R/r
2	Starchy:Sweet	Su/su

For each of the **four ears** you should provide the following information:

1. Description of the phenotypes
2. Total count for each phenotype (the whole cob)
3. Actual phenotypic ratio and the Mendelian ratio inferred
4. The dominant and recessive alleles with a support statement
5. The genotype of the parents and Punnett square of the cross
6. The possible genotypes for each phenotype
7. A Chi-square analysis of your data sets as illustrated in Appendix C of the lab manual.

B. Dihybrid Cross

In dihybrid crosses the inheritance patterns of two genes are followed at the same time.

Case One: Challenging

Examine two ears of corn (purple:yellow:starchy:sweet). One cob should approximate a 9:3:3:1 ratio with the other ear showing a 1:1:1:1 ratio. For each ear of corn

1. Report the total count for each phenotype.
2. Calculate the actual phenotypic ratio from each ear and compare it to the theoretical value.
3. Give the genotypes of the parents of each ear of corn showing the Punnett square for each ear.
4. Show the possible genotypes for each of the four phenotypes.
5. A Chi-square analysis of your data sets as illustrated in Appendix C of your lab manual.

Case Two: More challenging

The endosperm is not always visible through the aleurone, but if it is the aleurone may be yellow or white. (Y) is used to identify the gene that controls endosperm color. Pick up a Yellow:White ear of corn and answer the following questions:

1. Which allele for the Y gene is dominant, yellow or white? Construct a Punnett square as proof.
2. What is the phenotype of the aleurone?
3. With respect to the endosperm, what is the genotype of the aleurone?

Case Three: Most Challenging

The aleurone has at least two genes that control color. One is the Purple (R) gene described above and the other is a dominant color inhibitor gene (C1). For example, in the genotype C1CRR, there would be no pigment in the aleurone because the dominant allele C1 blocks pigment formation. Locate the Yellow:Purple ear that claims a phenotypic ratio of 13:3.

1. How close is the phenotypic ratio to the claimed ratio.
2. The parents of this ear were C1CRr X C1CRr. Explain the ratio using a Punnett square.

Activity 3
Genetics Problems

Check with your laboratory instructor to see if you are to work these problems.

The best way to understand genetics is to work genetics problems. The easier problems listed first will follow simple Mendelian ratios. Some of the later problems will be variations on the Mendelian theme. You will have to draw from lecture information and also from the text book to answer these questions.

In solving genetic problems, you should also follow the same format in order to facilitate someone following your work. Remember to always:

1. Define the alleles that will represent each characteristic.
2. Write down the suspected genotypes of the parents involved.
3. Work through the problem showing the Punnett square employed.
4. Write down and label your answer to the problem. Be sure and determine whether the problem is asking for a phenotypic or genotypic ratio.

1. In man, normal pigmentation is due to a dominant allele (A), albinism to the recessive allele (a). A normal man (if there is such a thing) marries an albino female. Their first child is an albino. Is the man homozygous or heterozygous for the pigmentation gene?

2. In man, the allele for brown eyes is dominant to that for blue eyes. A man with blue eyes marries a brown-eyed woman whose mother had blue eyes. What proportion of the children would be expected to have blue eyes?

3. All of the offspring produced by crossing red four o'clock flowers with white four o'clock flowers have pink flowers. What proportions of what flower colors will be found among the offspring of each of the following crosses?

   a. pink X pink

   b. pink X red

   c. pink X white

   d. red X white

4. In mice the genotype yy is gray, Yy is yellow, and the YY genotype dies as a small embryo.

   a. What offspring would be expected from a cross between a yellow mouse and a gray mouse?

   b. What offspring would be expected from a cross between two yellow mice?

   c. In which of the two above crosses would you expect the larger litter size?

5. The ability to taste phenyl-thio-urea is dominant to the inability to taste the chemical. In a second nonlinked gene, the brown eyed allele is dominant to the blue eyed allele. The parents of a cross are both heterozygous for these two genes.

   a. What proportion of their offspring would be blue eyed tasters for phenyl-thio-urea?

   b. What proportion would be blue eyed nontasters?

   c. What offspring would be produced from a male heterozygous for the above two traits and a nontasting blue eyed female?

6. In cocker spaniels, the dominant allele for hair color (B) produces a black coat. The recessive allele produces a red coat. A second nonlinked gene possesses a dominant allele (S) that produces a solid color pattern. Its recessive allele produces a white spotting effect.

   a. What offspring are to be expected from a mating of two spotted red dogs?

   b. What offspring are to be expected from a mating of two dogs heterozygous for both of the above two traits?

7. In man, either the homozygous recessive state of gene A or one dose of the dominant allele from gene B produces a form of blindness called "retinitis pigmentosa." Only the genotypes of AAbb or Aabb produce a normally sighted individual. A woman with "retinitis pigmentosa," whose parents both had normal vision, marries a man with the genotype of AaBb. What proportion of their children would be expected to be blind?

8. In tomatoes, round fruit shape (O) is dominant over elongate (o) fruit. Smooth fruit skin (P) is dominant over peach skin (p). Test crosses of F_1 individuals heterozygous for these pairs of characters gave the following results:

smooth and round	12
smooth and long	123
peach and round	133
peach and long	12

Explain the inheritance seen in the above results. Are the genes controlling fruit shape and fruit skin linked? What percent of recombination do the data indicate? How far apart are the two genes expressed in map units?

9. Four pairs of genes are known in cucumbers. One controls leaf color [green (G) and yellow-green (g)]; a second gene controls fruit skin texture [spiny (S) and spineless (s)]; a third gene controls fruit shape [elongate (E) and oval (e)]; and a fourth gene controls plant height [tall (T) and short (t)]. Crosses involving these four characteristics indicate that all four pairs of genes must be on the same chromosome. Indicate the order of the four genes on the chromosome:

   G and S = 28% recombination
   S and E = 24% recombination
   G and E = 4% recombination
   G and T= 35% recombination
   S and T = 7% recombination

10. Two genes in onions [(P = purple bulbs, p = white bulbs), (T = tall plants, t = short plants)] are known to be 5 map units apart. A purple bulb tall plant was crossed with a white bulb short plant. Both lines were pure. The resulting F_1 plants were then test crossed. What will be the expected percentages of phenotypic classes in the offspring?

11. In cats, the genotype BB is black, Bb is tortoiseshell, and bb is yellow. The gene is on the X chromosome. Sex determination in the cat is the same as it is in humans. A tortoiseshell female is crossed with a black male. What offspring and in what ratios would be expected from this mating? Would you see any tortoiseshell male cats?

12. In humans, hemophilia is inherited as a sex-linked recessive gene. Queen Victoria, who did not have hemophilia, had nine children: five normal daughters, three normal sons, and one son with hemophilia. Two of the daughters produced one or more sons with the disease. What was the genotype of Queen Victoria with respect to this gene?

13. In the human, one type of toothlessness is due to a recessive gene carried on the X chromosome. What offspring would be expected from a toothless woman and a normal man?

14. Color blindness in man is sex-linked and recessive to normal vision.

    a. A normal woman whose father was color blind marries a color blind man. What types of children may be expected, and in what proportion?

    b. What types of offspring may be expected if she marries a normal man instead?

15. A woman with type A blood accuses Bob of being the father of her child. The child has type B blood. Bob furnishes proof that he possesses blood type O. Is Bob the father or not? Explain your reasoning.

16. A woman (blood type A) and a man (blood type B) produce a child with blood type O.

    a. Show the genotypes of both parents.

    b. What types of offspring and in what proportions would you expect this couple to produce?

Activity 4
Pedigrees

Much of our knowledge of human inheritance is the result of long studies of family trees or pedigrees. It is customary in pedigrees to represent males with squares, females with circles, individuals showing the particular trait by filled-in squares or circles, and individuals not showing the trait with open squares or circles.

Could you tell if a trait were dominant or recessive strictly from looking at a pedigree? Could you determine the genotypes of the individuals represented in a pedigree? Try to construct supported inferences for the following three human pedigrees.

In working with pedigrees, first determine which is the dominant allele. This can be done in two ways. First, search for parents who are identical in phenotype who produce an offspring different from themselves. The child's trait must be recessive since it did not show up in either of the two parents. A second situation showing dominance exists where two dissimilar parents produce only offspring like one of the two parents. What dominance relationship does this suggest?

Construct three inferences about the inheritance of the traits in the following pedigrees. Give supporting evidence from the pedigree which supports your inferences. These inferences can deal with dominance relationships, or genotypes of specific individuals in the pedigree.

Pedigree #1: Dimpled Cheeks

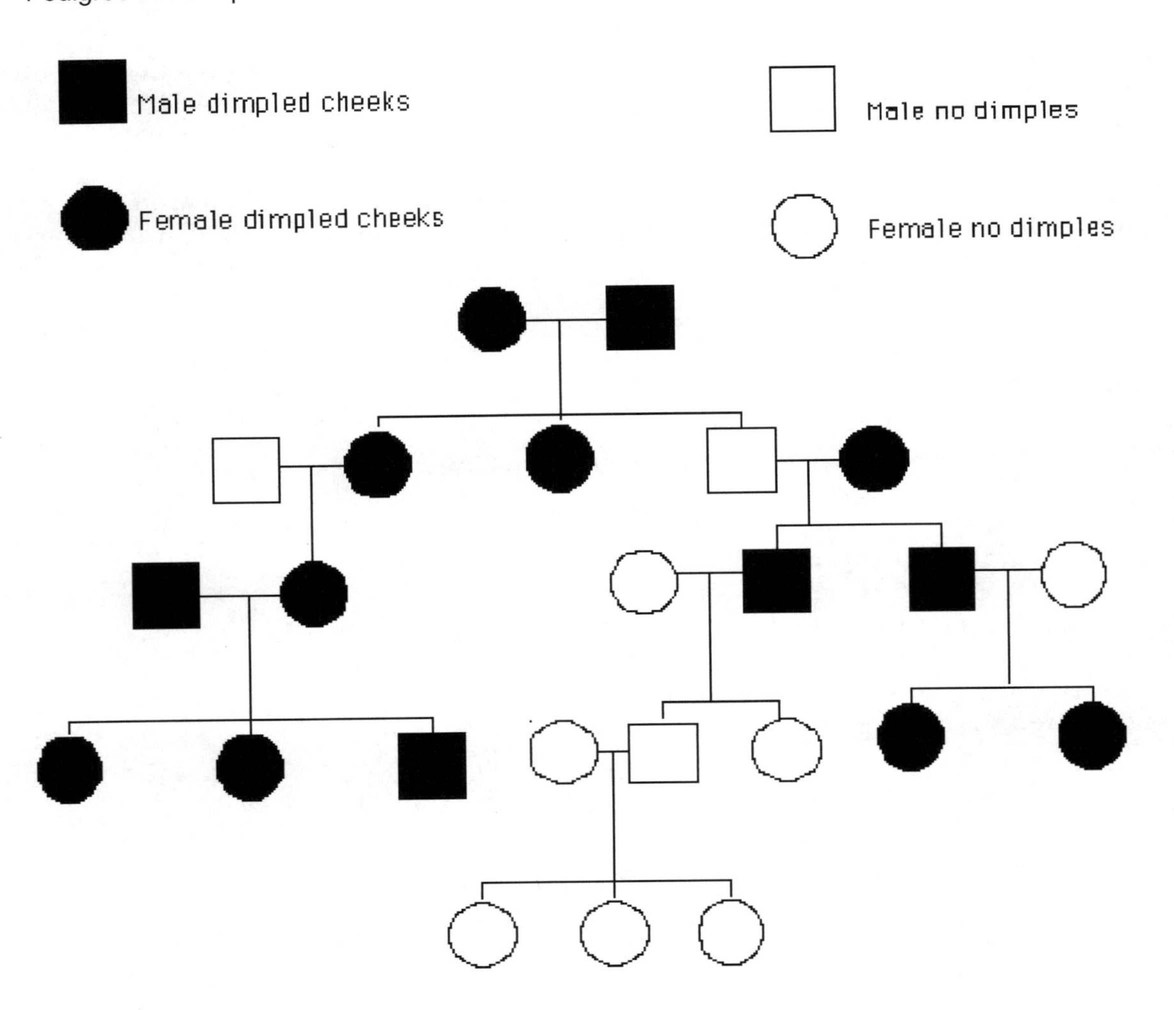

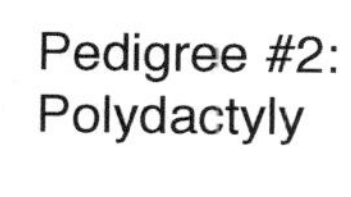

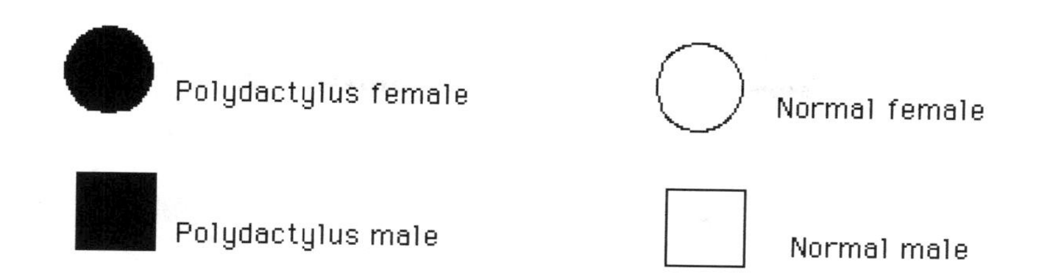

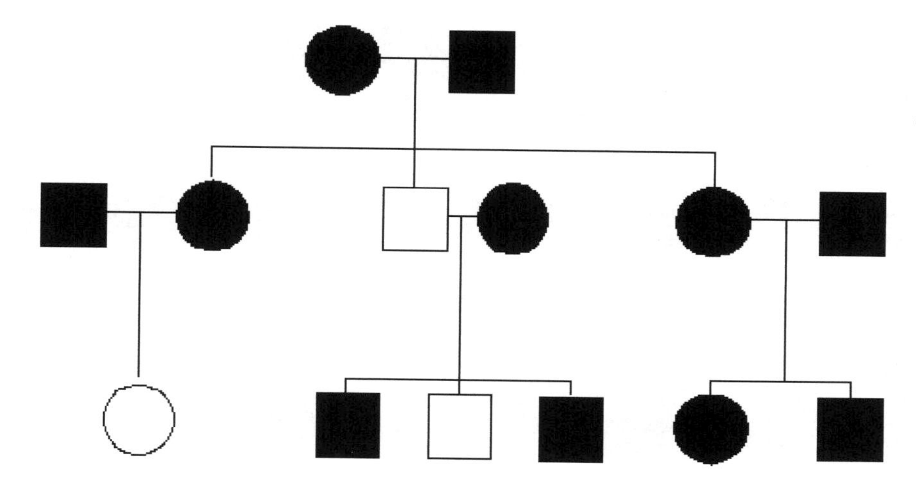

Pedigree #3:
Palmar Muscle

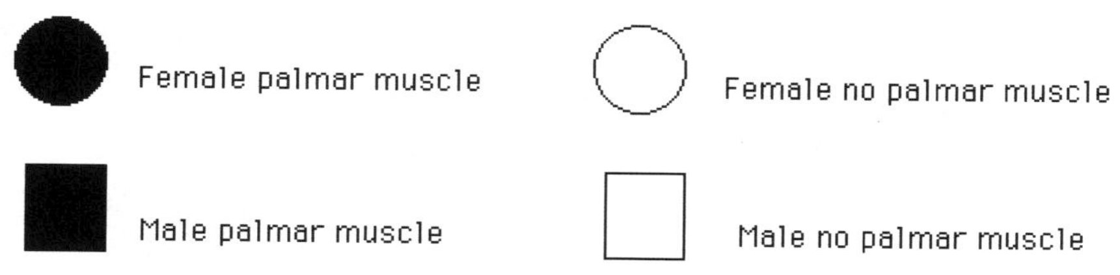

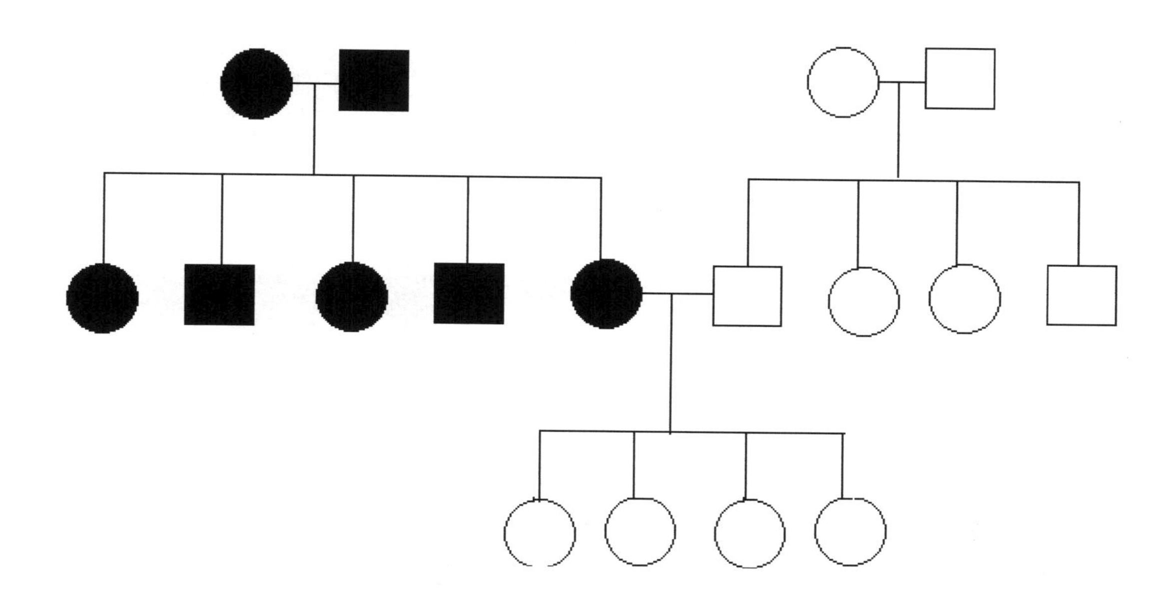

Questions for Further Thought and Study

1. People heterozygous for a trait are often called carriers of that trait. What does this mean?

2. What determines how often a phenotype occurs in a population?

3. Is it possible to determine the genotype of an individual having a dominant phenotype? If so, how? If not, why not?

4. What blood type(s) are not possible for children with parents both having AB blood? both having O type blood?

5. Are dominant characteristics always more frequent in a population than recessive characteristics? Why or why not?

AN INTRODUCTION TO THE FUNGI

21

Objectives

1. To become familiar with the general structure of a fungus.

2. To make labelled sketches of fungi representative of three different fungal divisions.

3. To place unknown fungi into their correct division based upon structural characteristics learned in the drawing of detailed sketches.

Introduction

Fungal organisms are very unusual creatures which exhibit some plant-like and some animal-like characteristics. They also have some traits that are unique to their own group. The fungi possess cell walls and are usually attached to a substrate (being nonmobile). Fungi however lack chlorophyll characteristic of plants. They are heterotrophic, depending upon pre-formed organic matter as animal cells do.

The organization of a fungus is hard for the average person to grasp. A fungus is composed of a tangled mass of individual hypha (pl. hyphae). A **hypha** is a slender filament of protoplasm which grows into the substrate. The hyphae may or may not be divided into individual cells. Some fungi have hyphae possessing septa which divide the filament into individual cells while others do not. The septa of some hyphae possess perforations which allows cytoplasmic strands to join adjacent cells. Some fungi are characterized by having incomplete septal walls, while other types of fungi lack septa altogether. The lack of partitions in these hyphae promotes cytoplasmic streaming which is very important in the distribution of nutrients to the fungal mass. The term **mycelium** is used to describe the cotton-like mass of hyphae composing an individual organism. Mycelium may extensively invade soil, water, and living, or dead tissue.

Hyphae are enclosed by a cell wall composed of chitin. The cell walls of plants are composed of cellulose. Chitin is a substance not found in the Monera, Protista, or Plant kingdoms, but is unique to the Fungi and Animal groups. The presence of the rigid chitinous cell wall prevents fungi from taking food into their cytoplasm by phagocytosis. This necessitates the release of digestive enzymes into the medium through which the hyphae are growing with the subsequent extracellular digestion of surrounding nutrients. The products of this digestive process can then be absorbed across the hyphal membranes. The fungi which digest the remains of dead organisms are called **saprophytes.** Other fungi depend upon living organisms for their sustenance and are therefore called **parasites.** The parasitic fungi may have slender hyphae which penetrate living cells and absorb nutrients directly.

The general life cycle of the fungi is different than that of the plant and animal world. Almost all of the nuclei of a fungal mass possess haploid nuclei. **Asexual reproduction** is very characteristic of fungi involving mitotic production of haploid spores produced in structures called **sporangia.** The sporangia are usually borne on stalks that facilitate the dissemination of the spores. **Sexual reproduction** occurs often toward the end of a growth season. Two dissimilar cell types of the same species (+, - mating types) have a fusion of their hyphae. In the chamber formed by the fused hyphae, haploid nuclei acting as gametes come together to form the only diploid portion of the typical fungal life cycle. Very shortly after "fertilization," the zygote undergoes meiosis returning the fungus to the haploid condition. The fungal kingdom is separated into three different divisions (phyla) based upon the different reproductive structures formed in this life cycle.

You will be making detailed sketches of the fungi discussed in the following sections labelling all the indicated structures in **bold type.**

Activity 1
Making Sketches of Representative Fungi

A. Division Zygomycota-Bread Molds

Using a slide of ***Rhizopus,*** locate the following structures. When looking at this slide, your first impression might be that it is nothing more than a tangled mass of cytoplasmic threads that have been smeared across the slide. By careful examination and imagination, you will be able to be see and sketch the following structures in some detail. You should take artistic license and make your sketches realistic. Simplify the mass of threads and have the **stolons** pass horizontally across your sketch with the **rhizoids** branching vertically downward from the stolons. For best results, stay near the edges of the hyphal mat. Each threadlike strand that you see is an individual **hypha** of the *Rhizopus*. The hyphae of the zygomycetes lack septa (cross walls). Cytoplasm, nuclei, and organelles freely move through the **mycelium** (total collection of hyphae seen) by cytoplasmic streaming. There are different types of hyphae which can be distinguished in the slide of *Rhizopus sp.* Horizontal hyphae which grow across the surface of the substrate are called **stolons.** The stolons tend to be of a larger diameter and straighter than the two other categories of hyphae. The more slender and shorter hyphae which branch from the stolons to penetrate the substrate for anchorage and absorption of nutrients are called **rhizoids.** Upright stalks which arise from the stolons carrying **sporangium** are called **sporangiophores.** The sporangiophores elevate the sporangia from the surface facilitating the dispersal of released spores. In your sketch note that some of the sporangia are immature and do not yet carry spores.

Place a slide of *Rhizopus* conjugation on your microscope. This slide was made from a culture containing the two different mating types of Rhizopus (+,-). The most prominent structures seen on the slide, called **zygosporangia,** are the result of the sexual reproductive effort between the two different mating strands. Zygosporangia are formed when **gametangia** of opposite mating types fuse at their end walls. Within the common cytoplasm, two opposite nuclei come together to form the zygote. The zygote develops a large, dark, and granular protective **zygosporangium** around itself. The complex almost resembles two ice-cream cones which have been mashed together at their ice-cream ends. The ice-cream would represent the darkened zygosporangium with the tapering wafer cones being the two gametangia which originally fused in this reproductive effort. Meiosis takes place within the zygosporangia returning the fungus to the haploid state. By careful scanning and observation, you should be able to locate **gametangia** that have just contacted one another prior to the formation of a zygosporangium. You should also be able to find immature zygosporangium which have not yet developed their characteristic black coloration as well as a few senescent zygosporangia which have ruptured releasing the products of the reproductive effort.

B. Division Ascomycota-Sac Fungi

The Ascomycetes sexually reproduce by forming a sac-shaped structure called an **ascus.** Different mating strains of the same species fuse forming a class of hyphae referred to as dikaryotic hyphae. These hyphal stands possess complete septa partitioning the cytoplasm into cells containing two haploid nuclei each. These dikaryotic hyphae form a fruiting body known as an **ascocarp.** Within the ascocarp, specialized sacs called **asci** (sing. ascus) develop where two nuclei of the different mating types fuse to form a zygote. The zygote immediately undergoes meiosis within the ascus to form four **ascospores.** These four undergo a mitotic division to produce the final eight ascospores located within the ascus.

Obtain a slide of *Morchella* and view it under low power. Relate its overall structure to the photograph on p.577 of your text book (Campbell, *Biology*). The fruiting body (ascocarp) shows many fenestrations (windows) which appear as channels in your slide which communicate with the outside of the ascocarp. Lining these channels are the numerous **asci** which house the ascospores. How many ascospores do you see within each ascus? How are the ascospores arranged? Make sure that these details are included in your sketch.

The ascomycetes reproduce **asexually** by producing spores called **conidia.** Obtain a slide of *Penicillium* conidia. Within the mass of hyphal fibers find the structures which resemble "hands" in a rough way. The palms of the hands are called **conidiophores.** The fingers of the hands represent the rows of nuclei which will be packaged into membranes to form the conidia (spores). Conidiophores grow upright elevating the conidia above the substrate to which the fungus is attached. This aids in the dispersal of mature condia.

C. Division Basidiomycota: Club Fungi

The Basidiomycetes are probably the most familiar fungi and include the mushrooms, puff balls, and shelf fungi which we have all seen. The above ground visible portions of these fungi are actually just reproductive structures (fruiting bodies) which produce millions of spores for the fungal organism. During sexual reproduction, cells within mycelia of different mating strains touch and fuse. During growth, the underground mycelium eventually protrudes above the substrate as tightly bundled groups of hyphae called basidiocarps (mushrooms, puff balls, or shelf fungi). Within the basidiocarp, there are numerous basidia which will produce and release the sexual spores.

Obtain a slide of *Coprinus* and observe it under low power. This slide is made from a cross section through a common mushroom at right angles to it stem. In a sketch made under low power, label the **cap** (pileus), with its long finger like **gills** extending inward toward the **stalk** (stipe) of the mushroom. This fruiting body **(basidiocarp)** is an aggregate of tightly bound hyphae. Move to high power to observe and sketch the surface of the gills. **Basidia** are club shaped cells where union of haploid nuclei occurs forming the zygote of the life cycle. This zygote immediately undergoes meiosis to produce the numerous dark staining **basidiospores** attached to the basidia. Try to imagine the number of spores which a single mushroom can produce. This is just one very thin horizontal slice through the mushroom. Multiply the number of spores on this section by four hundred or more possible cuts that could have been made on the same mushroom. Obviously one mushroom is capable of producing many spores.

Activity 2
Classification of Unknowns

There are four slides of fungi at the front of the room which you are going to observe and sketch. Based upon the work that you have previously done, you are going to classify each one of the four unknowns into the correct fungal division based upon the criteria above. I expect you to justify your classification of each of the four unknowns in a written sentence.

You should have sketches for *Boletus,* Apple scab *(Cleistothecia), Xylaria,* and *Mucor.*

Thought Questions

1. Mushrooms often sprout in circles which increase in radius year after year. Can you explain this phenomenon?

2. What advantage does asexual reproduction have over sexual reproduction?

3. Animals and most higher plants have a dominant diploid stage in their life cycle as compared to the dominant haploid pattern seen in the fungi. Does the haploid condition offer an adaptive advantage to the fungus? Why or why not?

AN INVESTIGATION OF NUCLEIC ACIDS

22

Laboratory Objectives

1. To understand the structure of the nucleic acids.
2. To recognize the structural similarities and differences between DNA and RNA.
3. To understand the copying mechanism (replication) by which DNA molecules are produced.
4. To understand how RNA is synthesized from DNA (transcription).
5. To understand the nature of the DNA code and the methods of transcription and translation that convert these instructions into polypeptide sequences.

Introduction

The nucleic acids include DNA (deoxyribonucleic acid) and RNA (ribonucleic acid). In most organisms, hereditary information is coded in the structure of DNA. This hereditary information has been passed on to the current cell from its predecessors through the processes of meiosis and mitosis. In most cells, this family legacy (DNA structure) is protected from the everyday activities occurring in the cell cytoplasm by the nuclear membrane. However, since it is the job of the DNA of the nucleus to direct this cytoplasmic activity, a link must exist between these two compartments of the cell. It is the job of RNA to copy the genetic instructions present in the DNA molecule (transcription) and to carry these instructions to the cytoplasm for protein synthesis (translation). Proteins are the primary structural and regulatory material of the organism. DNA directs the cell's activities by directing the types and amounts of protein synthesized. It is this link that allows DNA to direct the development, growth, and repair of a cell.

Stressed throughout this course and your text book has been the very important theme of "structure determining function." Keep this theme in mind as you study nucleic acid structure.

As you study the structure of the nucleic acids, keep in mind that:

1. The DNA molecule must be capable of replicating itself in the process of cell division.
2. The structure of the DNA molecule must contain a code capable of determining protein structure.
3. The DNA molecule must be capable of transmitting its coded information to the RNA molecule.
4. The RNA must be capable of making this code available to direct protein synthesis in the cytoplasm.

Activity 1

DNA and RNA Molecular Components

You will use model kits to study the structure and properties of DNA and RNA. The following list presents the basic components of both:

Table 1. Components of DNA and RNA

Types of molecules	DNA	RNA
Sugar	*Deoxyribose*	*Ribose*
Phosphate	Phosphate	Phosphate
Purine bases are larger	Adenine Guanine	Adenine Guanine
Pyrimidine bases are smaller	Cytosine *Thymine*	Cytosine *Uracil*

As you study the list, notice that most of the components used for building DNA and RNA are the same. The exceptions are shown in italics. The sugar of DNA lacks an oxygen when compared to the RNA sugar. The pyrimidine bases of thymine and uracil are also different.

Components are represented by the kit pieces shown below. Identify each of the pieces and arrange them in stacks for use in Activity 2.

Sugars

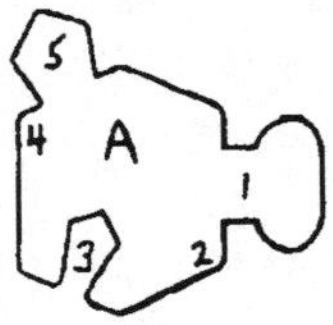

Deoxyribose
DNA only
color code: blue

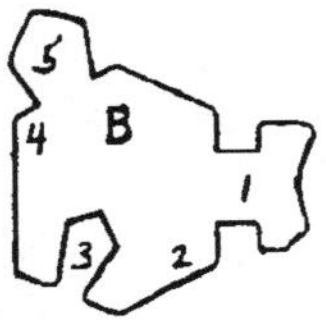

Ribose
RNA only
color code: red

The carbons of the sugars can be numbered from 1 to 5. This helps to identify where the other components will attach to the sugar.

Phosphates

both DNA and RNA
color code: green

Bases

Purines (larger bases possessing two rings)

Adenine
yellow color

Guanine
red color

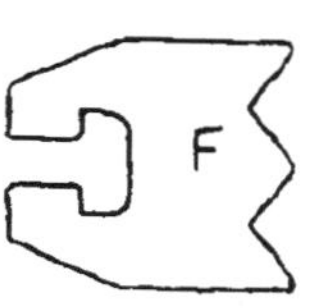

Pyrimidines (smaller bases possessing one)

Thymine
DNA only
red color

Uracil
RNA only
blue color

Cytosine
DNA and RNA
green color

Activity 2
DNA Nucleotides

The basic monomer of the DNA molecule is a unit called a nucleotide. A DNA nucleotide is made up of a sugar, a phosphate and one of the four nitrogen bases (adenine, guanine, thymine, or cytosine). All of the bonds within a nucleotide group are strong covalent bonds.

Construct the four nucleotides of DNA. Check your work with the illustrations below and at the top of the next page. Notice that the nitrogen bases attach to carbon #1 of the sugar, while the phosphate component attaches to carbon #5.

deoxyadenosine phosphate

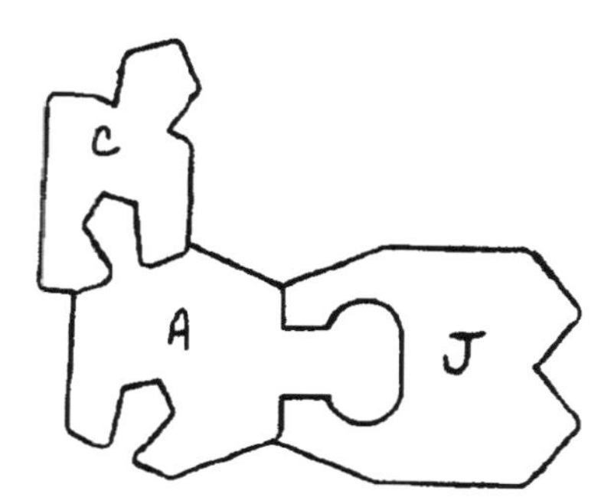

deoxythymidine phosphate

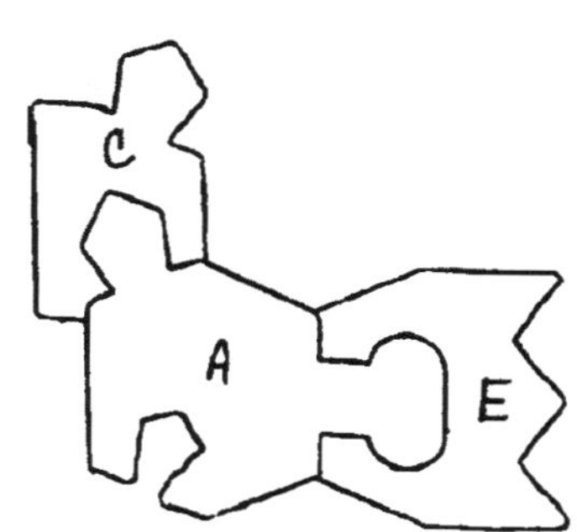

deoxyguanosine phosphate deoxycytidine phosphate

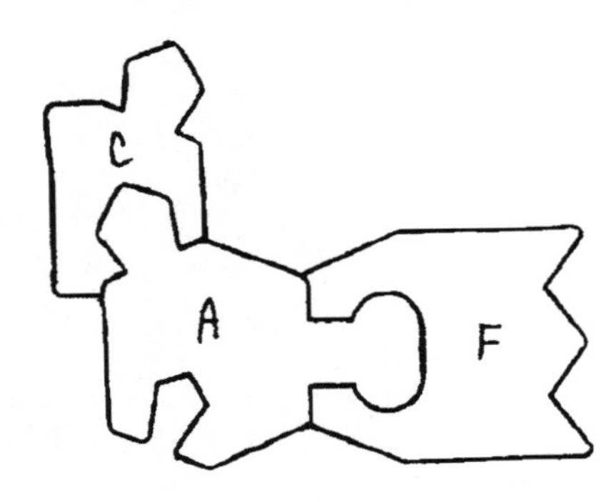

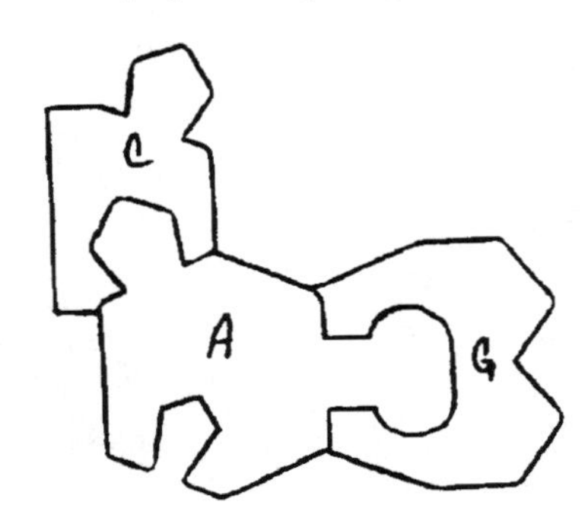

Notice that each molecule has been tagged with a cumbersome name. Luckily it is common practice to refer to the entire nucleotide by just the base name as the other components of the nucleotides are identical.

Activity 3

Two Dimensional DNA Structure

In order to build DNA, we must first join the nucleotides by strong covalent bonds. Join the four nucleotides you have just made in the following order.

5'

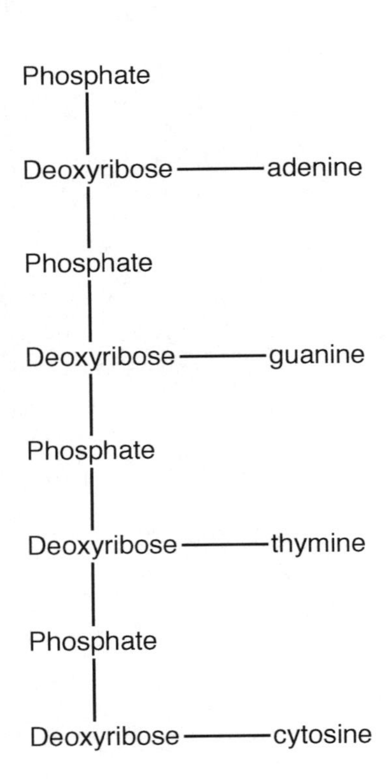

3'

You are linking the nucleotides together with 5' to 3' ester linkages. The phosphate on the 5' carbon of one nucleotide molecule has a carboxyl group (-COOH) which undergoes a dehydration synthesis with a hydroxyl group (-OH) attached to the 3' carbon of the second nucleotide molecule. Notice that the molecule produced possesses a polarity (or a top and a bottom). The 5' carbon at the top of the chain possesses a free phosphate group while the 3' carbon of the sugar at the bottom of the string has a free hydroxyl (-OH) group.

The DNA molecule is made up of thousands of nucleotide units arranged in two antiparallel chains. You have just completed a small segment of one chain.

To build the complementary chain that runs in the antiparallel direction to the short segment which you just built, it is important to be aware of a discovery made by a biochemist named Henri Chargaff. In investigating the DNA of many different species, he found that there was a constancy in the numbers of adenine and thymine nucleotides present in the organisms' DNA. These numbers would be different for the different species investigated, but within any of the species investigated, the amount of adenine always equalled the amount of thymine. Conversely, the amount of guanine always equalled the amount of cytosine. A careful examination of the structure of these four bases provides the chemical basis for this observation. Adenine and thymine fit together in a way that allows them to share two hydrogen bonds. Guanine and cytosine are capable of forming three hydrogen bonds. The specificity of these weak H bonds restricts the base pairing to

$$A{=}T \quad \text{and} \quad G{\equiv}C$$

Recognizing that adenine pairs with thymine and guanine with cytosine, construct a nucleotide chain with a sequence of bases which will base pair with the sequence in the chain that you have already made. Remember that the two ribbons of DNA should be antiparallel to each other.

What is the sequence of nucleotides in this new chain?

Join the two chains by pairing the appropriate bases. The bases should be directed toward the inside of the molecule with the sugars and phosphates toward the outside. Your model should resemble a ladder with the uprights consisting of sugars and phosphates and the rungs consisting of base pairs.

Answer the following structure/function based questions:

1. To be a "good" candidate for the job of carrying genetic information, DNA's structure must be capable of coding information. Where is the code stored in the molecule which you have just produced?
2. If building onto your model in a "downward" direction, what would be the next base pair in the sequence?
3. What might be the significance of having the base pairs on the inside of the molecule with the sugar-phosphate chain on the outside?
4. In looking at each base pair, what generalization can you make about the nature of the pairing? Is the chain always the same diameter? Why does this always occur?
5. How does the model illustrate the differences in bonding strengths of the base-base bonds as compared to the sugar-base or sugar-phosphate bonds? This difference in bond strengths allows DNA to unzip to replicate and also to serve as a template for the synthesis of RNA.

Check the diagrammatic representation on the next page to see if your work is correct.

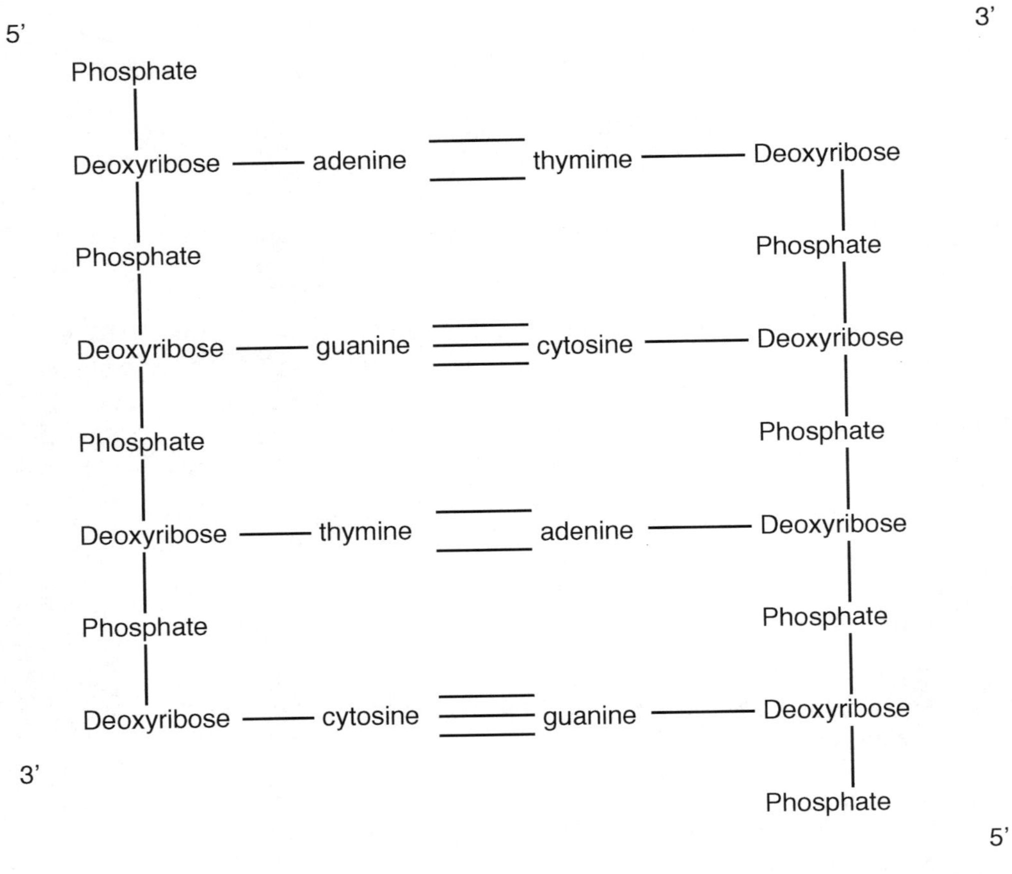

Activity 4

Three Dimensional DNA Structure

The model just created is only two dimensional. The actual molecule is more like a ladder twisted in the form of a spiral staircase. Examine the models on display in the laboratory. These show the three dimensional form of DNA. Be able to recognize the appropriate components which make up DNA structure. You should recognize and be able to find:

nucleotides	adenine	periodicities
hydrogen bonds	guanine	2.0 nm
deoxyribose	thymine	.34 nm
phosphate groups	cytosine	3.4 nm
3' end	5' end	
covalent bonds	H bonds	
pyrimidines	purines	

Activity 5

DNA Replication

Before cells undergo mitosis, DNA must be synthesized with great accuracy insuring that each new daughter cell acquires a complete set of genetic instructions. This occurs during the "S" phase of the cell life cycle. The introduction of a few genetic mistakes (mutations) might result in the death of the new cells. Duplication (in the sense of making a "Xerox" copy of the DNA) of the molecule is not enough. You have all seen what happens to the quality of a copy of a copy of a copy. It rapidly deteriorates. The replication process that we will investigate below is almost like the "rebirth" of the DNA structure which occurs every time the cell prepares for mitosis.

Structure is again related to function. At the time of replication, DNA polymerase unzips (splits) complementary base pairs separating the two strands of the original DNA molecule. This would not be possible if the base pairs were connected by covalent bonds. Each parent strand then acts as a template to which free complementary nucleotides attach. The result is the formation of two double-stranded molecules which are identical to the original parent DNA.

Using the small segment of DNA that you have just manufactured, split the molecule into its two strands. Do this by separating the molecule between the base pairs. What type of bonds are you breaking?

Now use each strand as a template to produce a small DNA segment. Remember to follow the rules for base pairing. If you complete this task correctly, you should end up with two segments of DNA exactly like the original molecule.

What does the term semi-conservative mean with reference to DNA replication?

Activity 6

DNA Transcription

The second quality that a genetic molecule must possess is an ability to put its information into physiological action-to put words into deeds. DNA is capable of doing this through its interactions with the RNA molecule.

Just as DNA served as a template for its own duplication, DNA can also serve as a template to determine RNA structure. All three types of RNA are made in the nucleus under DNA direction.

Review the basic RNA components given in Activity 1. Construct the four nucleotide units found in RNA. They are:

Adenosine phosphate
Guanosine phosphate
Uridine phosphate
Cytidine phosphate

RNA is a single stranded molecule made up of a single nucleotide chain. The nucleotide sequence of RNA is determined by the nucleotide sequence on one side (sense strand) of the DNA molecule. The opposite side of the DNA molecule which does not act as a template is referred to as the "anti-sense" strand.

In forming RNA, DNA is used as a template and the rules of base pairing are followed. Since thymine is not present in RNA, uracil takes its place. In the synthesis of RNA, **RNA polymerase** plays a similar role to DNA polymerase as it unzips the DNA molecule and assists the RNA bases in their complementary base pairings. Unzip one of your DNA molecules. Take a single DNA strand and using the RNA nucleotide units you have already constructed, make the complementary RNA strand. The DNA strand is acting as a template. You should be forming a double-stranded hybrid molecule made up of one strand of DNA and one strand of RNA. See your instructor if you are having difficulty.

Is the base sequence on the RNA identical or complementary to the DNA base sequence?

Do you suppose the double-stranded molecule just made should separate if RNA is to do its job? Why?

What is the function of the anti-sense strand of DNA that does not act as a template? What is its fate after RNA synthesis is completed?

There are three types of RNA made by the above process. They include *mRNA* which is a complementary messenger of a structural gene, *tRNA* which transports appropriate amino acids to the growing polypeptide chain, and *rRNA* which is a component of the protein factory (ribosome).

Activity 7

Translation and Protein Synthesis

The information of a gene is carried in the nucleotide sequence of the DNA molecule. Based on extensive research, biologists believe that a three base sequence (coding unit or codon) on one DNA strand determines the location of an amino acid in a polypeptide chain of a protein.

If the sequence of nucleotides along a DNA strand can be read off in groups of three, each triplet will specify the location of a specific amino acid in a linear sequence of amino acids making up the primary structure of a polypeptide. In other words, the nucleotide sequence of a gene has to be translated into the primary structure of a protein. The three types of RNA described above are necessary for this translation process. Messenger RNA (mRNA) detaches from its DNA template in the nucleus to move into the cytoplasm where protein synthesis occurs. A ribosome composed of ribosomal RNA (rRNA) attaches to the beginning of the mRNA molecule exposing successive codons to tRNA. Appropriate tRNA's complementary base to the exposed codons on the ribosome carrying with them specific and unique amino acids. With two amino acids lying side by side on the ribosome, the ribosome assists (catalyzes) the formation of a peptide bond between the two amino acids. With the completion of the peptide bond, the ribosome moves along the mRNA exposing the next codon. This cycle is continued until the entire nucleotide sequence has been read by the ribosome and the resulting polypeptide has been released.

It is difficult to understand the highly choreographed dance which must occur between the three types of RNA to produce a completed polypeptide. View the animated sequence present in lab to gain a better understanding of this intricate process.

Name ______________________________

Section ______________________________

Post-lab Questions

These questions may be assigned for the following week or used as quiz material at the discretion of your laboratory instructor. You will have to go to your textbook for some of the answers to these questions. Other questions may not have answers found in the book. You will have to use some imagination in these responses.

1. In the translation process, is the code present in mRNA read in an overlapping, partially overlapping, or nonoverlapping manner?

2. Relate the following statement to the material which we covered in today's laboratory: "What is true for *E. coli* is true for the elephant." What relevance does the statement have in today's world?

3. What is a mutation? What effect does a mutation have in the production of proteins?

4. Distinguish between frame shift and substitution mutations. Which of these is more likely to be harmful and why?

5. Why do most organisms utilize DNA as the "stuff" they make their genes out of instead of proteins: Instead of RNA?

STARFISH DISSECTION 23

Laboratory Objectives

1. To carefully observe some of the external landmarks of the starfish.
2. To perform a gross dissection of the star fish to see the "structure-function relationship" between an animal's organ systems and its way of life.
3. To examine a microscopic section of a starfish ray to more closely observe the structures discovered in the gross dissection of the starfish.

Evolutionary Relationships

In the evolution of the animal kingdom, a major evolutionary event led to the bifurcation of the kingdom into two major subsets: the **protostomes** (arthropods, annelids, and mollusc lines) and the **deuterostomes** (echinoderms and the chordates). The names "protostome" and "deuterostome" refer to the fate of each group's blastopore. Recall from the developmental biology lab that the process of gastrulation commences at the blastopore. The blastopore develops into the mouth of protostomes while it develops into the anus of deuterostomes.

Other developmental differences separate these two lineages of animal evolution. Recall that early in the development of the zygote, cleavage produces many smaller cells from the original zygote as mitosis occurs without cell growth. One cell divides into two, two cells form four, and the four cells form an eight cell stage. At this point the eight cells are arranged differently in the two animal groups. The two tiers of four cells lie directly above each other in the deuterostome lineage. In the protostomes, the cells of the second tier lie within the grooves separating the cells of the first tier. Each new successive tier of cells spirals around the main axis of the body producing the spiral cleavage pattern found in the protostomes.

Deuterostomes exhibit **indeterminant divisions** in the early cleavage stages. A cell can be removed from the organism at this point, and the animal will proceed along its normal developmental pathway. Surrounding cells are flexible enough in their developmental potentials to fill in the void. In the protostome lineage, the early divisions are **determinant.** The developmental fate of the early cells produced by cleavage is fixed. A cell's removal at this time will result in missing parts in the adult.

The final developmental difference between these two groups is the origin of the mesoderm germ layer and coelom (body cavity). In the protostomes, the mesoderm arises from cells splitting off at the base of the gastrula (schizocoelous development; split coelom). Deuterostomes have the mesoderm arising as outpocketings from the endoderm (enterocoelous development) at the end of the developing archenteron.

The real outcome of all of the above means that vertebrates (animals with backbones) are more closely related to echinoderms (starfish and the like) than to worms (annelids), clams (molluscs), and grasshoppers (arthropods). On the surface we appear to have little in common with the echinoderms (starfish). The flattened radial bodies of starfish certainly do not resemble our own, but the developmental similarities mentioned above place us in the same camp.

Introduction

Sea stars (class Asteroidea) have arms or **rays** which radiate from a **central disc.** They belong to a phyla which exhibits radial symmetry in the adult form, although as larvae they are bilaterally symmetrical. They have a well-developed coelom (body cavity) lined by ciliated peritoneal cells, an **endoskeleton** of calcareous ossicles and spines (echinoderms = spiny skin), a complete digestive tract possessing both a mouth and an anus, and a **water vascular system** which aids in their locomotion. Due to their low metabolic rate, these animals lack specialized respiratory surfaces or circulatory systems. They simply do not require these systems which can be expensive to operate and maintain.

Starfish are active predators. This seems to contradict the paragraph above where mention of their low metabolic rate was made, but when you consider that their prey are animals such as clams, activity levels can be seen to be relative. Starfish feed in a unique manner. Using chemoreceptors to locate a prey item such as a clam, the starfish surrounds its prey, clamping onto it with its tube (sucker-like) feet. In the case of the clam, the muscle holding the valves of the clam shell shut tires at some point producing a small gap or opening between its valves. The starfish takes advantage of this small opening by everting its stomach through its own mouth into the mantle cavity of the clam. A starfish stomach can squeeze its way through a 1 mm opening. Once inside of the clam, the starfish secretes copious amounts of digestive enzymes into the clam's mantle cavity beginning the process of digestion. The digested materials then enter the digestive system of the starfish to be absorbed. As will be seen, this ultimate in "dining out" strategy influences the entire structure of the starfish's digestive tract.

Activity 1
Starfish External Anatomy

Obtain a preserved starfish from the front of the room. Place it in a dissection tray and cover it with water. This will help to "float" some of the delicate structures that you will be attempting to find on the exterior of the animal.

A. Observe the aboral surface (upper side opposite the mouth) of the starfish.

The most prominent landmark that you will find is the button-like **madreporite** of the water vascular system. Using the dissecting microscope, notice the sieve-like holes in its surface. The madreporite is the opening into the animal's water vascular system that we will be examined in more detail later. The madreporite restricts the entrance of debris and other contaminants while allowing fluid to either enter or leave the water vascular system.

Try to locate the **anus** of the starfish on the **aboral** surface. It should be found in the central portion of the disc. It may be difficult to locate on a preserved specimen, but should be observable on a freeze dried animal. Recall that the starfish digests its prey item "outside" of the starfish body, and then absorbs the products of its efforts. Little nondigestible material enters the digestive tract of the starfish. Does this help to explain the apparent insignificant size of the starfish anus? Who would have ever thought that you would be looking for a starfish anus in the first place?

Observe the blunt **spines** on the animal's aboral surface using the dissecting microscope. Remember the animal should be under a thin layer of water at this time. The spines seen are the characteristic from which this phylum gets its name. Echinoderm means "spiny skinned." At the base of these spines are two different types of small appendages which aid the animal in maintaining homeostasis. The "pincher-like" **pedicellariae** help to groom the animal's exterior keeping it remarkably free from encrusting growth. The polyp-like **dermal branchiae** are extensions of the body wall which aid in gas exchange and waste excretion. Recall that the sedentary life of a starfish allows it to get along quite well without a specialized respiratory, circulatory, or excretory system. Sketch the pedicellariae and dermal branchiae in the space provided at the top of the next page.

Scrape the animal's fleshy outer covering to reveal the internal skeleton of the starfish. It is make up of interlocking **calcareous plates and ossicles.**

B. Turn the animal over to observe the oral surface of the animal.

The mouth is surrounded by a soft membrane called the **peristome** and is guarded by **oral spines.**

Five **ambulacral grooves** radiate from the area of the mouth down the midline of each arm. Each groove is filled with **tube feet,** extensions of the water vascular system to be studied in more detail later.

Although difficult to see, a sensory tentacle and a light-sensitive eyespot are located at the tip of each arm.

Activity 2
Internal Anatomy

A. Getting Started

We will get to the internal anatomy of the sea star by removing the animal's aboral surface. This is a delicate procedure. Begin by snipping off the tip of each ray. This gives you an entry point for your scissors' tip. Avoid using a scalpel at this point. Snip along the side of each ray toward the animal's central disc. Keep the tips of the scissors pointed upward to avoid damaging internal structures. When the cuts are done, gently lift the aboral tip of each arm toward the central disc. The **hepatic ceca** may remain attached to the aboral surface that is being removed. If so, gently replace it in the lower portion of the coelomic cavity.

Carefully snip around the periphery of the central disc avoiding the madreporite. Leave this structure intact as you remove the remainder of the aboral wall. As you lift the wall gently up from the central disc, note the point of attachment of the **small intestine to the anus.** The junction is found near the fairly prominent **rectal gland.**

B. Digestive System

The entire structure of the digestive system reflects the feeding strategy of the starfish. Filling most of the volume of each of the starfish's arms are pairs of greenish colored **hepatic ceca (digestive glands).** They are responsible for producing the digestive enzymes which are used by the starfish's stomach in the digestion of its prey. Recall that the stomach of the starfish is everted through its mouth as it snakes its way into the mantle cavity of a clam. The digestive enzymes are not retained in a closed cavity like your stomach, but they are released into the mantle cavity of the clam. As the digestive enzymes are essentially diluted in this process, tremendous quantities of digestive enzymes must be manufactured and released. The products of digestion are also stored in the hepatic ceca.

Each hepatic cecum has a delicate **pyloric duct** which attaches it to the **pyloric stomach** of the starfish. A duct is a small tube which carries secretions from one organ to another. In this case the pyloric duct carries the digestive enzymes of the hepatic ceca to the stomach and the products of digestion from the stomach to the hepatic ceca for storage.

The **pyloric stomach** connects to the short inconspicuous **intestine** on its aboral surface and to the **cardiac stomach** on its oral surface. The cardiac stomach connects to the mouth through the short **esophagus.** The cardiac stomach and esophagus are the portions of the gastrointestinal tract which are everted during feeding. Carefully remove the hepatic ceca from one ray. Running from the **cardiac stomach** along the **ambulacral ossicles,** find the two slips of **retractor muscles** which pull the stomach back into the starfish body after feeding. The muscles seem to be very delicate considering their important function.

C. Reproductive System

Starfish are typically dioecious, meaning that there are two separate sexes. It is nearly impossible to determine the sex of the individual starfish unless you look microscopically at the products of the gonads.

Remove the hepatic ceca from two or three rays to observe the **gonads** lying below (more toward the oral surface of the starfish). Depending upon the time of year these animals were collected, the gonads may nearly fill the entire ray at the expense of the hepatic ceca. When the gonads are actively producing sperm and eggs, stored energy reserves are withdrawn from the hepatic ceca to be invested into reproductive products. Starfish release incredible numbers of eggs and sperm into their surroundings where external fertilization takes place. If collected out of the breeding season, the gonads are fairly inconspicuous. **Gonoducts,** not visible in the gross dissection, will be seen later in the microscopic examination of the starfish ray. Gonoducts lead from each gonad to very small genital pores located at the periphery of the central disc on the oral surface between the arms.

D. Water-vascular System

The water vascular system is unique to echinoderms. The **madreporite** (sieve-like plate observed earlier) is attached to a short **stone canal** which leads downward to the **ring canal.** The stone canal is named for its walls made of calcium. The circumoral ring canal connects and supplies fluid to the five **radial canals** that lead into the ambulacral grooves. The radial canals will be seen later in the microscopic examination of the starfish ray. The five radial canals run the length of each arm and are connected by short side branches to the **ampulla** of the **tube feet.**

Remove a few of the tube feet to examine them more closely. Each tube foot has an ampulla that serves as a hydraulic reservoir and a sucker-like base that is the key to its operation. The tube foot works on the same principal as does a "clamp-on" pencil sharpener. If water is "pulled out" of the sucker-like foot and stored in the ampulla, the lower surface of the tube foot becomes concave creating a suction force between it and the substratum. If fluid moves from the ampulla into the tube foot, its lower surface becomes convex and the grip upon the substratum is released. One tube foot alone does not produce a strong force, but many tube feet acting together strike fear into the world of the bivalve clams. See Figure 1 for an illustration of how a tube foot operates.

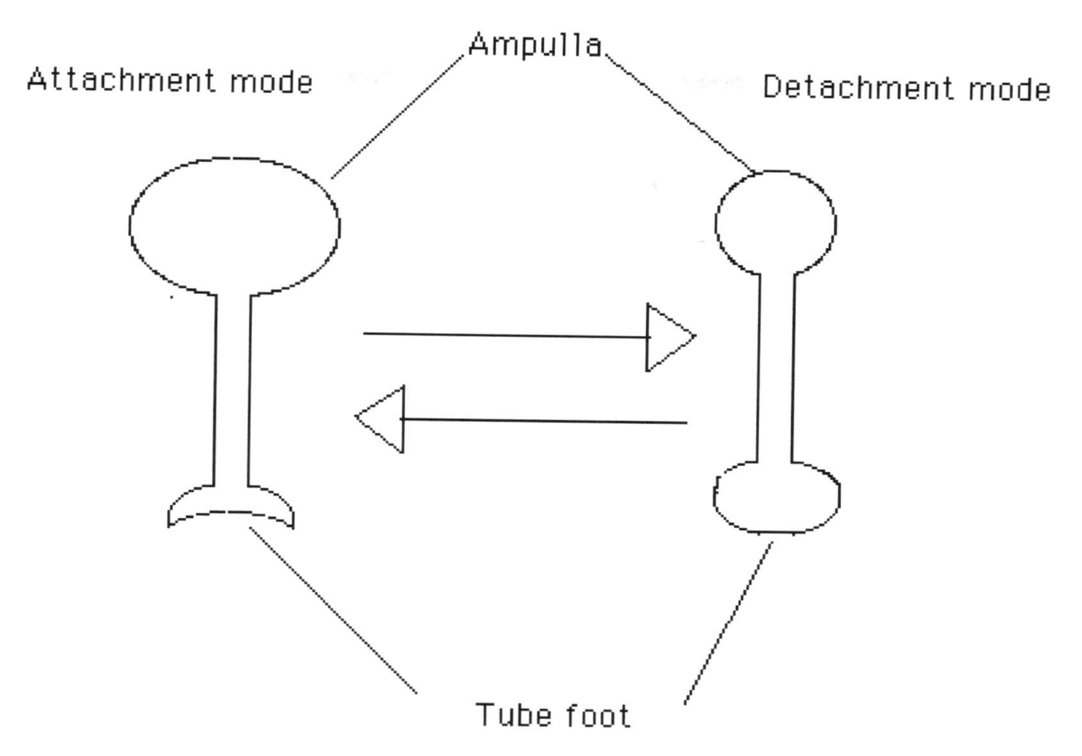

Figure 1. Operation of a Tube Foot

Lying along the inner wall of the cirumoral ring canal are nine **Tiedemann bodies.** Some biologists believe that they are important in the formation of amoebocytes which roam around the coelomic fluid cleaning up debris.

E. Nervous System

Echinoderms do not possess any cephalization. The nervous center is composed of the somewhat pentagonal circumoral nerve ring which lies within the peristomial epidermis. Recall that the peristome is the "skin" surrounding the mouth. The nerve ring will be next to impossible to see. A radial nerve extends from each angle of the nerve ring down each arm along the bottom of the ambulacral groove. If all of the tube feet are removed from one arm, the radial nerve may appear as a yellow to brown ridge running along the bottom of the ambulacral groove.

Activity 3

Microscopic Section of a Ray

Make a sketch of this entire slide illustrating the microscopic structures that we have seen in our gross dissection. Work under the 4X objective lens making sure that the aboral surface is toward the top of the field of view. Provide labels for all of the following terms seen in bold face print. Your instructor may have you hand in your sketch as a graded exercise.

Begin your examination along the aboral surface of the ray. Along this margin you will see **dermal branchiae** extending through the **calcareous plates** of the endoskeleton. The dermal branchiae resemble large polyps whose lumen is continuous with the **coelomic cavity** (body cavity) of the ray. Recall that the starfish lacks circulatory, respiratory, and excretory organs due to their sluggish life style. The dermal branchiae provide surface area for the exchange of gases and waste products with the surrounding environment. The calcareous plates of the endoskeleton are made up of porous looking roundish cells.

Also along the aboral surface you will be able to find the pincher-like **pedicellaria** which keep the surface of the starfish free from the encrusting growth of other organisms. These will be cut in different sections so it will take some patience to find an entire pedicellaria. With a good section, move to the 40X objective lens. The "beak" of the pedicellaria seem to be re-enforced with calcareous ossicles. Just below the beak notice the "V-shaped" bands of muscle that insert well up into the tips of the pedicellaria beak. This certainly looks like a nasty pinching machine.

The most prominent organ occupying the majority of the coelomic cavity immediately below the aboral surface will be the **hepatic ceca.** Recall that this large organ had to be peeled off the upper aboral surface and replaced into the coelomic cavity as you were removing the aboral wall from your starfish. Notice that this organ has a multibranched lumen into which the digestive enzymes are secreted. After entering the lumen the digestive enzymes would then proceed into the pyloric ducts and finally into the pyloric stomach of the starfish.

In your gross dissection, the hepatic ceca had to be removed to expose the starfish **gonads** that lie further down in the coelomic cavity toward the oral surface. The gonads are seen in the same relative location in the cross section of the starfish ray. They appear as very densely staining purple bodies containing abundant gametes. Based upon your earlier examination of starfish gonads in the developmental laboratory exercise, is your specimen a male or female? Recall that a macroscopic determination of sex is almost impossible. Inferior and medial to (below and toward the center line) the gonads you will find the **gonoducts.** The gonoducts carry the sperm or egg cells from the gonads to the external genital pores. In a male starfish scattered sperm cells may be found in these genital ducts.

In the midline of the starfish ray very close to the oral surface, find the **ambulacral ridge** in the coelomic cavity. It resembles a saddle or a two-humped camel's back. The "humps" of the camel's back possess **calcareous ossicles** providing the rigid structural support exhibited by the ambulacral ridge in the gross dissection. Immediately below the trough of the ambulacral ridge you can see the **radial canal** of the water vascular system. Remember that the radial canal connects to the ampulla of the tube feet.

Below the radial canal will be the large and obvious sections of **tube feet.** Survey the area for a tube foot that has been cut longitudinally exposing most of its structure. You should be able to find the stalk of the tube foot as well as its sucker-like end. Recall that the sucker-like end can either exhibit a concave or a convex inferior surface. The outside walls of the stalk exhibit a scalloped appearance that would suggest the ability to change size and shape. The empty lumen (space) in the center of the stalk allows the fluid of the water vascular system to enter or leave the sucker-like disc at the end of the tube foot. The portion of the tube foot closest to the starfish ray exhibits abundant muscle fibers. These muscle fibers allow the tube foot to be actively moved around.

DEVELOPMENTAL BIOLOGY

24

Laboratory Objectives

1. To learn the terminology of early developmental stages.
2. To compare the events of starfish development to some of the patterns seen in the development of the frog.
3. To be able to describe the patterns of early cleavage in the starfish and the frog.
4. To be able to describe the process of gastrulation in the starfish and the frog.
5. To know the origin and fate of the three embryonic germ layers.

Introduction

The development of a new individual organism from a one-celled zygote is a truly amazing feat. From a simple diploid cell arise the billions of cells making up the adult of the species. Each of these cells will have a complete copy of the genome that was present in the initial zygote. More amazing than the number of cells produced are the specializations which occur during the development of the individual organism. Which cell will become a skin cell or a liver cell or a nerve cell depends upon which of its genes are turned on and which remain silent. The sequence of gene stimulation is very important in determining the eventual fate of the cell. What stimulates certain genes to be active and others to remain quiet is not well understood. Some of the factors involved in determining the future fate of a cell include cytoplasmic conditions within the cell itself, the cell's location within the developing individual, the effects of neighboring cells, and hormones produced in some distant part of the developing organism.

The factors which influence development may be poorly understood, but there are common events in diverse life forms which can be followed. The development of an individual involves five separate stages. Gametogenesis has been investigated previously in the meiosis lab of General Biology I. The remaining four stages include: (1) fertilization, or the union of the sperm and egg beginning the development of the zygote, (2) cleavage, a rapid series of mitotic divisions, without growth, partitioning the egg's cytoplasm into progressively smaller cells, (3) gastrulation, a morphogenic movement of cells establishing the adult (or larval) body plan, and (4) organogenesis, the process whereby the various organ systems of the body are formed.

The mechanics of development are affected by the amount of yolk that is present in an egg cell. The basic events of development are best studied in starfish as these eggs possess the least amount of yolk to complicate matters. The eggs of starfish have yolk evenly distributed throughout their cytoplasm. This condition gives rise to an **isolecithal** egg.

Activity 1
Starfish Development

A. Obtain a slide labelled "Starfish testis sec." Using the scanning lens notice that the testes are composed of several lobes. Focusing on one of the lobes, observe that the lobe is built more like a tube. The opening of the tube (lumen) is filled with mature sperm cells waiting to be released into the waters of the sea. The walls of the tube (composed of finger-like columns of cells) are producing sperm cells to be released into the lumen.

Using the oil immersion lens, focus on the sperm in the middle of the lumen of the testes. Look for flagella on the sperm. Do you think that the sperm is motile or do they just drift around moving with the ocean currents?

With the oil immersion lens still in place, move to the cells forming the walls of the testes. What process is occurring here (meiosis or mitosis)? Can you see any indication that this process is occurring?

B. Obtain a slide labelled "Starfish ovary sec." Observe the slide first with the 10X objective lens. Notice as with the testes slide, the ovary appears to be composed of several lobes. Focusing on one of the lobes, notice the many immature eggs found along the periphery of the ovary.

Using the 40X lens, focus on an immature egg. The younger and smaller eggs are concentrated around the periphery of the ovary. Notice that the egg cells are cut in different planes. A few will be found showing the prominent **nucleus** with a smaller more condensed **nucleolus.** The densely staining cytoplasm is packed full of oil droplets which will serve two basic functions when the mature eggs are released. The oil droplets will serve both as an energy source as well as a flotation device. Each starfish produces about 250 million eggs during a single reproductive season.

Find and sketch a mature egg in the ovary labelling the nucleus, nucleolus, and cytoplasm. Place your sketch in Figure 1. Notice how the yolk is evenly distributed throughout the entire cytoplasm of the egg.

C. Obtain and view a slide labelled "Starfish ova unfertilized w.m. This is a slide that has been prepared without any cutting to reveal internal structures. The eggs were just transferred to the slide and then a coverslip was put on the surface. Examine several under high power.

Again you can see the **nucleolus** within the larger nucleus. What is the significance of this large and distinct nucleolus within the nucleus? What cell activities have been going on to prepare the egg for maturity?

Does the nuclear area appear to be on the egg's surface or deep within the cytoplasm? What might be an advantage to its location?

Upon fertilization, a fertilization membrane will form around the egg to prevent further sperm from entering the egg cell. The distinct nucleus also disappears following fertilization. Survey your slide to see if you can find any fertilized eggs.

D. Meiosis in the female is inherently different than is meiosis in the male. Rather than having equal cytoplasmic divisions during meiosis I and II, female meiosis exhibits unequal cytoplasmic divisions during both divisions. By partitioning the majority of the cytoplasm into one large egg cell rather than four smaller egg cells, more energy can be efficiently stored to sustain the future development of the egg cell (if fertilized). The excess genetic material of both meiosis I and II is enclosed in a polar body and released from the developing egg.

Obtain a slide labelled "Starfish eggs polar bodies w. m." Scan the entire slide looking for eggs that appear to have a polyp of extraneous material being budded from the side of the egg membrane. Sketch the egg cell with the attached polar body. Place your sketch in Figure 1.

E. Obtain a slide labelled "Starfish early and late cleavage w. m."

Cleavage is the very rapid division of the early zygote with little or no time between divisions for cell growth. The large mass of cytoplasm present in the mature egg is being partitioned into smaller and smaller portions as this rapid series of cell divisions occurs. Cleavage is important in the development of an individual organism for many reasons. A large cell has a small surface to volume ratio. This means that there is a shortage of membrane across which needed nutrients must pass to supply the metabolic needs of the cell. Secondly, mRNA diffuses from the nucleus carrying instructions to the cell's ribosomes directing the types of proteins the cell is to make. Since diffusion is a slow process, a large cell has trouble efficiently getting its mRNA to its ribosomes.

By surveying the slide, find and sketch zygotes that have divided into **2, 4, and 8 cells.** In the eight celled stage, the zygote appears to have two tiers of four cells each. The upper tier of cells is located directly over the lower tier. This pattern of cleavage (radial cleavage) is characteristic of the deuterostome lineage of animal evolution. Deuterostomes include the echinoderms and vertebrates (animals with backbones). In the protostome lineage (annelids, molluscs, and arthropods among others) the upper tier of cells is located in the furrows separating the cells of the lower tier. Each new successive tier of cells produced spirals around the main axis of the body producing a spiral cleavage pattern. The individual cells are now called **blastomeres.**

When the cell has divided four times (producing 16 blastomeres), the embryo has reached a stage called the **morula** stage. A morula is a solid ball of blastomeres.

Compare the size of the 2, 4, 8, and morula stages of development. Note that they are approximately the same size. No growth in total cytoplasmic mass has occurred during this period of cleavage.

F. The product of cleavage is a stage of development called the **blastula** stage. In the starfish, the blastula resembles a hollow ball of cells with a large symmetrical central cavity called the **blastocoel.** The **blastomeres** are the cells surrounding the blastocoel. In the starfish the blastula possesses cilia on its exterior surface. The cilia allow the embryo to swim out of the fertilization membrane which has surrounded it up to this stage.

Obtain a slide labelled "Starfish blastula, w. m." Under high power sketch a blastula placing it in Figure 1. You can gain an appreciation of how small the cells have become. By careful manipulation of the fine focus, notice the three dimensional nature of the blastula. The blastocoel is more easily seen in this slide.

G. **Gastrulation** is the next significant event in an animal's development. Gastrulation is a *morphogenic cell movement* that begins to establish the anterior-posterior symmetry of the animal. Look carefully at the phrase . *morphogenic cell movement*.. The prefix "morpho" refers to morphology or shape. The root "genic" comes from "genesis" which means "to create." Gastrulation involves the creation of the future shape of the organism through the movement of cells to their adult locations. The animal's digestive tract begins to develop and the animal's germ layers (ectoderm, mesoderm, and endoderm) are also laid down during gastrulation. The germ layers are important in later development. The **endoderm** for example develops into the lining of the gut while the **ectoderm** becomes part of the skin and nervous system. The **mesoderm** develops into muscle, bone, and some of the internal organs.

If the starfish blastula is viewed as a balloon, gastrulation is analogous to sticking a finger into one end of the inflated balloon pushing the one side in toward the other. This **invagination** occurs at the **blastopore** as cells in this region begin to reproduce faster and migrate inward. This invagination of cells destroys the old blastocoel as a new cavity is formed, the **gastrocoel.** This cavity communicates to the outside through the blastopore, which in deuterostomes will form the animal's anus. The blastopore in protostomes will form the mouth. The cells lining the gastrocoel make up the animal's **endoderm** which will form the digestive system and its accessory glands. The outer cells of the gastrula are called **ectoderm** and will give rise to the exterior coverings of the animal as well as its nervous system.

Figure 1. Starfish Developmental Stages

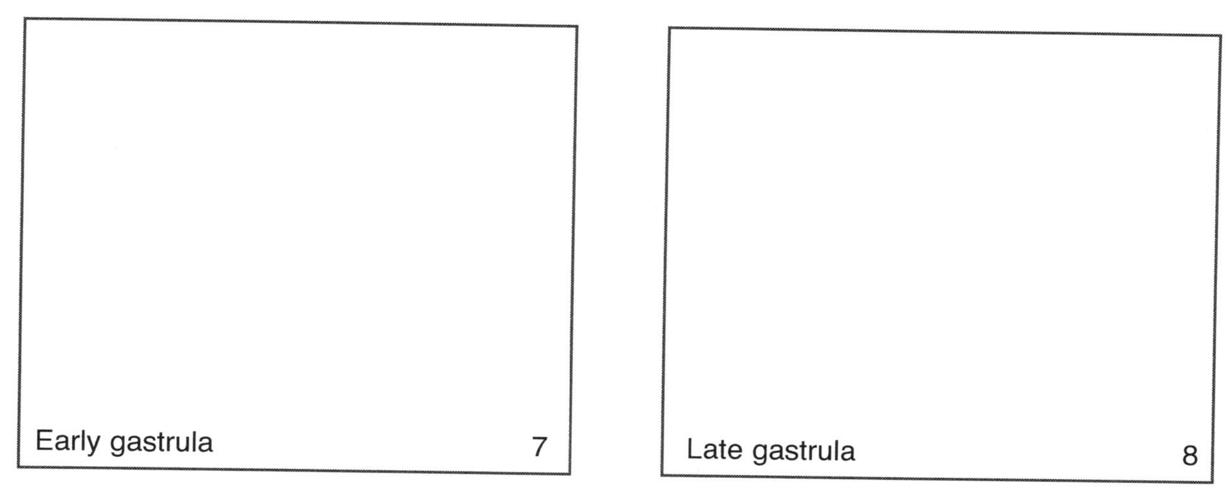

Figure 1. (con't.)

Place a slide labelled "Starfish gastrula w. m." under the 10X objective lens. Look at and sketch several different gastrula stages. Place your sketches in Figure 1. Be sure and label the gastrula's **ectoderm, endoderm, gastrocoel, blastopore,** and disappearing **blastocoel.** Early gastrulation looks as if a dimple had formed in one end of the blastula. Later in the process, the forming gastrocoel has progressed approximately 1/2 the way toward the opposite pole of the animal.

H. As the gastrula matures, it elongates. The elongated gastrocoel is now termed the **archenteron** or primitive gut. The inner end of the archenteron eventually develops two **lateral pouches** that will grow outward and pinch off, forming the third germ layer called the **mesoderm.** The musculo-skeletal system, circulatory system, and gonads develop from the mesoderm. The animal's **coelom** (internal body cavity in which the internal organs are suspended) is also created in the process of mesoderm formation. The outside of the late gastrula begins to assume the shape of the first larval stage of the starfish, the bipinnaria larvae.

Obtain and sketch a representative gastrula from a slide labelled "Starfish late gastrula w. m." How does it differ from the previous slide? Look for the lateral sacs which develop at the inner end of the archenteron. These sacs upon further development will produce the animal's coelomic cavity and mesoderm germ layer.

I. Survey the slide labelled "Starfish bipinnaria larvae w. m." What type of symmetry do you see? Does the larvae exhibit bilateral or radial symmetry? What adaptive advantage might this symmetry offer the animal?

J. Observe the slide labelled "Starfish brachiolaria larvae." At this stage appendages have appeared. They appear to have "sucker-like" discs at their tips. These appendages are the predecessors to the tube feet of adult echinoderms that we will study in the next laboratory exercise.

K. Observe the slide labelled "Starfish young w. m." What type of symmetry do you see? Is symmetry of this type conducive to greater or lesser mobility than seen in the brachiolaria larvae?

L. Study the "Starfish development composite section." This slide has a variety of stages of development. Be able to identify the stages of development seen on this slide. Use it as a review and check to see if you are proficient at using the terminology of development.

Activity 2
Frog Development

Frog eggs possess more yolk than do starfish eggs. The distribution of the yolk produces a polarity to the frog egg. The **vegetal pole** of the frog egg possesses a larger concentration of yolk than does the **animal pole.** When the eggs are deposited into the pond by the female frog, the animal pole is oriented toward the sun with the vegetal pole below. If you have ever observed freshly laid frog eggs, you might have noticed that the top of the egg (animal pole) is darker than the bottom of the egg. This pigmentation protects the frog egg from the damaging ultraviolet rays of sunlight.

The presence of more yolk in the vegetal pole of the frog egg makes the cytoplasm in this region more viscous than in the animal pole. Molasses is more viscous than is water. This polarity in yolk distribution has an influence on early cleavage divisions in the frog embryo. Recall that cleavage is a rapid series of mitotic divisions without overall growth in size. Each mitotic division is accompanied by a cytokinetic division partitioning cytoplasm to the two forming daughter cells. The cleavage furrow of cytokinesis has a much tougher time forming in and dividing up the yolky cytoplasm of the vegetal pole in comparison to the less viscous cytoplasm of the animal pole. As a result, cleavage divisions occur more quickly and easily at the animal pole of the original frog egg producing more and smaller cells in this area.

A. Obtain a slide labelled "Frog, early cleavage sec." This slide represents an early frog embryo which has been cut to reveal its inner structure. Under 10X survey the periphery of a sectioned frog embryo. You are observing cleavage at the 8-16 cell stage. One side of the frog embryo will possess a thin darker covering of melanin that protects the organism from the damaging ultraviolet rays of the sun. This represents the animal pole described above. The opposite side of the embryo is the vegetal pole.

   Notice that the cells **(micromeres)** located just under the pigment layer are smaller and more numerous than the cells of the vegetal pole. The **macromeres** contain more yolk than do the smaller micromeres. You will not see a nucleus in each of the cells as they are sectioned in different planes. Sketch this stage of the frog embryo's development showing the animal pole, vegetal pole, pigment layer, micromeres, and macromeres.

B. Obtain and sketch a slide labelled "Frog blastula, sec." In this slide you can see that the frog's **blastocoel** is displaced toward the animal end of the developing embryo. This displacement is again attributable to the unequal distribution of yolk within the original egg. The larger yolky macromeres are found below the blastocoel. Be sure and label the more numerous micromeres under the pigmented animal pole.

C. **Gastrulation** occurs next in the developmental sequence. Recall that gastrulation is described as a morphogenic movement of cells. In the case of the frog, the macromeres (heavily laden with abundant yolk) do not become involved in the migration of cells to their adult location. The smaller micromeres continue to divide much more rapidly and, as a result, grow downward and over the larger macromeres. This gastrulation process resembles an oceanic wave cresting as it heads toward a beach. The smaller micromeres sweep across and cover the larger macromeres as they spread toward the opposite pole of the frog embryo. At the end of the morphogenic movement of micromeres, the only external evidence that can be seen of the macromeres is a plug of the nonpigmented cells sticking out of the animal's blastopore. This stage is often referred to as the **yolk plug stage.**

   1. Examine and sketch the yolk plug stage that can be found under a dissecting microscope set up by your instructor.

   2. Obtain and sketch a slide labelled "Frog yolk plug sag. sec." with your 10X objective lens. Place your sketch in Figure 2. This structure can be compared to your hand when your curled index finger is almost

touching the tip of your thumb. The C-shaped structure opens to the outside world through the **blastopore** (the space remaining between your index finger and your thumb). Recall that the frog's blastopore will become its anus with further development. The open space inside the "C" is the animal's **gastrocoel (archenteron)** or forming gut. The gastrocoel is lined with larger yolk laden macromeres that are destined to become the **endoderm** germ layer. The outside of the gastrula is covered with micromeres that will develop into the animal's **ectoderm.** The mesoderm is not easily visible. Be sure and label the **yolk plug** that is sticking out of the animal's blastopore. A small portion of the original **blastocoel** may still be evident just under one of the two "fingers" which defines the opening of the blastopore. Be sure and label the ectoderm, endoderm, gastrocoel, blastopore with yolk plug, and old blastocoel if still evident.

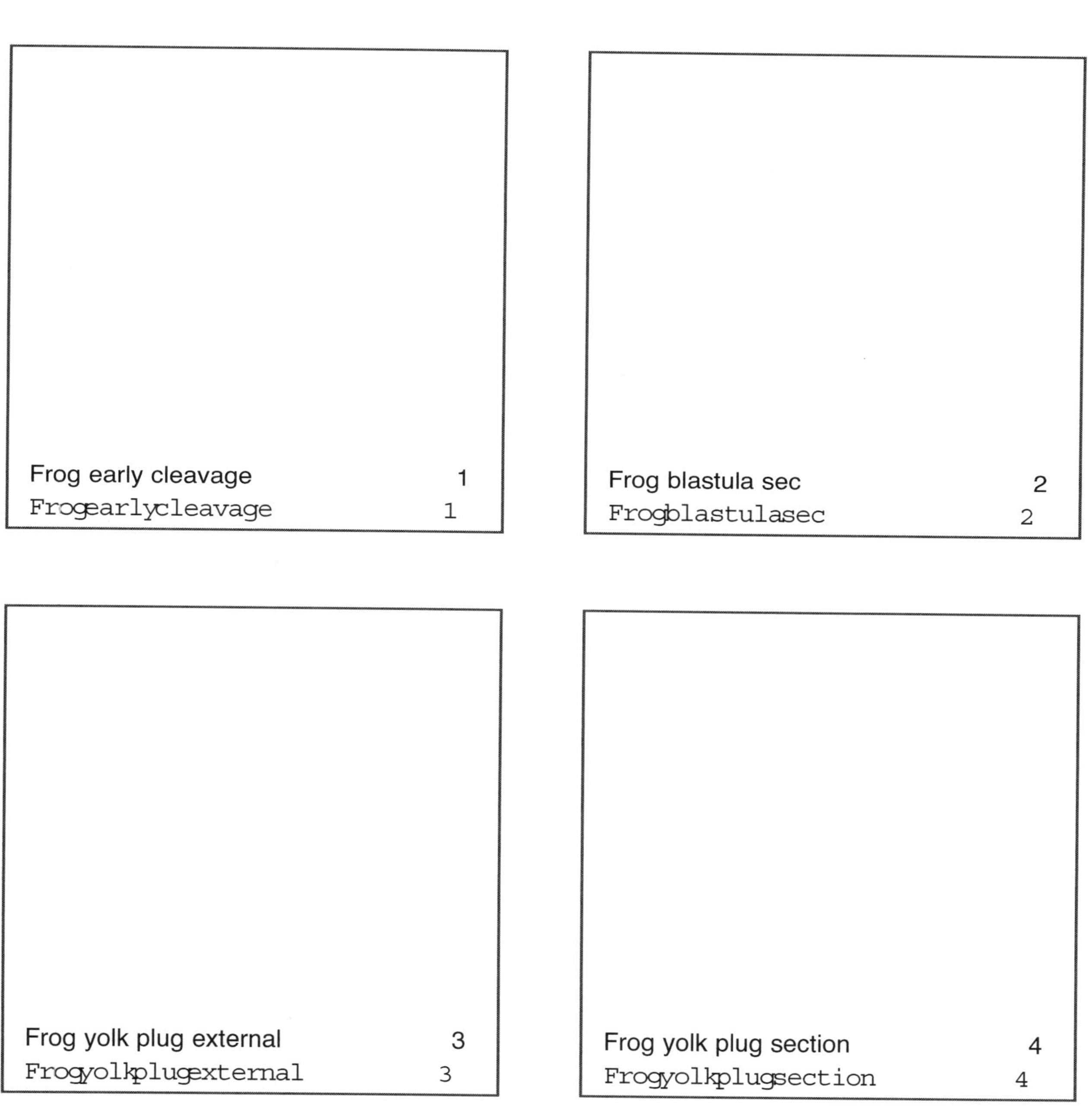

Figure 2. Frog Developmental Stage

AN INTRODUCTION TO RECOMBINANT DNA TECHNOLOGY

25

The aim or goal of genetic engineering is to combine the DNA of different organisms into new patterns to produce useful products. Sometimes these products are the DNA molecules themselves that scientists wish to investigate in more detail. The DNA sample to be analyzed might need to be amplified to produce an adequate sample to work with. Desired products may also be protein molecules coded for by the DNA. Human growth hormone is a protein molecule that can be used to treat different growth disorders in young children. Prior to the advent of genetic engineering, the only supply of this hormone was the pituitary glands of cadavers. In the last few years, genetic engineers have been able to isolate the gene for human growth hormone and splice the gene into the cytoplasmic machinery of bacterial cells to produce large quantities of the hormone relatively inexpensively. Genes for pest resistance can be inserted in the genome of crop plants. Microbes have been produced which are capable of digesting oil spills. These "new" life forms have even been granted a patent. In the coming years, the genetic revolution should rival or surpass the effects that the Industrial Revolution had on human history.

To understand how genetic engineers manipulate the DNA of different sources, it is necessary to understand the tools that these scientists work with. Most of these "tools" happen to be actual products of living cells.

If genes (DNA fragments) are going to be moved from one genomic environment to another, a mechanism is needed to cut DNA into smaller fragments. Nature has supplied the molecular geneticist with very precise molecular scissors called restriction endonucleases first isolated from bacterial cells. Bacterial cells often fall victim to "attack" by viruses. The viruses inject viral DNA into the cytoplasm of the bacteria in hopes of reproducing themselves. The bacteria have evolved "restriction endonucleases" as a defensive mechanism against this viral attack. Dissecting the name "restriction endonuclease," some of the characteristics of these enzymes become clear. Restriction endonucleases (restriction enzymes) cleave or cut DNA at very specific base sequences within the confines of the DNA molecule. EcoR1 is a restriction endonuclease which was the first isolated from *E. coli.* It cuts at a very specific 6 nucleotide base sequence which occurs (on the average) every 1000 base pairs. Table 1 lists two examples of restriction endonucleases and the sequences at which they cut. Notice that each sequence of nucleotides is a palindromic sequence that reads the same in the forward as in the backward direction ("dog" and "god" in the English language). The resulting staggered cuts produce "sticky ends" or exposed bases to which other foreign molecules of DNA, cut with the same restriction enzyme, can be attached.

If fragments of DNA cut by the same restriction enzyme did come together, their union would be short lived. With only four hydrogen bonds holding them together, the fragments would soon split apart. DNA fusions can be made permanent with an enzyme called **ligase.** DNA ligase catalyzes the formation of phophodiester bonds that links together nucleotides in a DNA chain. You might recall that ligase is the enzyme which joins the Okazaki fragments together which are formed on the lagging strand of the DNA template during DNA replication. Again, the second tool of the genetic engineer is a naturally occurring chemical.

Once hybrid molecules are synthesized it is necessary to move them from the test tube back into an environment where they can be reproduced. **Cloning vectors** can be used for this purpose. One such vector is the **plasmid** of bacteria. A plasmid is a small round extranuclear piece of DNA which is not necessary for the survival of the bacterium carrying it. The plasmid may carry genes, however, that allow the bacterium to prosper in environments that are lethal to other bacteria lacking the plasmid. Often plasmids carry genes that allow the bacterium to survive in the presence of antibiotics. This is useful for the biologist as the antibiotic resistant gene can serve

Table 1. Restriction Enzymes

Restriction Endonuclease	Bacterial Source	Restriction Site
EcoR1	*Escherichia coli*	↓ - G - A - A - T - T - C - - C - T - T - A - A - G - ↑ - G A - A - T - T - C - - C - T - T - A - A G -
Bam H1	*Bacillus amyloliquefaciens*	↓ - G - G - A - T - C - C - - C - C - T - A - G - G - ↑ - G G - A - T - C - C - - C - C - T - A - G G -

as a tool to help isolate cells which have successfully taken up the recombinant plasmid. A plasmid may be opened up with a restriction enzyme, have a foreign piece of DNA match up with its sticky ends, and then be closed with the enzyme ligase.

Plasmids can move from one bacterial cell to another by the process of **transformation.** In the textbook, you were introduced to the process of transformation in the experiments of Frederick Griffith working with mice exposed to *Streptococcus pneumoniae.* The uptake of manufactured plasmids by bacterial cells can be enhanced if the cells have been treated with calcium chloride. After exposure to the salt, the permeability of the competent cells increases greatly. If the plasmid was engineered to contain an antibiotic resistant gene, cells which were successfully transformed can be separated from those cells which did not take up the chimera by exposure to the antibiotic. Once a cell is transformed, rapid division of the bacterial cell produces a clone of cells possessing the engineered plasmid. Large amounts of the plasmid DNA are now available.

Often the entire genome of an organism is stored in a genomic library. To create such a library, the entire genetic complement of an organism is cut with a restriction endonuclease and all of the fragments are connected to plasmids cut with the same restriction enzyme. Clones of the plasmids are made by transformation of competent bacterial cells. Imagine the complexity of the human genomic library produced in this way. Mammalian genomes contain approximately 4,000,000 kilobase pairs in a haploid cell. If the genome were cut with EcoR1 (which possesses a restriction site approximately every 4 kilobases of genomic DNA.) one million restriction fragments would be produced. Don't overlook the fact that restriction sites do not observe gene boundaries on the chromosome. It is as if someone went into the collections of a library and indiscriminately cut every book every 100 pages and scattered the sections of pages all over the library. The organization of the genomic library rivals the organization of the world wide web. Molecular geneticists can use radioactive probes to search for specific nucleotide sequences in the library. The mRNA from a tissue actively producing the protein of interest can be made radioactive and then set loose in the genomic library. The radioactive probe will complementary base pair with the DNA sequence responsible for the coding of the desired protein.

Not all of the "equipment" utilized by the genetic engineer comes from the natural world. It is often necessary to separate DNA fragments produced from an indiscriminant restriction enzyme digest. Restriction fragments can be separated based on their size and charge characteristics by using gel electrophoresis. Samples of DNA digests are placed into the wells of an agarose gel. Wells are depressions which have been manufactured into the agarose gel. The samples are loaded into the wells with the use of a micropipet. The loaded gel is placed on a platform in an electrophoresis chamber, and a buffer solution is added. An electric current, which causes the restriction fragment molecules to migrate, is turned on. Restriction fragments of DNA digests at the neutral pH of the buffer solution possess a negative charge. The negatively charged fragments migrate toward the positively charged pole. The agarose gel acts like a molecular sieve. The rate of migration of the different fragments is determined by their molecular size. The pore size of the gel is very close to the range of sizes of the molecules that are being separated. The smaller the fragment, the faster it will migrate through the maze. The smaller molecules can zip through the small pores while the movement of the larger molecules is more restricted.

The wells of the electrophoresis gels are not loaded with "naked" restriction enzyme digests. The sample buffer is mixed in with a 10-20% glycerol component making the sample more dense than the electrophoresis buffer. This greater density ensures that the sample will stay in the bottom of the well. The other addition to the sample buffer is the dye bromophenol blue. As the current is applied to the gel, bromophenol blue will migrate as fast as the smallest restriction fragment in the sample. The dye serves as an indicator to mark the progress of the wave front of the separation. This allows the power to the chamber to be shut off before some of the smaller restriction fragments are pulled off the far end of the gel.

Once the gels are finished in the electrophoresis chamber, the different bands of DNA need to be made visible for inspection. In research labs, ethidium chloride is used to stain the nucleic acid bands. The bands are then made visible with ultraviolet light. Both of these procedures carry some severe health risks. We will employ methylene blue to develop our agarose gels. The gel will be incubated with enough methylene blue to cover the surface of the gel in a freezer dish over night. The next day the dye will be carefully rinsed from the gel and the distances the bands are from the wells will be measured.

ESTIMATING LENGTHS OF DNA FRAGMENTS

EXERCISE 25A

Laboratory Objectives

1. To learn some of the techniques of genetic engineering.
2. To learn how to micropipet small samples onto electrophoresis gels
3. To run electrophoresis gels to obtain a separation of a restriction enzyme digest.
4. To estimate the sizes of DNA restriction fragments.

Introduction

Natural selection is often used to explain the unity and diversity seen among species of organisms that inhabiting the earth. The theme of unity and diversity carries over to the molecular world of the gene.

The term **genome** is often used to describe the total complement of DNA contained in a cell. The common denominator between all genomes is that information is stored in the genome in the nucleotide sequences of the cell's chromosomes. As dots and dashes contain the information in the Morse Code, the sequence of nucleotides determines amino acid sequences of proteins coded for by the genes of the genome. The nucleic acid alphabet is almost universal in the living world. The universal language of the genetic system is a strong piece of evidence suggesting that life arose once and that all living organisms share a common ancestor. Whether the nucleic acid is found in a bacterium, virus, plant, animal, or fungus, all employ the same common language of nucleotide sequences carrying information from one generation to the next. This common system of encoding information points to a common ancestor of all life forms existing today.

Even though all living things use the same genetic language, the sizes and complexity of genomes are very diverse as a result of evolution by natural selection. Table 2 lists the wide range of sizes that different genomes possess.

Table 2. Genome Sizes

Genome Type	Size of genome (Kilobase pairs)	Haploid number of chromosomes
Plasmids (pBR322)	4	1
Viruses		
Phage Lambda	50	1
HIV virus	10	1
Procaryotic cells (E. coli)	4000	1
Eucaryotic cells		
Yeast	14,000	16
Fruit Fly	170,000	4
Human	3,000,000	23
Corn	5,000,000	10

The smallest genome that we will investigate in this laboratory exercise is the bacterial plasmid. Plasmids are small circular pieces of extrachromosomal DNA that can be found in the cytoplasm of bacterial cells. They may carry genes that confer antibiotic resistance to the bacteria which house them. A map of the plasmid pUC18 is shown in Figure 1.

As discussed earlier, geneticists use molecular scissors called restriction endonucleases to cut the DNA of genomes into smaller pieces called restriction fragments. One example of a restriction endonuclease used in this lab is EcoR1 which was first isolated from the *E. coli* bacterium. The nucleotide sequence at which EcoR1 cuts occurs on the average of once in every 4 KB pairs of nucleotides. Since plasmids contain a total of 4 KB pairs (as seen in Table 1), you would expect to find one restriction site in each plasmid. The restriction site for EcoR1 is shown in Figure 1 for the plasmid pUC18. Upon digestion with EcoR1, a plasmid of this size would yield one restriction fragment in an electrophoretic analysis.

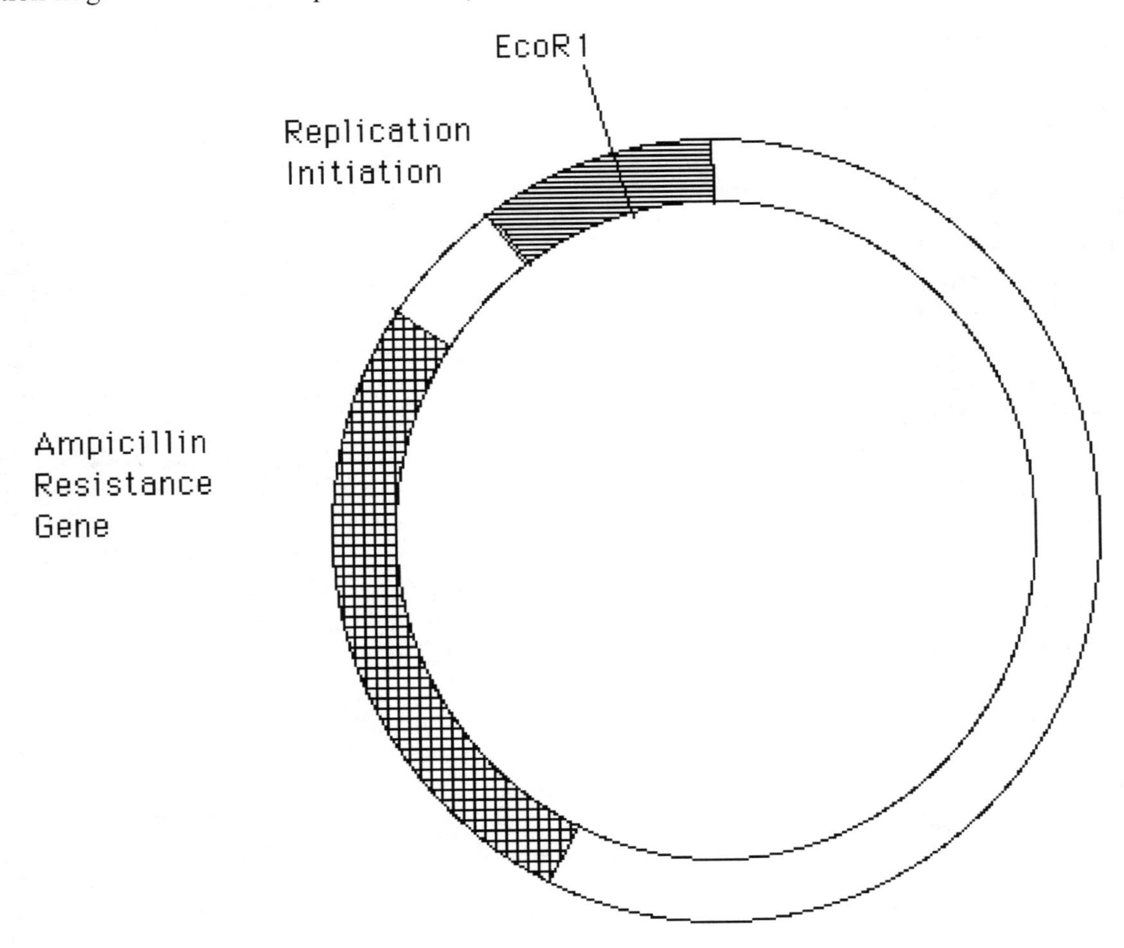

Figure 1. Plasmid pUC18

The biology of a virus is more complicated than is the biology of a plasmid and this difference is reflected in the size and complexity of its genome. Viral genomes are approximately 40 KB pairs in length (about 10X longer than plasmid sequences). Viruses are very small protein coated capsules (some resembling space craft) which contain viral DNA. The protein capsules possess head proteins which house the viral DNA and tail proteins which are capable of attaching to the cell membranes of host cells. Viruses lack some of the common characteristics often associated with living organisms. Viral particles don't actively move about. They do not excrete, metabolize, grow, nor do they respond to environmental stimuli. They do, however, reproduce within host cells to form new viral particles. Viruses inject their DNA into the host cell and turn it into a viral nursery. Here new viral proteins and nucleic acids are produced and assembled. Plant, animal, fungal, and bacterial cells are all susceptible to viral invasion. Viruses which prey upon bacterial cells are called bacteriophages (or "phages" for short). A bacteriophage that has been studied in much detail is **phage lambda.** See Figure 2 for a map of its genome with restriction sites for Hind III indicated.

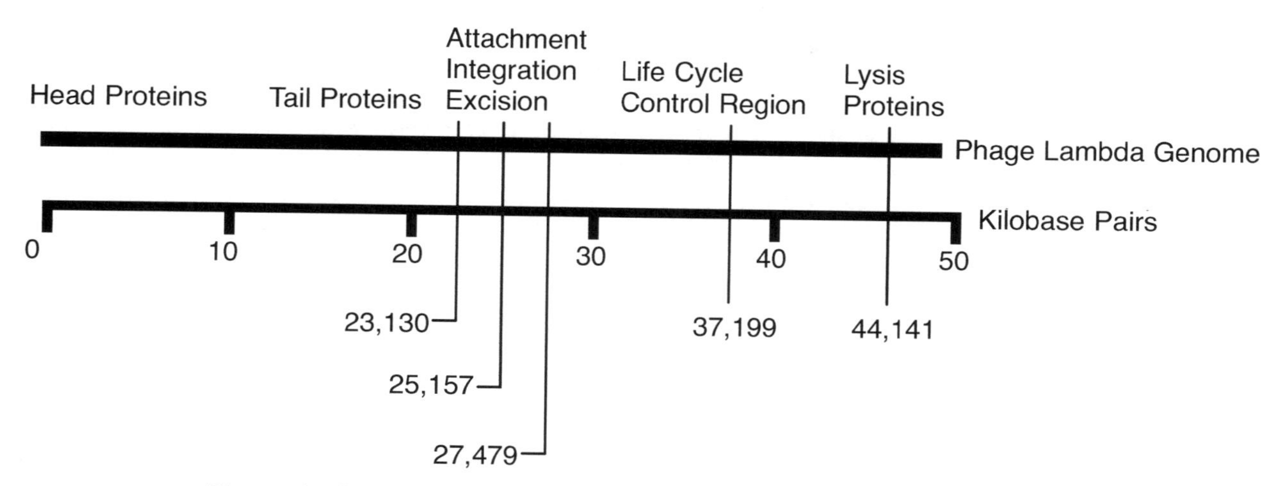

Figure 2. Genetic Map and Hind III Restriction Sites for Phage Lambda DNA

The upper line represents the linear sequence of the different genes controlling the biology of the virus. Notice that related genes are located next to one another. The genes which code for the structural head proteins are found upstream from the genes coding for the tail proteins. As indicated in the table, the entire genome is approximately 50 kilobases long. The lower scale (kilobase pairs) shows the location of restriction sites where the restriction endonuclease Hind III would cut the phage lambda genome. The fourth site indicated actually represents two different restriction sites that are very close together. They essentially function as one site. Upon digestion, Hind III produces six restriction fragments of varying lengths. The sizes of the restriction fragments produced in sequence of largest to smallest in base pairs is: 23,130 bp, 9,416 bp, 6,557 bp, 4,361 bp, 2,322 bp, and 2,027 bp. Often, as in today's lab exercise, a digest of Phage Lambda DNA is included in an electrophoresis run to serve as a standard against which other fragments can be compared.

The genome of bacteria is more complex than the relatively simple genome of phage lambda. The single circular chromosome of *E. coli* contains approximately 4.5 million base pairs of DNA (4,500 kilobase pairs) representing about 4,000 genes. Doing a little arithmetic tells us that each gene is approximately 1000 base-pairs long. The circular chromosome of the bacteria is very neatly organized. Genes are sequenced along the circular chromosome based upon function. Related genes are packaged close to one another in operational units called operons (Lac operon studied in class). Each gene is present in one (and only one) copy. There are very few noncoding DNA sequences between genes. This economy of space and efficiency in packaging allows bacteria to replicate very rapidly. The important thing to remember about the bacterial genome is that there are very few repetitious sequences of base pairs in the genome. When cut by a restriction endonuclease like EcoR1, approximately 1,100 restriction fragments are produced. Remember that the genome is 4,500 KB in length and (on the average) a restriction site for EcoR1 occurs about every 4 KB pairs. (4,500 KB/4KB=1100 restriction fragments.) Since there is very little repetition in the bacterial genome, no distinct bands should form on an electrophoretic run of an EcoR1 digest. The 1000's of restriction fragments will be equally dispersed forming a smear along the electrophoresis gel.

Referring to Table 1, it can be seen that the genomes of eucaryotic cells are much larger than the bacterial genomes. There is approximately 1000 times more DNA in a eucaryotic cell than in a bacterial cell. However the actual numbers of genes present in a eucaryotic cell is only about 10 times greater. This implies that eucaryotic genes are 100 times longer than procaryotic genes, but in reality that is not the case. Eucaryotic genes are a bit longer as they possess introns in the middle of their coding sequences. An intron is a sequence of DNA that does not code for amino acid sequences of proteins. Before RNA transcribed from a eucaryotic gene is sent to the cytoplasm, the intron sequences must be cut out of the RNA and the remaining coding sequences spliced back together. Introns do not account for all of the size difference between eucaryotic and procaryotic genomes. Eucaryotic genomes also contain multiple copies of one gene. For example, humans possess 10 to 20 copies of the gene for growth hormone. Remember that procaryotic cells possess one copy of each gene. There are also a lot of noncoding nucleotide sequences between the eucaryotic genes. Many of these sequences may be related to the control of gene expression (regulatory genes). Lastly, eucaryotic chromosomes have large numbers of highly repeated noncoding sequences of DNA whose functions are largely unknown. The same nucleotide sequences can be repeated a thousand or more times in the eucaryotic genome. Bacterial chromosomes lack these repeating segments of DNA. One example of these repeating noncoding sequences are the DNA satellite series found in eucaryotic DNA.

Satellite DNA may represent a good portion of an organism's genome. At the extreme, Kangaroo rats and meal worms may have 60% of their genome present as satellite sequences. Satellites are not transcribed into RNA. The satellite sequences also seem to be isolated in discrete regions of the chromosomes. Their function is not well understood, but some satellites are concentrated in the region of the chromosomes' centromeres. Recall that mitotic spindle fibers are associated with the centromeres of chromosomes during karyokinesis. However not all satellites are found in centromeric regions. There is also no widely accepted function for the highly repeated satellite sequences.

Possessing an average of 4,000,000 KB pairs of nucleotides in a eucaryotic genome, a restriction endonuclease digest by EcoR1 would be expected to produce about 1,000,000 different restriction fragments. This large number of different sized restriction fragments produces a smear in an electrophoretic run, but the highly repeated satellite sequences show up as distinct and heavier bands on the gel. See the sample gel below for clarification.

In the development of a multicellular organism from a single-cell zygote, every cell receives a full complement of genetic instructions during the development process. Even though multicellular organisms possess specialized tissues for specialized functions, all of the cells of an organism receive an identical genome. Muscle cells have the genetic blueprints necessary to produce insulin, even though insulin production is the job of certain pancreatic cells. Pancreatic cells possess the genes to produce hair follicles, but obviously they don't. Specialization depends upon what genes are turned on and off, not their retention or loss in specialized cells. The same is true for satellites. Their nucleotide sequence and length is retained in all of the cell's of an organism. This property will be demonstrated in today's lab.

In this exercise, you will learn how to load sample wells in electrophoretic gels with sample solutions. You will also separate the products of a restriction endonuclease digest using electrophoresis obtaining results similar to those shown below.

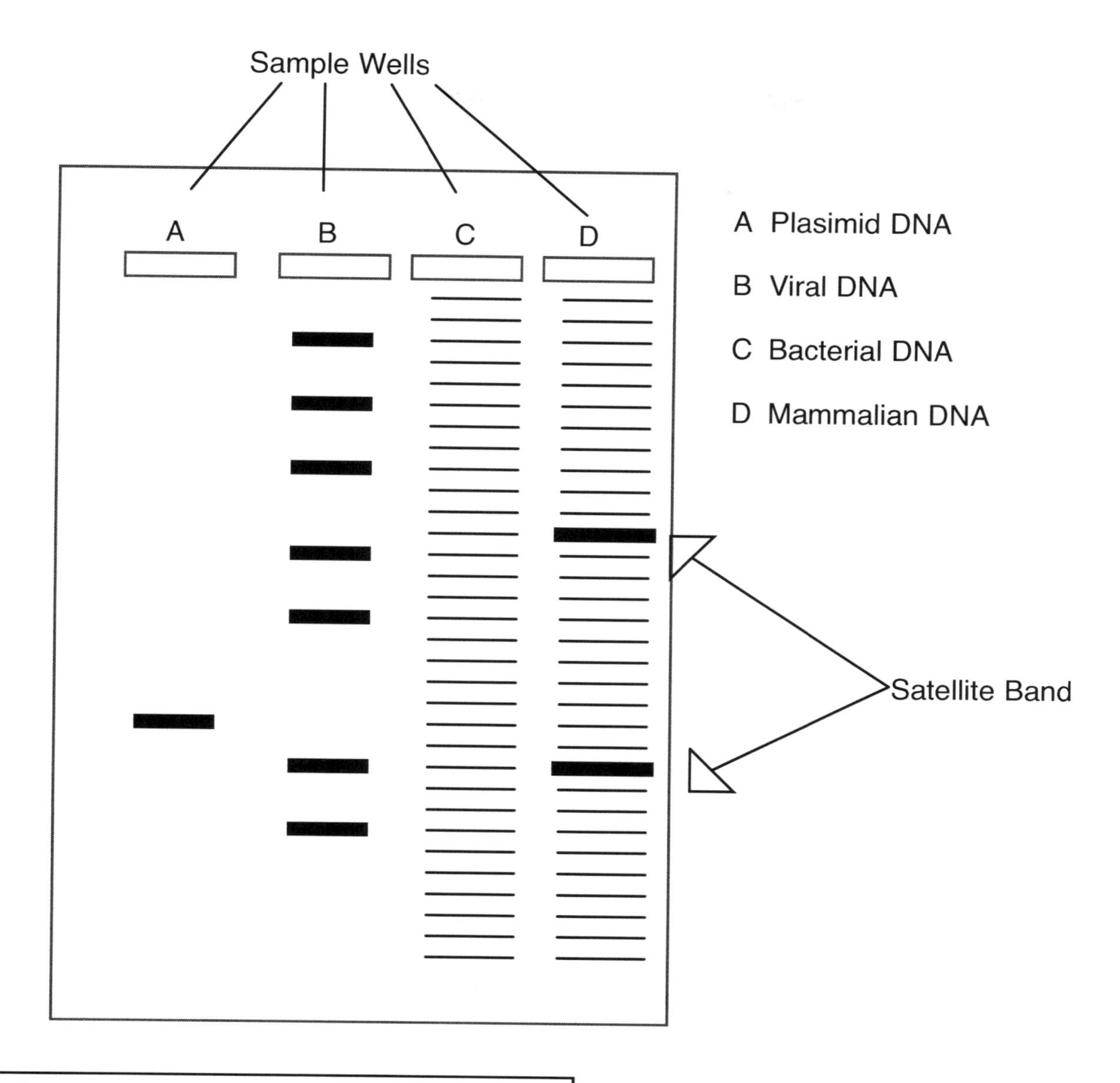

Activity 1

Micropipetting—A Standard Molecular Genetics Technique

The micropipet is an instrument used to transfer very small quantities of sample used in the genetic engineering laboratory. The apparatus that we are going to use is very simple in construction. The micropipetor is made of a stainless steel plunger used in conjunction with a micropipet. The 25 microliter pipet body is calibrated at five microliter intervals. Remember that one milliliter is equal to 10^{-3} liters while a microliter is equal to 10^{-6} liters. Therefore one microliter is equal to 10^{-3} milliliters.

To operate the micropipetor:

1. Insert the metal plunger into the micropipet at the end possessing the large white band. The micropipet can be held between the thumb and the middle finger. The index finger is now free to operate the metal plunger.

2. Gently push the plunger all of the way into the micropipet.

3. Holding the micropipet vertically, place its open end into the sample solution.

4. Draw up the specified sample amount by raising the plunger to the indicated calibration.

5. Carefully wipe off excess sample from the outside of the micropipet with a tissue or Kimwipe.

6. Place the end of the micropipet into the receptacle to receive the sample and slowly depress the metal plunger to eject the sample.

7. Rinse the micropipet between samples by drawing up and expelling water three times from the micropipet.

Even though difficult to see, electrophoresis gels have "wells" at one end to receive small quantities of sample to be analyzed. Your job is to get 15 microliters of practice sample into the wells of the practice gel as shown in Figure 3.

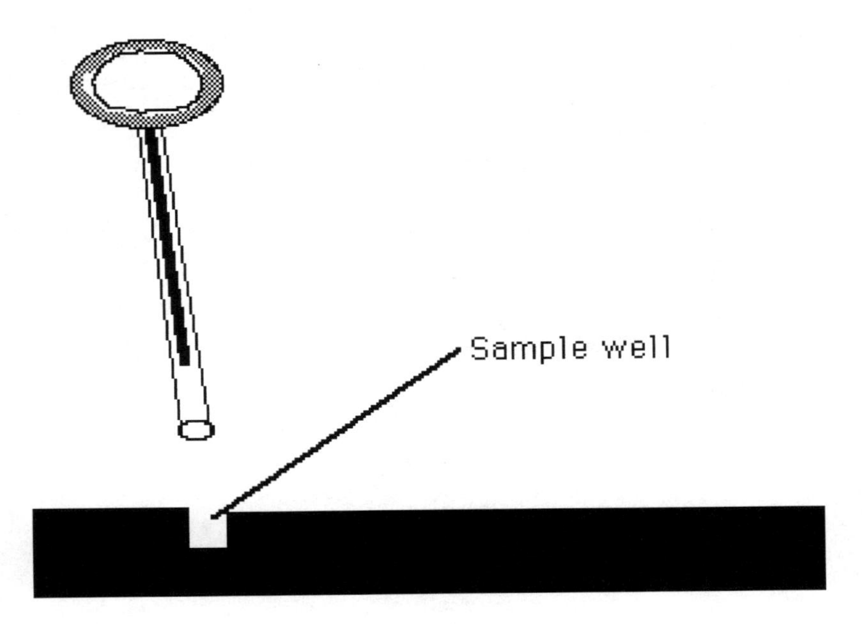

Figure 3. Miropipetting into a Sample Gel

Procedure

1. Obtain a practice sample gel from your instructor.

2. Place the practice gel into a freezer dish and cover the gel with one-half inch of water.

3. Draw up 15 microliters of the practice sample solution using the micropipet.

4. Fill one of the eight sample wells present in the practice gel. Be sure to have the tip of the micropipet in the well without going through the bottom of the gel.

5. Have everyone in your group practice charging the wells. When you are done, each of the wells should be visible due to the presence of the bromophenol blue dye. If the dye is floating around in the water above the sample well, you did not have the end of the pipet far enough down in the sample well.

Activity 2

Determining the Length of DNA Restriction Fragments

One of the first steps in the characterization of a restriction fragment is to determine the size of the fragment in nucleotide base pairs. This is done by running an electrophoretic separation of the unknown restriction fragment on the same gel with DNA markers of known length. Recall that DNA fragments migrate in an electrophoretic field according to their respective sizes. The smaller the DNA fragment, the further the fragment will travel through the agarose maze. A linear relationship exists when the logarithms of the sizes (in base pair units) of the DNA fragments are plotted against the distances which they migrate on the electrophoresis gel. A calibration curve can be produced by using the known marker DNA fragments. The length of the unknown DNA fragment can then be estimated from this calibration curve. See Figure 4 for an example of the procedure.

The box on the right of Figure 4 represents an electrophoretic gel with two sample wells. The well on the left was filled with marker DNA of known lengths. The gel on the right was filled with a sample of a restriction enzyme digest of unknown length. The gel was put in an electrophoresis chamber for a period of time allowing migration of the restriction fragments to occur. The gel was stained to reveal the presence of the different migration bands. The distances of the bands to the wells was measured.

The graph appearing on the left of Figure 4 is a plot of the logarithms of the base length in KB of the marker DNA fragments against the length traveled in millimeters. To determine the length of the unknown DNA fragment, find the distance it migrated on the X axis of the graph and run a perpendicular line up to the standard curve. At the point of intersection run a second line (perpendicular to the first) toward the Y axis. Read the length of the unknown segment at the point of intersection of the second line with the Y axis. The unknown fragment is about 1.4 KB long.

Procedure for Activity 2

1. Wear gloves to protect your hands from the electrophoretic buffer. Secretions from your hands also contain "nucleases" which would digest your electrophoretic samples.

2. Obtain an agarose gel from the front of the room and place it on the tray of the electrophoresis chamber. Make sure that the wells are over the **negative (black) electrode** of the electrophoresis chamber as the current travels from the "-" to the "+" electrode. Also make sure that the black wires from the power source are connected to the "-" terminals and the red wires are connected to the "+" terminals.

3. Fill the electrophoresis chamber until the gel is just covered with approximately 1/2□ of buffer.

4. Load 15 microliters of each sample into the wells as indicated below.

Sample Well Number	Sample
1	Plasmid DNA - EcoR1 cut
2	Phage Lambda - Hind III cut
3	Calf Thymus DNA-EcoR1 cut
4	Calf Kidney DNA-EcoR1 cut
5	Plasmid DNA - EcoR1 cut
6	Phage Lambda - Hind III cut
7	Calf Thymus DNA-EcoR1 cut
8	Calf Kidney DNA-EcoR1 cut

Notice that there is repetition of the samples. Lanes 1-4 are duplicated by lanes 5–8.

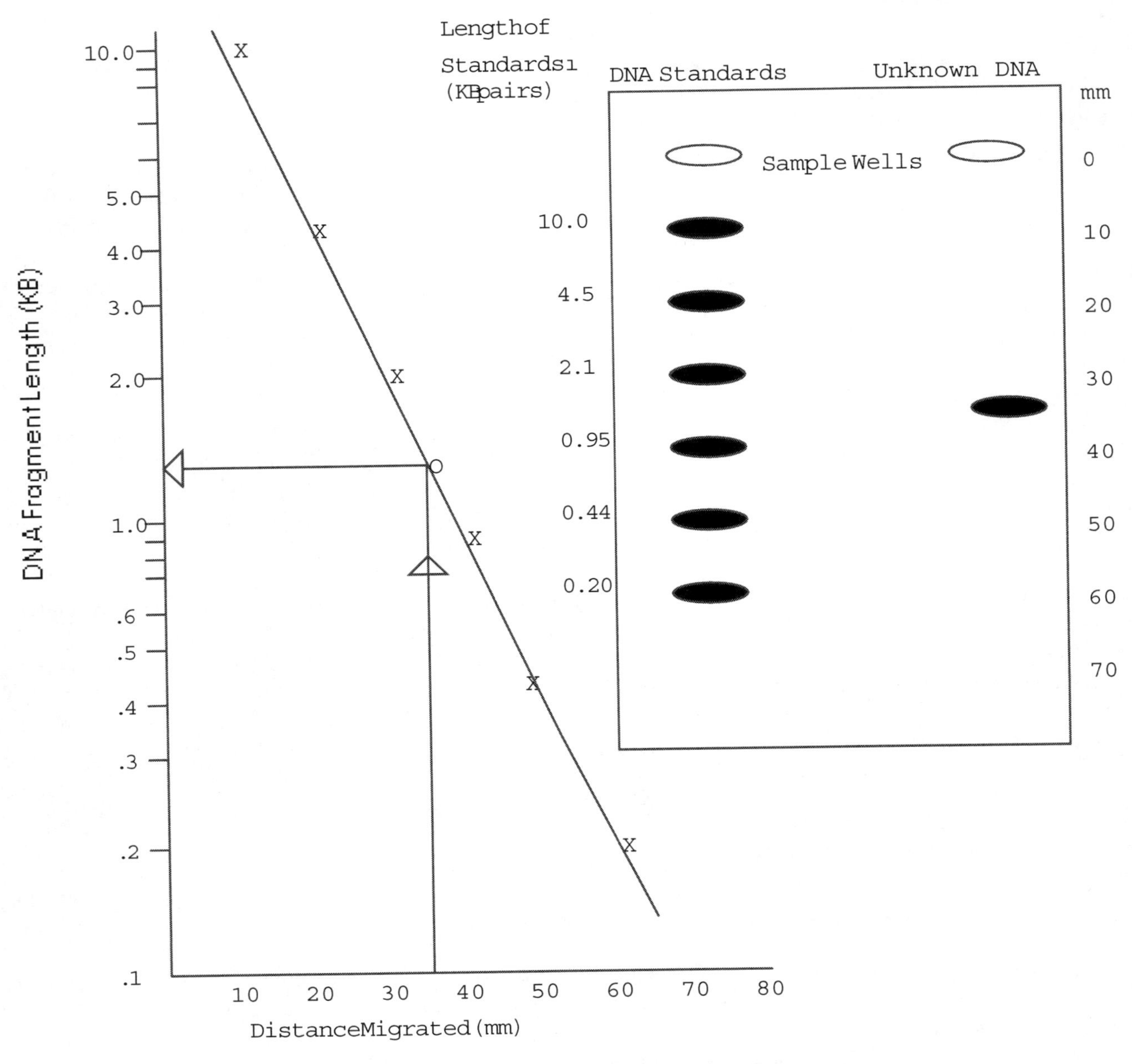

Figure 4. Determining the Length of DNA Fragments

5. Electrophorese until the bromophenol blue (blue dye) in the DNA samples has migrated to within 1 mm of the positive electrode end of the gel. This should take about 50-60 minutes.

6. After the electrophoresis is completed, remove the gel and place it in a **labelled** freezer container. Cover the gel with methylene blue and store in the refrigerator over night.

7. The next day, destain the gels by lightly running water over their surface. Be careful, the gels are delicate.

8. Place your gel over a light source and measure the distances of all DNA bands (in mm) from the wells of the gel. Record the values in Table 3 and make a careful sketch of the gel with its bands.

Table 3. Results of Electrophoretic Run	
	Band Distances (mm)
Lanes 1, 5	
Lanes 2, 6	
Lanes 3, 7	
Lanes 4, 8	

9. After observing your gels over the light boxes, wipe the light boxes clean with alcohol pads and return them to their storage area.

10. Store your gels in the refrigerator in **labelled freezer bags** for later observation by the rest of your group.

Activity 3

Lab Report

Follow **Appendix A** of your lab manual in the writing of your lab report. Be sure to address the following points in the report in the appropriate sections. Do not number your responses to the points below as they are only intended to help guide you in the writing of a successful report.

1. On semilog paper, plot the distance migrated by the DNA bands from the phage lambda (lanes 2 and 6) as a function of the lengths of these fragments. The lengths of the fragments are 23.1 KB, 9.4 KB, 6.6 KB, 4.4 KB, 2.3 KB, and 2.0 KB (see Figure 4).

2. Determine the length, in KB, of plasmid pUC18 by the method described in Figure 4.

3. Estimate the length of the major satellite band from the cow digests (lanes 3, 4, 7, 8). Why does the band appear in both the thymus and kidney samples?

4. The length of DNA helix occupied by one nucleotide pair is 3.4 A (angstrom). What is the length of pUC18 in mm?

5. A human has 10^{14} cells and each human cell has about 6.4 X 10^9 nucleotide pairs of DNA. What is the length of the DNA double helix from one human cell (in meters)? What is the length of the double helix if all of the DNA of all of the cells of a human are combined (in kilometers)? Why is there more DNA in a human cell than in a plasmid?

Clean Up Procedures

1. When done, place used gels and gloves in the orange biohazard waste bag on the side table.
2. Place the gel combs, pipets, plungers, casting trays, and scoops in the metal cans found on the side table. Please replace their metal lids.
3. Don't discard the electrophoresis buffer from the chambers unless indicated by your instructor. Return the electrophoresis buffer to its original glass bottle and refrigerate.
4. Make sure that the electrophoresis chambers are turned off and are unplugged.

DNA FINGERPRINTING

EXERCISE
25B

Laboratory Objectives

1. To prepare electrophoresis gels.
2. To perform a restriction endonuclease digest and separation of restriction fragments by electrophoresis.
3. To interpret banding patterns on an electrophoretic gel to determine the source of unknown DNA samples.

Introduction

The DNA that makes up a mammalian genome can be divided into three different classes depending upon the amount of repetition present. Approximately 50% of the mammalian genome occurs in only 1–20 copies per haploid genome. This nonrepetitive DNA contains almost all of the coding portion of the genome and tends to be consistent in sequence between organisms of the same species. The same gene in closely related species also tends to be fairly constant in nucleotide sequence. Changes in the sequence of nucleotides in a gene, either by base substitution or addition or deletion of bases, may produce vast changes in structure of the protein coded for. A random change in protein structure usually disrupts the functioning of that specific protein. Such mutations are not tolerated by natural selection and therefore are not passed on to descendents.

The second class of DNA contains moderately repeating DNA sequences which contain a few genes coding for ribosomal RNA. The remainder of this second class of DNA contains long sequences of noncoding nucleotide sequences (sequences that don't code for protein structure) that separate many of the coding genes in the eucaryotic genome. These moderately repeating sequences may play a role in the regulation of transcription of the structural genes of the genome.

The highly repetitive DNA sequences are composed of units which are found 10^5 to 10^7 times per haploid cell and are frequently referred to as **satellite DNA.** As has been mentioned before, satellite DNA is not transcribed into RNA. Often the satellite DNA is found in isolated patches along the chromosomes. No known function exists for the satellites, but some satellite DNA is concentrated in the centromeric regions of the chromosomes. The centromere is the portion of the chromosome which apparently attaches to the spindle fibers during mitotic division. Since satellites do not code for protein structure, mutations are more easily tolerated along their length than in the coding portion of the genome. Satellite sequences therefore tend to be more variable between closely related species than do coding sequences of DNA.

Sheep, cows, and goats are closely related mammals. These species belong to the order Artiodactyla (even-toed mammals) and comprise the family Bovidae. As mentioned above, the bulk of the coding sequences of each of their genomes is fairly similar. The nucleotide sequences of most genes that have been examined in these three species are more than 90% identical. The genomes of sheep and cows also contain satellite DNA which comprise about 15% of the total DNA. The satellites of each are found in the centromeric regions of the chromosomes. However, the sequences of nucleotides in the two species' satellites are very different. When satellites are digested by the same restriction endonuclease, different lengths of restriction fragments are produced. In Figure 5, a portion of each species' satellite genome is shown with the distances between restriction sites for two different restriction endonucleases indicated.

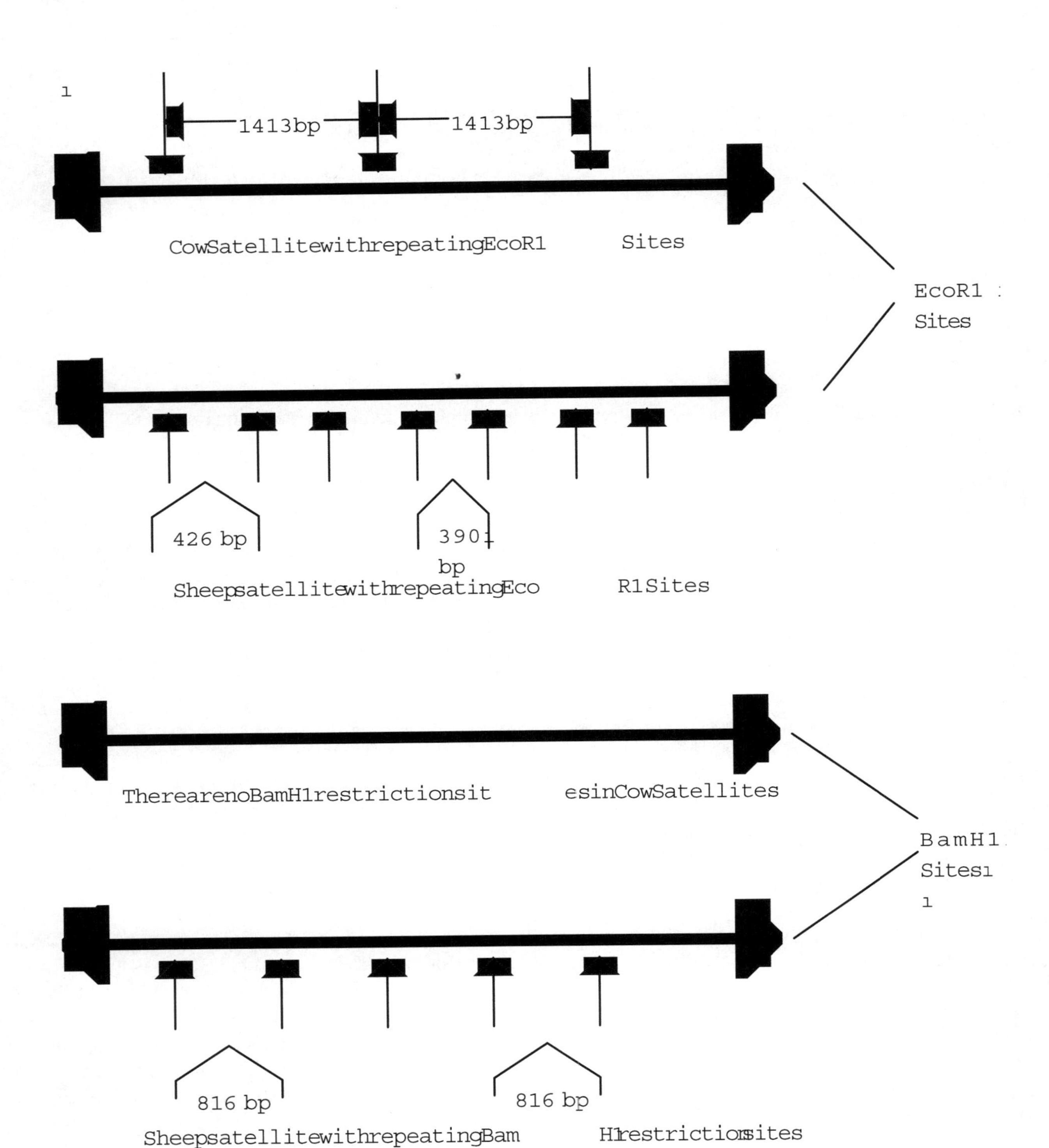

Figure 5. EcoR1 and BamH1 Restriction Sites in Cow and Sheep Satellites

EcoR1 cuts cow satellite DNA every 1413bp. EcoR1 restriction sites in the sheep satellites produce two shorter restriction fragments of 390 and 426 bp. BamH1 produces restriction fragments of 816 bp in the sheep satellites, while cow satellites lack BamH1 restriction sites.

Recall from a previous lab that the large numbers of nonrepetitive DNA nucleotide sequences of a eucaryotic genome when cut with a restriction endonuclease produce a smear when separated with gel electrophoresis. Superimposed upon this smear are the satellite band restriction fragments. Differences in satellite banding patterns are going to allow us to identify unknown samples of DNA as belonging to either a sheep or a cow.

Procedures for Activity 1 (Group of four students):

A. Gel Preparation (two students of the four)

1. **Wear gloves** to protect your samples from "nucleases" contained in secretions from your hands.
2. Place the gel support deck on a level work surface.
3. Wrap tape around the gel support deck. The tape must be firmly pressed against the edges of the deck to to insure a tight seal.
4. Add the premeasured electrophoresis buffer and agarose together and heat over a hot plate. Swirl with a glass stirring rod until the agarose forms a suspension.
5. Heat the suspension, stirring gently with a glass rod, and allow it to come to a near boil. Remove the flask from the heat source, stir gently with a glass rod, and cool at room temperature for about 2–3 minutes. At this time the agarose solution should be **absolutely clear** with no particulate matter observable.
6. Pour the melted agarose into a test tube with the indicated desired volume marked on its side.
7. Pour the contents of the test tube onto the gel support deck in a sinusoidal pattern. You are trying to get an even depth of coverage of agarose solution across the casting tray with no trapped air bubbles.
8. Insert the comb into the casting tray slots. Push down gently on the top of the comb until resistance is felt. The teeth of the comb will form the wells of the electrophoresis gel.
9. After the gel has cooled for 15-20 minutes it will have changed from a clear to an opaque whitish appearance. The tape and comb can be removed. Lift the comb gently straight up so that you do not destroy the preformed wells. The gel is now ready for use.

B. Sample Preparation (two students):

Table 4 shows the sequence of steps outlined below in a flow chart format.

1. Wear plastic gloves.
2. Number six 1.5 ml tubes with the numbers 2, 3, 4, 6, 7, 8 with a water-proof marking pen.
3. Add 40 microliters of EcoR1 in tubes 2 and 6.
4. Add 40 microliters of BamH1 in tubes 3 and 7.
5. Add 10 microliters of water in tubes 4 and 8.
6. Place 20 microliters of unknown DNA #1 into tubes 2 and 3. Place 5 microliters of unknown DNA #1 into tube 4.
7. Place 20 microliters of unknown DNA #2 into tubes 6 and 7. Place 5 microliters of unknown DNA #2 into tube 8.
8. Cap the tubes and place them into the microcentrifuge for not more than 10 seconds to insure complete mixing.
9. Place the tubes into a flotation device and incubate them at 37° C for forty minutes.
10. After the incubation period, add 10 microliters of buffer to each of the tubes and mix well.
11. Load **15 microliters** of the following samples into the sample wells of your agarose gel.

Sample Well	Sample	
1	DNA markers of known length	
2	Tube 2	
3	Tube 3	Tubes 2, 3, 4 contain
4	Tube 4	unknown DNA #1
5	DNA markers of known length	
6	Tube 6	
7	Tube 7	Tubes 6, 7, 8 contain
8	Tube 8	unknown DNA #2

Table 4. Flow Chart of Sample Preparation Procedures

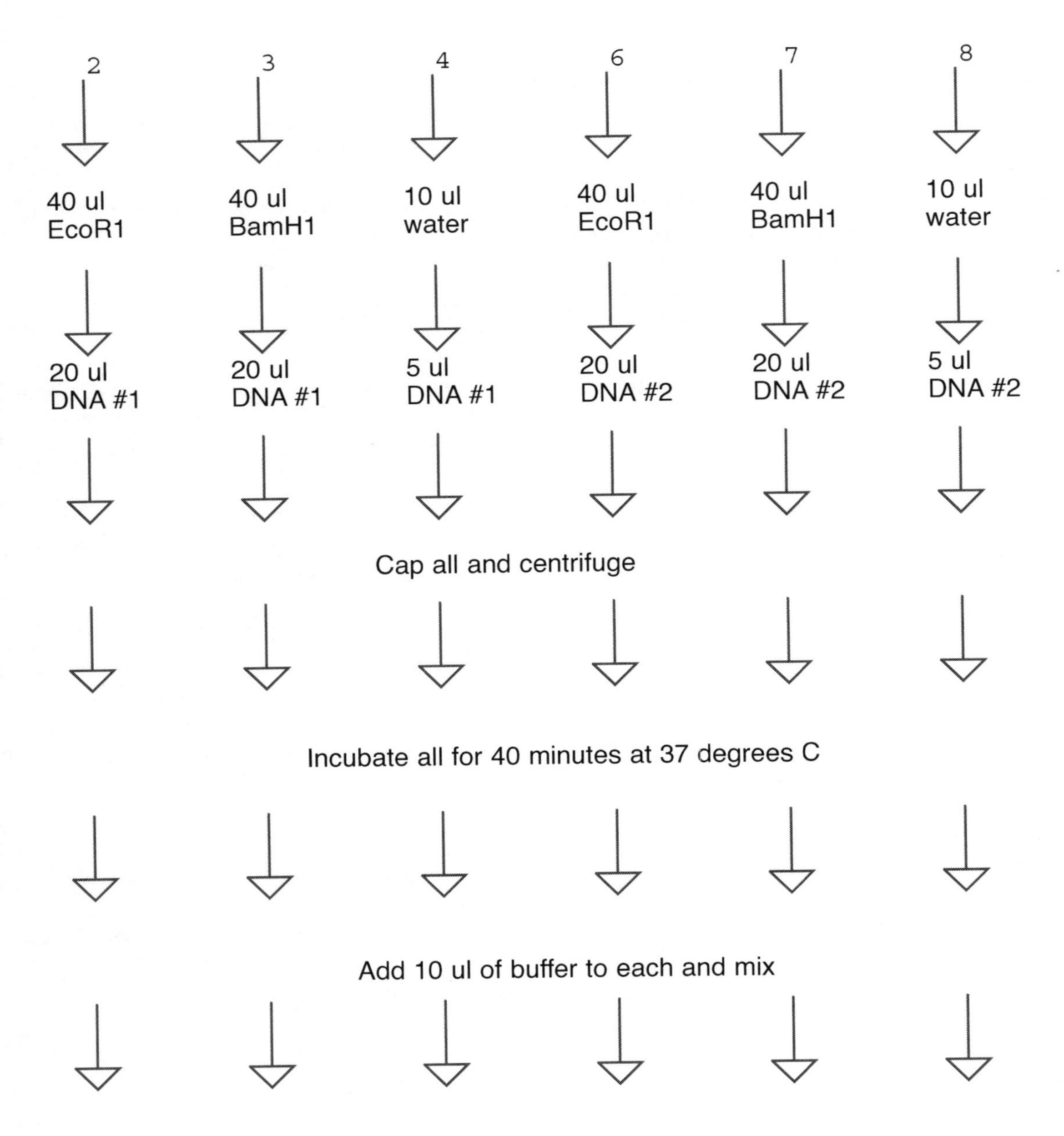

12. Electrophorese until the bromophenol blue has migrated to within 2 mm of the positive electrode end of the gel. Make sure that the sample wells of the agarose gel are placed over the "-" black electrode at the beginning of the electrophoresis run.

13. Remove the gels from the electrophoresis chamber and stain overnight at 4° C with methylene blue as done last week.

14. The next day, rinse the gels off gently with water. Measure the banding patterns revealed over the light boxes provided. Record the distances migrated in Table 5 below. Make a sketch of your gel with its banding patterns for inclusion in your lab report.

15. After observing your gels over the light boxes, wipe the light boxes off with alcohol pads and return them to their storage area.

16. Store your gels in the refrigerator in labelled zip lock freezer bags for later observation by the rest of your group.

Table 5. Restriction Band Lengths

Lane #	Distance migrated (mm)			
1				
2				
3				
4				
5				
6				
7				
8				

Activity 2

Laboratory Report

Follow the format presented in Appendix A of the lab manual. Include the following points in the appropriate sections of your report. Do not number your responses. Don't necessarily limit yourself to the points below as they are included only as a guide as to what should be in a successful report.

1. Introduction

   a. Clearly state the purpose of the experiment.

   b. Explain why two restriction nucleases were used and explain what a satellite is.

2. Results.

   a. Include a sketch of your gel in your lab report complete with labels and measurements of each detectable band.

   b. On a piece of semilog graph paper, plot the electrophoretic mobilities of the standard DNA's (lanes 1 and 5) as a function of their lengths. The lengths of these standards are 3621, 2040, 1120, and 784 base pairs.

c. Determine the lengths of all detectable DNA bands in the EcoR1 and BamH1 digests for both cow and sheep using the standard curve produced above. Record the results in a data table.

3. Discussion

a. A comparison of your results to the restriction maps of the sheep and cow satellites shown earlier will enable you to identify which sample belongs to which organism. Provide a rational explanation for each identification.

b. How close did your approximations of satellite band lengths correspond with the known restriction fragments lengths provided? If different, what factors might account for the differences noted?

c. What was the function of lanes 4 and 8?

Clean Up Procedures

1. When done, place used gels and gloves in the orange biohazard waste bag on the side table.

2. Place the gel combs, pipets, plungers, casting trays, and scoops in the metal cans found on the side table. Replace the lids on the metal cans when done.

3. Don't discard the electrophoresis buffer from the chambers unless indicated by your instructor. Return the electrophoresis buffer to its original glass container and refrigerate.

4. Make sure that the electrophoresis chambers are turned off and are unplugged. Also unplug the water baths.

5. Return the sponge flotation devices to the side bench to drain. Do not stack them on top of one another.

6. Refill the ice trays giving the class that follows yours an adequate supply to complete their lab.

BACTERIAL TRANSFORMATIONS

EXERCISE 25C

Laboratory Objectives

1. Students will demonstrate careful sterile microbiological techniques.
2. Students will learn how to transform bacteria with plasmids.
3. By analysis of colony phenotypes, students will deduce the nature of three unknown plasmids.

Background Information

Plasmids are useful tools for the molecular biologist serving as gene-carrier molecules. They are sometimes referred to as cloning vectors. To create a recombinant plasmid (a plasmid containing foreign DNA), both the plasmid and foreign DNA of interest are cut with the same restriction endonuclease. Recall that many restriction endonucleases produce staggered cuts at their restriction sites resulting in "sticky ends." Sticky ends are exposed bases on the restriction fragments of DNA which can base pair with other complementary sticky ends of other restriction fragments. If a piece of foreign DNA and an opened plasmid complementary base pair, the fragments can be joined together with the enzyme DNA ligase producing an intact recombinant plasmid. It must be noted that few foreign DNA fragments and opened plasmids ever come together. A number of things can go wrong. Plasmids may not be universally cut with the restriction endonuclease. If a plasmid is cut and opened, it may spontaneously reseal and return to its normal configuration. Foreign DNA restriction fragments may join with other foreign restriction fragments and never get inserted into a plasmid. The molecular biologist needs mechanisms to identify those plasmids which have been successfully engineered to contain the foreign DNA segments.

Once the recombinant plasmid has been manufactured (the vector loaded with foreign DNA), it must be transferred to other bacterial cells for reproduction and subsequent cloning. Bacterial cells in nature take up free plasmids from their environment by the process of transformation. This was first demonstrated by Frederick Griffith in 1928 in the study of *Streptococcus pneumoniae*. The phenotype of nonvirulent R strains of bacteria was altered by the uptake of plasmids from dead S virulent strains. After incubation with dead S bacteria, the live R strains were transformed and assumed the pathogenicity of the S strain. The transforming agents turned out to be plasmids which had been housed in the cytoplasm of the dead S strain of bacteria.

To increase the efficiency of transformation, bacterial cells can be treated with calcium chloride ($CaCl_2$) to increase the permeability of their membranes to the plasmids. Cells subjected to the $CaCl_2$ are said to be **competent** as this treatment increases the numbers of bacteria successfully assimilating the engineered plasmid. Again, the process of transformation is not 100% efficient. Not all of the bacterial cells in the presence of the engineered plasmid will be transformed. The molecular geneticist needs a second tool to distinguish between those bacteria which have been transformed and those which have not.

Plasmids themselves have been altered to increase the efficiency of recombinant DNA technology in an attempt to solve some of the problems mentioned above. An example of such an engineered plasmid is illustrated by pUC18 as seen in Figure 6. This is a relatively small plasmid containing only 2686 base pairs. This small size makes it less susceptible to physical damage during handling. In addition, smaller plasmids generally replicate more efficiently in bacteria and produce larger numbers of plasmids per cell.

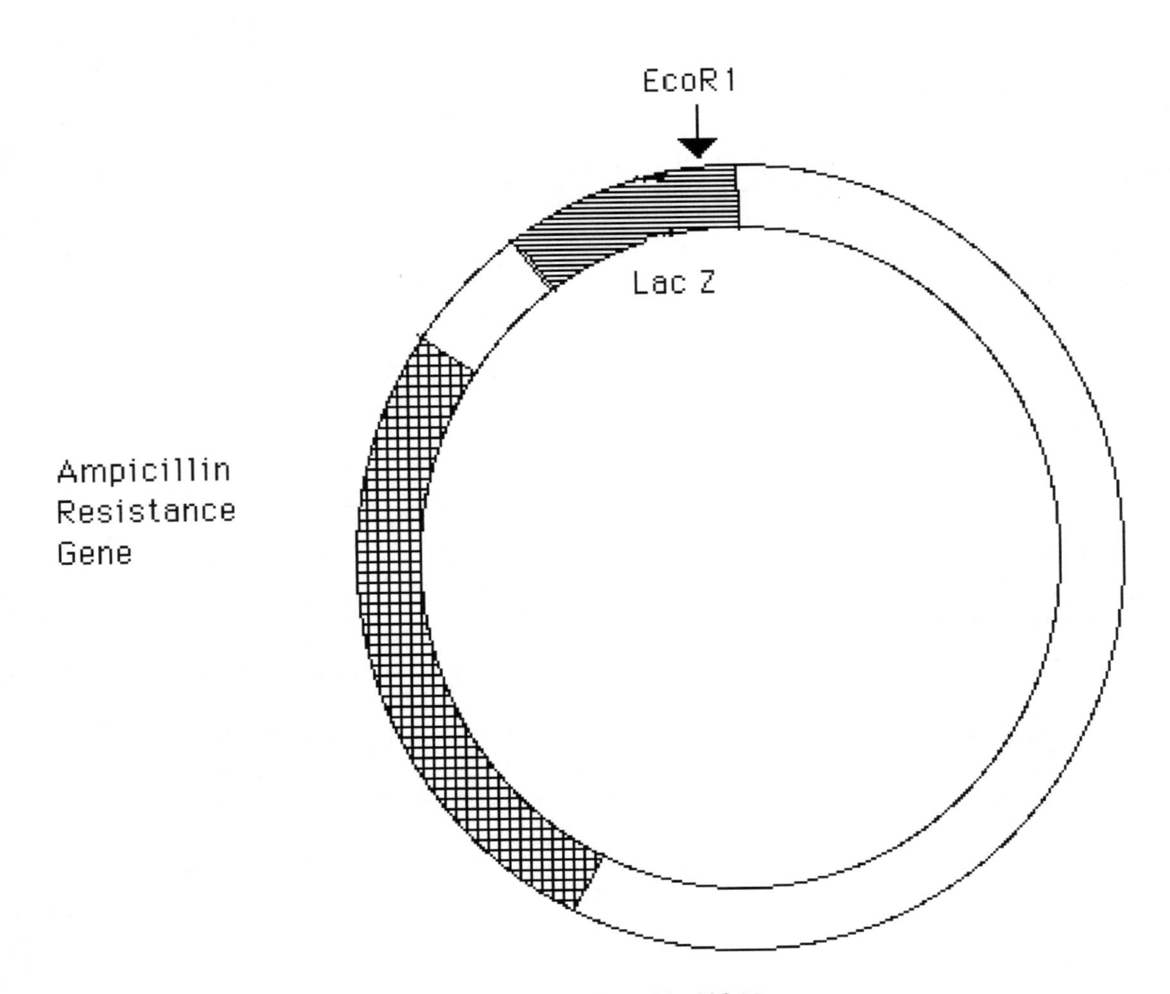

Figure 6. Plasmid pUC18

To be useful as a cloning vector, a plasmid must possess at least one restriction endonuclease cutting site for the insertion of foreign DNA. The pUC18 plasmid has been made more versatile with the insertion of a small piece of synthetic DNA containing closely spaced restriction sites for 13 different restriction endonucleases. This synthetic sequence is called a "polylinker." Plasmid pUC18 contains a 54 base-pair polylinker with unique sites for 13 different restriction endonucleases including a restriction site for EcoR1.

The plasmid has been designed such that the polylinker is in the middle of a segment of nucleotide sequences making up the **Lac Z gene.** The Lac Z gene, normally a component of the lac operon of *E. coli,* has been inserted into the pUC18 plasmid. Lac Z codes for an enzyme (B-galactosidase) which hydrolyzes lactose and related sugars into their component monosaccharides. One such related disaccharide is the synthetic substrate "Xgal" (5-bromo-4-chloro-3-indolyl-BD galactoside). When digested by functional B-galactosidase, "Xgal" is converted into two monosaccharides while releasing a distinctive blue dye. Bacterial colonies possessing pUC18 plasmids with intact Lac Z genes turn this distinctive blue color when grown on substrates containing "Xgal." However, the insertion of foreign DNA into the polylinker region of the pUC18 plasmid destroys the activity of the LacZ gene by interrupting its nucleotide sequence. In this case, the LacZ gene is no longer transcribed nor translated and no B-galactosidase is produced. Bacterial colonies which have been transformed with a recombinant plasmid that has incorporated a foreign piece of DNA into its polylinker sequence remain a whitish color.

Finally, notice that pUC18 possesses a gene conferring resistance to the antibiotic **ampicillin.** Ampicillin is a semi-synthetic penicillin that will kill a number of different types of bacteria. It works by interfering with the synthesis of the bacterial cell wall. The lack of a cell wall makes the affected bacterial cells susceptible to osmotic fluctuations in their environment. If a bacterium lacking a cell wall finds itself in hypotonic surroundings, water gain by osmosis will cause the cell to burst. Bacteria which have been transformed by pUC18 plasmids possess the ampicillin resistance gene. In a mass transformation with a cloning vector like pUC18, many of the bacterial cells will not take up the plasmid. By plating the bacteria onto a medium containing ampicillin, the bacteria that have not been transformed will be killed by the ampicillin leaving only bacteria that have assimilated the pUC18 plasmid.

Activity 1

Plasmid Analysis

Overview

In this exercise you will be given three tubes labelled Plasmid A, Plasmid B, and Plasmid C. One tube contains no plasmids, one tube contains pUC18 plasmids which have not received foreign DNA, and the third tube contains recombinant plasmids with foreign DNA inserted into the polylinker site of pUC18. You will transform *E. coli* using these three different sources of plasmids. Following the transformations, you will plate the *E. coli* onto nutrient agar containing Xgal and ampicillin. In a few days you will return to examine the phenotypes of the resulting bacterial colonies produced. The phenotypic analysis will allow you to characterize the nature of the plasmids in tubes A, B, and C.

Precautions

You will be working with bacteria in this laboratory exercise. Review the following general rules to ensure your safety.

1. Wash your hands before and after completion of the exercise. Disinfect your laboratory bench before and after use with the concentrated bleach solution provided. If you spill even small amounts of culture on the lab bench, wipe off the area with a disinfectant, wash your hands, and inform the instructor who will supervise the cleanup.

2. When you expose the media of a petri dish for the transfer of bacteria, keep the dish's lid over the bottom of the plate at a 45° angle. When pipetting from a test tube, keep the test tube at about a 45° angle to reduce the possibility of air-borne contamination.

3. Never place sterile objects on the lab bench and never pipet cultures by mouth.

4. Place used cultures and contaminated instruments in designated receptacles. These receptacles will be sterilized in an autoclave.

5. Think about what you are going to do before you do it and then think again!! Laboratory accidents are almost always due to stupid mistakes or students in a hurry.

Procedure

1. Label three sterile tubes (1.5 ml) A, B, and C.

2. Using a sterile micropipet, add 10 ul of plasmids A, B, and C to the corresponding tubes. Use a different sterile micropipet for each transfer and alcohol-flame the plunger each time.

3. Place tubes A, B, and C in an ice bath.

4. Gently tap the vial of competent bacterial cells provided to ensure that the cells are in suspension. With a sterile plastic transfer pipet add 5 drops of the bacteria to each of the 3 plasmid tubes. When transferring the competent bacteria to the plasmid tubes, do not touch the side of the vial with the plastic transfer pipet. You do not want to transfer plasmids inadvertently from one tube to the next.

5. After adding the five drops of competent bacteria to each tube, tap each of the tubes with the tip of your finger to mix the solutions. Place the well-mixed tubes on ice for 30 minutes.

6. Transfer the 3 tubes to a water bath set at 37° C for 5 minutes. This heat shock facilitates the uptake of plasmids into the bacterial cells.

7. Using a sterile transfer pipet (plastic) add about 0.7 ml of nutrient broth to each of the 3 tubes and incubate them for another 30 minutes at 37° C. This incubation allows the bacteria time to recover from the $CaCl_2$ treatment and to begin expressing the ampicillin-resistant gene in the plasmid.

8. Label 3 ampicillin-nutrient agar plates A, B, and C. Maintain sterility.

9. Using a new sterile plastic transfer pipet for each sample, drop 0.25 ml (2 drops) of bacterial suspension onto the surface of the corresponding agar plates. Discard the pipet in the beaker of disinfectant after each transfer maintaining sterile conditions. Use a sterile plastic inoculating loop to spread the bacteria evenly over the entire surface. Discard the loop in the same beaker of disinfectant as you used for the plastic transfer pipet.

10. Allow about 10-15 minutes for the liquid to be absorbed by the agar and then invert the plates. The plates will be incubated for 24 hours at 37° C.

11. Return to lab the next day to characterize the phenotypes of the colonies produced. Record your results in Table 6.

Table 6. Phenotypes of Bacterial Transformations

Plasmid	Colony Growth (yes or no)	Colony Color (blue or white)
A		
B		
C		

Activity 2
Lab Report

It is time for you to solo. No hints are going to be provided as to what you should include in your report. Be sure to follow the format of Appendix A. Your lab report is due the week after this lab exercise is completed.

EVOLUTION: HARDY-WEINBERG EQUILIBRIUM

26

Laboratory Objectives

1. To understand the mathematical principles of population genetics by the use of a simple model.
2. To calculate allelic and genotypic frequencies of a human gene.
3. To understand how natural selection can operate to change the allelic frequencies of a population as evolution occurs.
4. To graph collected data and make inferences about future occurrences from the graph.

Introduction

Population genetics is a marriage of Mendelian genetics and Darwinian evolutionary theory. The mathematical principles of Mendel are applied to the genome of a population to allow quantification of Darwinian evolutionary change. Earlier in the semester, we used a rather vague definition of evolution as being change in a lineage over time. This definition makes it difficult to measure evolutionary change. What do you measure? What are the units of measure? With the introduction of population genetics, a more mathematical model allows us to follow changing allelic frequencies of a population over time. This will be our new and more specific definition of evolution.

Some terms of population genetics must be introduced. The **gene pool** represents all of the alleles found in a population of organisms. It is as if the population represents a "superorganism," and its genome is made up of the alleles of its individual organisms. When dealing with a gene possessing two alleles, the gene can be characterized by its **allelic frequencies** present in the gene pool. The value "p" is equal to the frequency of the dominant allele in the gene pool with the value "q" representing the frequency of the recessive allele. Since the gene has been defined as only possessing two alleles, $\mathbf{p + q = 1}$. If the value of one of the variables is known, the other value is known by default.

If alleles are picked at random from a large gene pool to produce the offspring of the next generation, the relative proportion of each genotypic class can be predicted using the reasoning of Gregor Mendel. Using the Punnett square for a simple monohybrid cross of heterozygotes (Rr ∞ Rr), it is possible to calculate the expected frequency of the homozygous dominant genotype in the next generation as being equal to p^2. The expected frequencies of the homozygous recessive and heterozygous genotypes are q^2 and 2pq respectively. Logically $p^2 + 2pq + q^2$ must equal 1 since no other genotypes are possible in this simple example.

	R (p) .7	r (q) .3
R (p) .7	RR (p 2) .49	Rr (pq) .21
r (q) .3	rR (pq) .21	rr (q^2) .09

An example might help with this concept. Hypothetically, the gene for the color of flowers in the pea plants might be controlled by two alleles. The dominant allele "R" produces red flowers. The recessive allele "r" produces the white variety of flower with complete dominance being shown. In this situation, we will assume that p = .7 and q = .3. If the gene pool is very large and random pollination among plants occurs, the chance of two dominant alleles coming together in the next generation to form a homozygous dominant genotype is p^2 (or .7 ∞ .7 = .49). Calculate the expected genotypic frequencies for the other two genotypes using this example.

Some students have some difficulty understanding why the heterozygous genotype is represented by the algebraic expression of "2pq." They often ask "Why the "2" in the expression?" Looking at the Punnett square above, recall that there are two "ways" to be heterozygous. Pretend that you are "fishing" in the gene pool trying to "catch" alleles to form a heterozygous individual (Rr). Your first cast might get you a "R." To create a heterozygous individual, the next cast into the gene pool would have to snag a recessive allele "r." The chance of catching the recessive allele in the pool is equal to "q" giving a probability of "p ∞ q" for "catching" a heterozygous individual in this sequence. But what would the scenario be if you had "hooked" a recessive allele on your first cast into the pool? You would need to hook a dominant allele on your second cast. There are two ways to be a heterozygous individual, either "Rr" or "rR." Either way, the phenotype will look the same, thus the probability of the heterozygote condition in the next generation is equal to "2pq."

The above reasoning led to a model in population genetics known as the Hardy-Weinberg equilibrium theory. This idea states that the frequencies of alleles and genotypes within a population will remain constant from generation to generation if a number of conditions are met:

1. The population must be large and made up of diploid sexually reproducing individuals.

2. Mating must be random.

3. There must be no selection.

4. There must be no migration.

5. There must be no mutation.

If these conditions are met, the allelic and genotypic frequencies of the population do not change over time. With our new definition of evolution, a population which exhibits these Hardy-Weinberg characteristics will be in a state of stasis with no change occurring over time (no evolution).

Are these conditions ever found in nature? Probably not. Many students wonder why it is useful to study a situation that does not and probably never will exist. There are many good answers to this criticism.

1. The frequencies of alleles that result in serious genetic diseases vary widely among various ethnic groups. Calculations based on the Hardy-Weinberg principles allows these factors to be quantified.

2. The H-W model provides a standard against which natural populations can be compared. The model is almost like a "control" in an experimental situation. The forces driving evolutionary change can be identified.

3. Migration of populations as well as changes within populations can be traced by comparing allelic frequencies in reproductively isolated populations.

Activity 1

Estimating p and q Values in the Human Population

In actual populations it is usually not possible to tell the genotypes of all the individuals. If the "R" allele for a certain trait is dominant over the "r" allele, the "RR" and "Rr" genotypes will have the same phenotype. Recall that an individual's phenotype is the individual's outward physical appearance. It is therefore difficult to estimate the p and q values of a population by dealing with the dominant phenotype.

If, however, the frequency of the recessive phenotype (and therefore genotype) is known, q^2 becomes known. Likewise, if the the frequency of the dominant **phenotype** is known, q^2 can be calculated because the dominant and recessive phenotypes must add up to 1:

$p^2 + 2pq + q^2 = 1$ or rearranged $q^2 = 1 - (p^2 + 2pq)$

This translates to the fact that the frequency of the recessive phenotype is equal to 1 minus the frequency of the dominant phenotype. Once the q^2 value is known, the other allelic and genotypic frequencies can be calculated.

In this activity, you will determine your phenotype for a certain trait and use the results from the entire class to calculate the approximate number of individuals in the class who are homozygous dominant, heterozygous, and homozygous recessive for the trait, assuming that the class is a Hardy-Weinberg population.

Phenylthiocarbamide (PTC) is a chemical that tastes bitter to some people (tasters) but not to others. For the nontasters, paper impregnated with PTC simply tastes like paper. The ability to taste the PTC chemical is inherited as a dominant trait. "T" signifies the dominant allele while "t" represents the nontaster condition.

What are the genotype(s) for tasters of the chemical?

What are the genotype(s) for nontasters?

During man's evolution, how could this variation for the ability to detect a bitter taste be associated with survival potential? Think about food supply.

Procedure for Activity 1

1. Place a piece of PTC impregnated paper on your tongue. Since "bitter" taste receptors are located near the back of the tongue, roll the piece of paper around a bit. Chew the paper gently to release some of the chemical. This experiment will not harm you.

2. Classify yourself as a taster or a nontaster. Put your result on the board at the front of the room.

3. Calculate the percentage of your class who are nontasters? This value represents the value "q^2."

4. The square root of "q^2" represents the allelic frequency for the recessive allele. What is the value of q? the value of p?

5. Calculate the percentage of homozygous dominants and heterozygous individuals in your class.

The values calculated above, given that you have a reasonably large lab class, have a fairly good chance of being accurate. The human population is probably close to H-W equilibrium conditions with respect to the gene for tasting PTC. The breeding population in the Utica area is certainly a large one. Reproductive patterns are probably random with regards to the ability to taste PTC. People are not likely to discriminate between potential mates based on their ability to taste PTC. Finally, PTC tasting has no obvious survival value in today's world.

Even considering the above paragraph, there is ethnic variation in allele frequencies between different groups. For example, 63% of Arabs are tasters for PTC, whereas 98% of Native Americans can taste it. At least one condition for the Hardy-Weinberg equilibrium was not met with respect to this gene in the world population.

How do you think that this ethnic variation could have come about?

Activity 2

Selection Against the Homozygous Recessive

The textbook definition of natural selection is "the differential success in the reproduction of different phenotypes resulting from the interaction of organisms with their environment." In other words, organisms possessing a certain "suite" of characteristics may be more successful at making a "living" in their present environmental setting than other organisms. Those organisms that are more "fit" will have more potential energy to put into their reproductive efforts. The more "fit" phenotypes will leave more offspring to the next generation. The proportion of the population exhibiting the more fit phenotype will increase over time.

We are going to investigate an extreme case of total selection against a recessive phenotype to demonstrate the effect that natural selection can have on the allelic frequencies of a population. You should realize that natural selection is usually more subtle in its actions and doesn't normally wipe out whole phenotypes in one swipe.

Background Information

We are going to follow the fate of an allele producing red hair in a fictional population of rabbits (or is it a population of hares?). The allele for red hair (r) is recessive to the allele for white hair (R). The rabbits live in the Arctic region on a perpetual sheet of snow providing a white background. The population of rabbits has been in H-W equilibrium for years. Just this year, a new predator has moved into this environment upsetting the previous H-W balance. Since this class is taking place on the M.V.C.C. campus, our predator will be a hawk. Hawks are very efficient visual predators. A red rabbit against a white background is a sitting duck for our very efficient predator. In our simulation, the population of hawks will completely eliminate any new red offspring produced. The rabbits reproduce yearly.

Procedure for Activity 2

1. Make up a gene pool containing 50 red and 50 white alleles. The alleles are represented the red and white plastic extrusions found at the front of the room. The red extrusions represent the recessive red allele "r." The white extrusions represent the dominant (R) white allele. The gene pool will be held in a paper lunch bag. What are the initial p and q values for your population?

2. Reproduction is simulated by blindly drawing out two extrusions at a time from the paper bag. Place the offspring into the three different categories produced by the drawing. You will have a group of white-white extrusions, white-red extrusions, and red-red extrusions.

3. Record the numbers of each category below:

   ____white-white ___white-red _____red-red

4. Which category of offspring is totally wiped out by the visual predator? Remove the individuals of this category from the arena in front of you.

5. Calculate new p and q values for the surviving gene pool and place these numbers in the data table below.

Data Table		
Generation	p	q
1		
2		
3		
4		
5		
6		

   There are many ways to calculate new p and q values. Count how many red and white extrusions are left in the population. Divide these two numbers by the total number of **alleles** remaining in the gene pool to determine new p and q values or follow the steps below.

   a. After removal of the red-red pairs, calculate the number of surviving alleles in the gene pool. For example, pretend that 10 pairs of red-red were destroyed by the hawk in the first generation. You would have 100– (2 ∞ 10 pairs of "r")= 80 remaining alleles in your pool.

   b. You know that the hawk did not kill any rabbit possessing a white allele as it was protected by its white background. You started this generation with 50 white alleles, so there must be 50 white alleles still in the pool.

   c. The value of p would therefore be equal to 50/80 or 0.63. Since "p + q = 1," the new value of q is equal to .37

6. **This next step is very important.** Reconstitute the pool back to its original size of 100 alleles using the new allelic frequencies. In following the example above, **you would place 63 white extrusions and 37 red extrusions into the bag.**

7. Repeat the process of drawing, selecting, and reconstituting the gene pool for six more generations. Record the values for p and q in the data table below.

8. It is important to mention that the numbers for the calculations done in step 5 above will be different for each generation. Following through with our example, let's assume that the next generation of young rabbits only contains 5 red-red pairs (homozygous red rabbits).

   a. After removal of the red-red pairs, you must calculate the number of surviving alleles in the gene pool. For example, pretend that 5 pairs of red-red genotypes were destroyed by the hawk. You would have 100–(2 ∞ 5)= 90 remaining alleles in your pool.

   b. You know that the hawk did not kill any rabbit possessing a white allele as it was protected by its white background. You started this generation with 63 white alleles, so there must be 63 white alleles still in the pool.

   c. The value of p would now be equal to 63/90 = 0.7. What is the value of q now?

Assignment to be handed in at the beginning of the next lab meeting:

1. Graph the values of p and q produced above against generation number. What do you think will be the eventual fate of the recessive allele. Back up your answer with sound reasoning. Be careful here!! The answer may not be as obvious as you think.

2. "The gene responsible for achrondroplastic dwarfism in the human population is a lethal dominant, yet it still is found in significant numbers in the gene pool." Explain why this statement is difficult to understand. How can you account for the continuing presence of this allele in the human population?

3. One in 10,000 births results in a child with phenylketonuria (PKU), a recessive condition. What is the frequency of heterozygous carriers in the population?

4. A liver disease is caused by a recessive allele. One person in one hundred possesses the condition. What are the values of p and q in this population? What is the percentage of carriers in the population?

5. Tall is dominant to short. A group of individuals left to colonize a desert island probably wishing to get away from the rat race. Two short women and a short man join three heterozygous tall men on the raft.

   a. What are the p and q values for the pioneering group?

   b. After several years of inhabiting the island, its population grows to 2500 people. If H-W conditions were in effect, how many homozygous tall people would you find in the population?

   c. A decree is handed down from the governing body which rules that all short people have to leave the island tomorrow because it is getting too crowded. What are the new p and q values of the population?

6. In a group of fish, p=.4 and q=.6. The recessive allele confers a cryptic coloration. The dominant allele produces a brilliant coloration. Assume a stable population of 200 individuals which is in H-W balance. Yesterday a new visually oriented predator was introduced to the local reef environment which ate all of the brilliantly colored fish.

   a. Before predation how many cryptically colored fish existed in the population?

   b. How many brilliantly colored fish were found in the population?

   c. What are the new p and q values produced after predation?

KINGDOM MONERA: BACTERIA AND CYANOBACTERIA

27

Laboratory Objectives

1. To learn the different morphological shapes of bacterial cells.
2. To classify bacterial colonies as to their growth pattern.
3. To microscopically analyze types of bacteria captured by randomly sampling the environment.
4. To perform a Gram stain to characterize the student's oral bacterial flora.
5. To examine specimen of the other major group of Monerans, the cyanobacteria.

Introduction

Members of the kingdom Monera are procaryotic. All other organisms are eukaryotic, including the protistans and fungi. Prokaryotes are much smaller than eukaryotes and lack the more complex membrane bound organelles possessed by the eukaryotes. For example, while prokaryotes do contain DNA, they have no defined nucleus surrounded by a membrane. Also, while some Monerans are photosynthetic and possesses thylakoids containing photosynthetic pigments, they lack the highly organized chloroplasts of plant cells. Mitochondria are also absent. Prokaryotes do possess ribosomes which function in protein synthesis, however the ribosomes are of a different structure than the ribosomes of eukaryotes. The differences in structure have been useful in the development of antibacterial drugs. Drugs can be targeted to affect moneran ribosomes while eukaryotic ribosomes remain untouched.

The Monera can be separated into two divisions, the bacteria and the cyanobacteria.

Bacteria are for the most part heterotrophic, with many of the heterotrophs being saprophytic. Saprophytic organisms secrete digestive enzymes into the surrounding environment and then "absorb" the digested organic materials as a source of nutrition. These bacteria are very important in the environment as decomposers. Some heterotrophic bacteria are parasitic. Some bacteria are autotrophic, but they lack "chlorophyll a" found in the cyanobacteria.

Bacteria are mostly unicellular or form chain-like or clustered groupings of cells. Their reproduction is primarily by a simple asexual process of cell division known as **fission,** in which a cell pinches into two without the complex movement of chromosomes seen in mitosis. Several bacterial species also engage in a type of sexual reproduction known as **conjugation,** in which all or part of the DNA of one bacterium is transferred to a second bacterium through a specialized cell projection known as an F pilus.

Most bacteria have a cell wall. Many bacteria are also surrounded by some sort of capsule of gelatinous material, which may play a role in pathogenicity. If you recall the work of Frederick Griffith cited in earlier labs concerning bacterial transformations, you will remember that the S strain of *Streptococcus pneumoniae* was the virulent strain while the R variety was nonvirulent. The S strain possesses a plasmid carrying a gene for the production of a capsule around the cell wall of the bacterium. The possession of this capsule gives the virulent bacterial colony a smooth (S strain) appearance in comparison to the rough phenotype (R strain) of the nonvirulent strain.

It is the cell wall which gives bacteria their characteristic shapes. The three basic shapes that bacteria can assume are:

spiral-spirillum
spherical-coccus
rod-bacillus

Spirilla are always single cells, but cocci and bacilli may show a variety of cell arrangements because their cells tend to stick together after fission. Both bacilli and cocci may be paired or in chains (filaments), and cocci may also be clustered in various ways.

Activity 1

Collecting Bacterial Cells (to be performed the week before)

Bacterial cells are considerably smaller than either plant or animal cells. In addition, they are usually colorless and cannot be seen (unless stained) with a light microscope due to lack of contrast. You will collect bacteria from different sources in the environment and then grow cultures for staining and viewing during your next laboratory period. Work in groups of two or three for this work.

Procedure

1. Obtain a nutrient agar plate and a sterile swab from the front of the room. Do not open either the plate or the swab at this time. The Jello-like material in the plate is called agar which is actually a polysaccharide extracted from marine algae. With some additional nutrients added to it, the agar becomes a good medium for growing bacterial colonies.

2. Bacteria are located nearly on every surface in your environment. Open the sterile swab and sample a surface that you believe holds bacteria. Any surface is acceptable. You might sample the mucosa of your nose, the handle of the door, the top of the lab bench, or some of the plumbing in the rest rooms. The swab should only come into contact with one surface. Do not run the swab along the seat of a toilet and then sample the mucosa from your nose. The swab may not look like it has picked up any bacteria, but you will find next week that you collected plenty of interesting samples.

3. Crack open the top of the agar plate at approximately a 30° angle to run the swab over the surface of the gel in a sinusoidal pattern. Do not gouge the surface of the gel. You do not have to "plant" the bacteria which you picked up in your sampling effort. Keep the cover over the plate while you inoculate it to avoid contamination from air-borne bacteria.

4. Turn the plate over to label the bottom with your name, lab period, date, and the source of your sample. Give the plate to your lab instructor. They will be returned to you for the next lab period.

Activity 2
Cell Shapes

In the following procedure you will look at prepared slides to distinguish the three basic shapes bacterial cells can assume.

Procedure

1. Locate the bacteria first at low power and focus.
2. Change the objective lens to high power (40X objective lens) and center several bacteria in the field of view. You may need to adjust the fine focus slightly, but don't turn the coarse focus knob. You might also adjust the iris diaphragm for greatest contrast.
3. To view with the oil immersion lens, turn the nosepiece so that no objective lens is directly over the slide. Place a small drop of immersion oil on the area of the slide that you were viewing. Now turn the oil objective lens into place and fine focus only. Do not adjust the coarse focus.
4. Check with your instructor to make sure that you are looking at the right thing, and then draw the bacteria in the space below.

Bacillus

Coccus

Spirillum

5. Can you see any internal structures in the bacteria? Explain.
6. If you were looking at a eukaryotic cell at high power on your microscope, what structures might you see?
7. Be sure to clean the oil off the lens and the slide with a clean piece of lens paper.

Activity 3

Diversity in Colonial Morphology of Bacteria

The shape of individual bacterial cells does not provide a lot of morphological traits useful for classification. Much higher magnification is needed to see any characteristic other than shape, though sometimes aggregations (the ways that cocci cluster together) can be useful. One way to "magnify" bacteria is to look at many millions of bacteria together rather than looking at a single cell.

Bacteria may divide as often as once every 10 minutes. If a single bacterium finds itself in a hospitable environment, it can give rise to a population of millions of cells in 24 hours. While the individual bacterium is not easily seen, the colony derived from one cell is visible to the unaided eye. The appearance of the colony, such as its color and shape, may be useful in describing the species of bacteria. See Figure 1.

Bacteria are found everywhere. Last week you exposed agar plates to several different environmental sources of bacteria. These plates were placed in hospitable environments to allow the bacteria to replicate and form colonies. Fungal spores, single cells that are capable of germinating and growing into colonies of fungus, are also found everywhere in the environment. Your plates may also contain fungal colonies. In general, the colonies that have a fuzzy appearance are fungi, while the rest are bacteria.

If bacteria and fungal spores are found everywhere, why don't we see colonies of bacteria and fungi everywhere?

Procedure

1. Get your agar plate from the instructor. Choose three colonies with differing morphologies and identify their positions (as "1," "2," and "3") on the bottom of the plate with a wax pencil. Remember, do not choose colonies that have a fuzzy appearance since they are most likely to be fungi. Yeast colonies may also have been collected sampled last week. They will appear to be a bright pastel-like color. Record the colony morphology (in Table 1) of your three colonies below using the criteria shown in Figure 1. Use a dissecting microscope to examine the colonies more closely. The internal consistency can be described using the terms:

butyrous	butter-like consistency
mucoid	runny or watery colony
leathery	dry surface

Table 1. Colonial Morphology of Bacteria

Colony Number	Colony Color	Internal Consistency	Colony Form	Colony Margin
1				
2				
3				

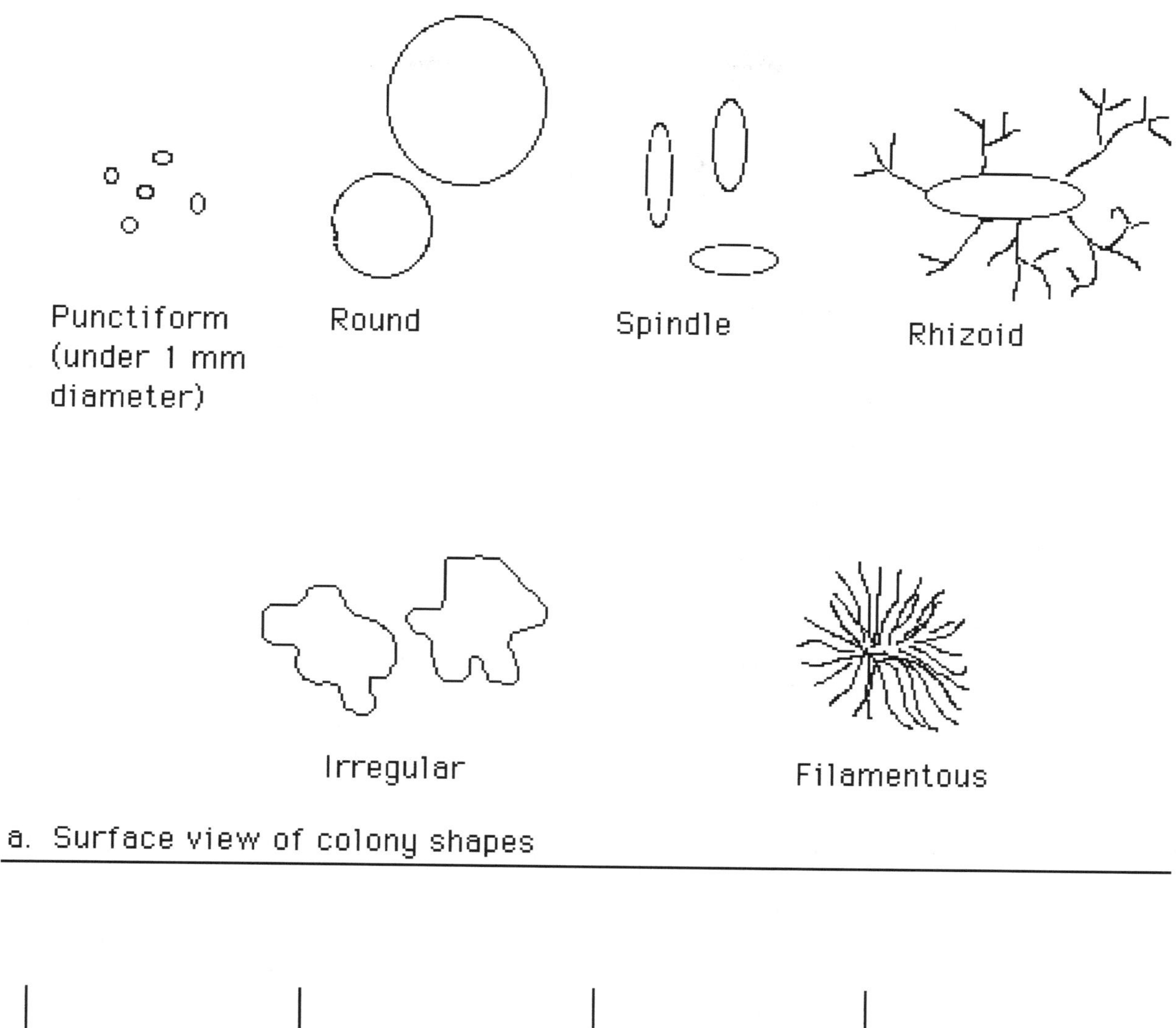

Figure 1. Colonial Morphology of Bacterial Colonies

2. Sterile technique to make slides to study cell morphology

   a. Although the bacterial and fungal cultures in your dish are probably not very dangerous, good sanitary technique should be observed.

   b. Prepare the work area by washing it down with disinfectant.

c. Sterilize an inoculating loop by holding the tip in a Bunsen burner flame until red-hot.

d. Allow the loop to cool, and then transfer a drop of water to the slide with the loop.

e. Sterilize the loop again, and let it cool. Do not set it down. With your free hand, lift the lid of the petrie dish slightly to gain access to the colonies on the agar. Be sure to hold the cover above the dish so that air-borne contaminants are less likely to fall on the plate.

f. Use the inoculating loop to transfer a small amount of bacteria from the first colony. Do not gouge the agar and do not set the loop down when you close the plate.

g. Mix the bacteria in the water droplet to form a dime-sized smear. Try to form a smear that is slightly cloudy. Smears that are too heavy will not stain properly. If too many bacteria were added, dilute your smear with a little water.

h. Sterilize the loop before setting it down.

i. While the first slide air dries, prepare the slides for the other two colonies.

j. Heat fix the bacteria to the slide by passing it through the flame two or three times with the **smear side up.** The slide needs to become only slightly warm to cement the bacterial cells to the slide. This prevents them from being washed off in the following staining procedure.

k. Allow the slide to cool and then set it on a staining dish. Flood your smear with several drops of crystal violet and let it stand for one minute. **Pay attention to the staining times.**

l. After the minute, rinse the stain off with water from a wash bottle. Allow your smear to air dry. Similarly stain your other two smears.

m. Examine the bacterial cells using the oil immersion lens of your microscope. Draw a few representative cells. Identify the cellular morphology of your bacteria as “bacillus,” “coccus,” or “spirillum.” Indicate if the bacterial cell arrangement is single, in pairs, clusters, or chains.

Activity 4

The Gram Stain

One of the most important procedures in identifying bacteria of medical importance is the Gram staining technique. Because of differences in cell wall composition, bacteria respond to this staining technique in one of two different ways: either retaining the initial violet stain when rinsed with alcohol, or bleaching out. Those cells which retain the stain are said to be Gram positive, while those that bleach out are Gram negative. It is believed that the two differ in their cell wall composition with Gram negative bacteria having thinner walls with a higher lipid content.

This difference is staining susceptibility corresponds to differences in susceptibility to different antibiotics, so it is often possible to begin treatment of an infection before the exact identity of the bacteria causing it is known.

To perform a Gram stain, bacteria are first stained with crystal violet. This is followed by treatment with an iodine solution which forms a complex with the stain. Alcohol is then used to wash the cells. The alcohol is thought to extract lipids from the cell walls of gram-negative bacteria. As the lipids are washed out, the crystal violet stain is carried along with them (making the Gram negative cells colorless). Finally, the slide is coated with the pink stain, safranin. Since Gram negative cells were decolorized by the alcohol treatment, they appear pink while Gram positive cells retain their violet color.

In this activity you are going to perform three Gram stains. Two of the procedures will be done on stock cultures of bacteria with known Gram stain responses. The two bacterial strains that you are going to work with are *Micrococcus luteus* and *Escherichia coli*. After determining what a positive and a negative gram stain look like, you are going to sample your oral cavity for a menagerie of bacterial types. The procedures for the Gram stain are given in detail below and are summarized in Figure 2.

A. Procedure for Staining Known Bacterial Types

1. Work in groups of two or three as before. Flame sterilize an inoculating loop and allow it to cool. Transfer a loop of *Micrococcus luteus* broth to the slide. Flame the loop again before setting it down.

2. While waiting for the first slide to dry, transfer a loop of *Escherichia coli* to a second slide. Remember to use sterile technique.

3. When your smears have dried, heat-fix the bacteria to the slides as you did before.

   Word of warning: In the next several steps, pay particular attention to the length of time that stains are left on the slides. Too much or too little time will affect the quality of the stain. Make sure that you have all of the staining solutions in front of you. You do not want to be looking for the next solution if you needed it twenty seconds ago.

4. Place the two slides over a staining dish. As you put several drops of crystal violet on the smears, make a note of the time. Wait one minute, and then gently wash the stain from the slide with water from a wash bottle.

5. Transfer several drops of Gram's iodine to the slide and wait another minute. Gently wash the slide with water again after the minute.

6. Decolorize with 95% ethanol for 10-20 seconds and then wash with water. At this point, the Gram-positive bacteria are still purple while the Gram-negative bacteria are colorless.

7. Apply the safranin (counterstain) for 30 seconds and then wash the excess stain off the slide with water.

8. Air dry the slides and then examine with your compound microscope using the oil immersion lens. Gram positive bacteria will appear blue or violet while gram-negative bacteria will appear pink.

9. Which was gram-positive? ____________________

   Which was gram-negative? ____________________

B. Characterization of Oral Bacterial Flora

1. Use the wide end of a toothpick to scrape your teeth near the gum line, then mix the scrapings thoroughly in a small drop of water at one end of a clean slide. The tartar from your back teeth often have a wider variety of bacterial flora than do the front teeth.

2. Allow the smear to dry completely and then heat fix your smear as before.

3. Follow the directions for the Gram staining procedure performed above.

4. Observe your slide under oil immersion characterizing the bacteria according to their morphological shape and reaction to the Gram staining protocol.

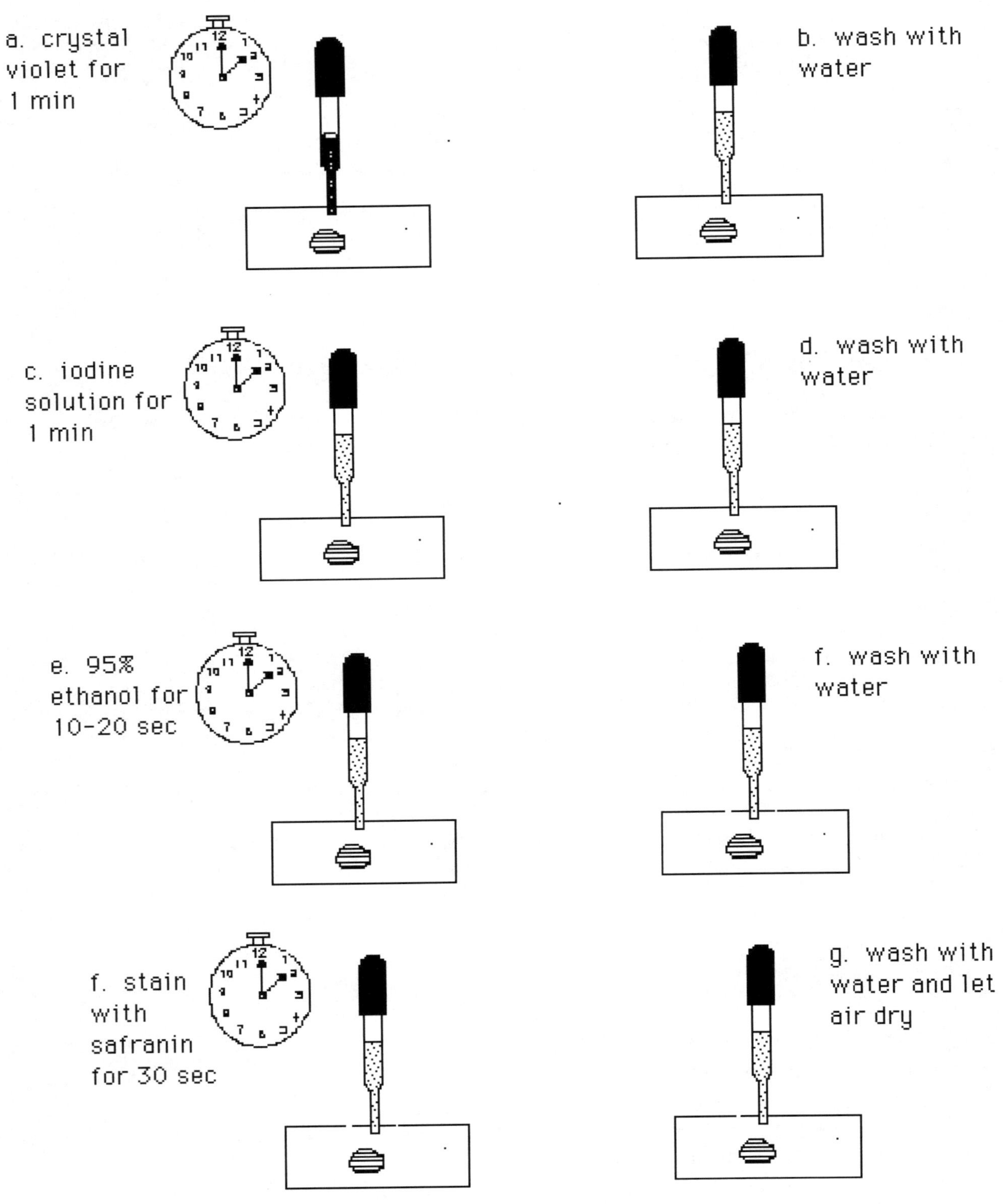

Figure 2. Gram Staining Procedure

Activity 4
Cyanobacteria

The division Cyanobacteria (formerly known as the blue-green algae) are always photosynthetic. Cyanobacteria are found in almost any environment including lakes, streams, soil, snow, oceans, rocks, ice, deserts, and even near boiling waters of hot springs. They reproduce by binary fission. They contain chlorophyll a, but the green color is often masked by other pigments. In fact, some cyanobacteria are red, brown, or even black. Many cyanobacteria reduce nitrogen gas, N_2, to a form usable by plants, NH_3 (ammonia). Recall that plants need a nitrogen source for the construction of proteins and nucleic acids. Cyanobacteria can be found as single cells, as colonies, or clustered in large filamentous mats. Cells in individual filaments are held together by outer cell wall layers and secreted materials.

A. Prepare a wet mount of an *Oscillatoria* culture, if available, or obtain a prepared slide. Note the long chains of cells called trichomes. This filamentous cyanobacterium with individual cells almost resembles a stack of pennies. This genus of blue-green bacteria gets its name from its oscillating movement through the water. Do not touch your slide, but carefully observe your living *Oscillatoria* to see if you can observe this motion. What might the function of this movement be? Diagram *Oscillatoria* below.

B. Prepare a wet mount of a *Nostoc* culture, if available, or obtain a prepared slide. *Nostoc* forms large, grape-like macroscopic colonies. These colonies contain two types of cells in their trichomes. The smaller are vegetative cells while the larger cells are known as **heterocysts.** Nitrogen fixation occurs within the larger heterocyst cells. Diagram cells of *Nostoc* below showing the two cell types.

C. Prepare a wet mount of a *Gloeocapsa* culture, if available, or examine a prepared slide, using oil immersion. *Gloeocapsa* is a cyanobacterium commonly found on moist rocks and on flower pots in green houses. The single cells adhere together because each is surrounded by a sticky, gelatinous sheath.

If viewing a living culture, place a drop of India ink to one side of the coverslip and wick the ink particles under the coverslip by touching an absorbent tissue to the opposite side. This technique allows you to see the gelatinous sheaths that surround the cells.

Describe the cells in terms of shades of green. Is the green localized in structures in the cytoplasm, or is it spread evenly throughout the cells? Are nuclei visible in the cells? How does the size of this blue-green algae compare to the bacteria that you were looking at in the last activity?

AN INTRODUCTION TO THE ANIMAL PHYLA

28

Introduction

The animal kingdom is composed of a wide variety of organisms differing in shape, size, structure, and way of life. They do, however, share many characteristics in common as they share a common ancestor. Animals are multicellular heterotrophic organisms composed of eukaryotic cells. They will differ in their shapes, sizes, structural plans, and how they move about in their environment. All of these attributes are inter-related as you will see. Animals are arranged into different phyla in which members of the phyla share common traits. Obviously animals of the same phylum tend to be more closely related to each other than to animals found in other phyla. Some of the traits that distinguish one phylum from another phylum are listed below. You should learn the definitions of the terms in **bold-faced print**.

A. Symmetry. Symmetry is one of the major characteristics used in discriminating between the major phyla. Symmetry involves the arrangement of body parts in relation to planes and central axes. There are four major types of symmetry found in the animal world.

1. **Asymmetrical** animals have no plane of symmetry that will consistently divide the animal into equal parts. Their growth pattern looks as if it is random. Only one of your specimens is asymmetric.
2. **Spherical symmetry** takes the shape of a ball. This type of organism can be divided into similar halves by a cut through the center of the animal in any direction. Most animals that possess spherical symmetry are free floating (planktonic) organisms living in the oceans.
3. **Radial symmetry** is shown by organisms that are cylindrical in form. They may be divided into a number of similar halves around a central longitudinal axis. This means that the animals have an upper (**dorsal**) and lower (**ventral**) surface, but no right and left sides. A can of soup or a whole pie possesses radial symmetry. Any cut along the diameter of the pie produces two identical halves of the pie. These animals tend to have no head or tail end. Their sensory organs are arranged around the periphery of the organism, and the animals tend to be relatively sedentary in their life styles.
4. **Bilateral symmetry** is the plan often associated with the more active and complicated animals. An organism that possesses bilateral symmetry has a right and a left side as well as an upper and lower surface. A bilaterally symmetrical organism also has an anterior and posterior end with sensory organs and associated nervous tissue concentrated at the front of the animal. An animal with this type of symmetry can be divided into mirror images by one, and only one, plane producing two equal halves. With the concentration of sensory structures at the head anterior end of the organism (**cephalization**), these animals tend to move actively in one direction.

B. Skeletal systems. Skeletons (presence, absence, and location) are also used as a criterion in producing animal taxonomic schemes. Skeletons function to protect inner soft body parts as well as serving as a place to attach muscles for locomotion.

1. **Endoskeletons** are rigid structures located inside of the body surrounded by fleshy parts of the animal. The layer of flesh present may be extremely thin in some animals. Even though they don't look like it, starfish possess an endoskeleton. If you were to bite into an animal with an endoskeleton, you would hit flesh before hearing the crunch of the skeletal parts.
2. **Exoskeletons** are worn on the outside of the body like a suit of armor. The exoskeleton may be jointed at places to allow the organism a certain amount of flexibility. If you were to bite an animal with an exoskeleton, you would hear a crunching noise before you hit flesh.

3. **Hydrostatic skeletons** are possessed by the soft-bodied invertebrate organisms (those lacking a backbone). These organisms possess a coelom (an enclosed body cavity filled with fluid) that acts as a hydrostatic skeleton. Muscles can work against the incompressible coelomic fluids as they contract. In biting an animal with a hydrostatic skeleton, there would be no crunch with your munch.

4. **Shells** possessed by some animals are not a true skelton as they serve only a protective role in the life of the animal. Shells are not jointed and don't have muscles attached to them for the purpose of producing locomotion.

C. Segmentation or metamerism is another criterion used by taxonomists to categorize organisms. The evolution of body segments has allowed the evolution of specialized feeding and locomotion structures. Segmented organisms are composed of many repeating units which may be structurally and functionally similar (as in the earthworm), or dissimilar in different regions of the body (as in arthropods and the vertebrates). Often segmentation is visible on the exterior of the animal even if the parts are different in appearance. In the vertebrate world, segmentation is seen internally in the arrangement of ribs, vertebrae, nerves, and muscle structure. Externally, however, the vertebrate integument hides these signs of segmentation.

D. Appendages are any projection from the body that may take various forms. These projections would include tentacles, antennae, legs, fins, wings, and arms. Three different types of appendages will be recognized in this exercise.

1. Tentacles are long, thin, and flexible any nearly every point along their length.
2. Spines, if they bend at all, only bend at the place where they attach to the body of the animal.
3. Jointed appendages bend only at special points along their length where two parts of the external or internal skeleton come together. Jointed appendages act as a lever system as the organism navigates through its environment.

Activity 1
Creating Natural Groupings

There are nine natural groupings representing nine distinct phyla present in your specimen set. Place the specimen on the table in front of you in ascending order.

Separate the group of organisms into smaller subsets following the directions supplied below. Keep track of the process of finer and finer subdivisions on a piece of paper including the specimen numbers as you proceed. You will be creating a multistage branching classification scheme as you proceed. This will assist you in writing a dichotomous key to the critters that you will be handing in to your instructor. This exercise is similar to the one done the first day of lab in General Biology I, except now we are using real organisms.

A. Separate the organisms into two smaller categories. One category will possess numerous external pores on their surface which will give the creatures a "spongy" appearance. The porous creatures may or may not be symmetrical. There will be two organisms in this category which we will not separate any further.

B. Place the larger subgroup of animals left from the above separation into two groups based on their symmetry. Examination of some of the smaller specimen will have to be done using the dissecting microscope. Place all of the animals with radial symmetry together in one group and those with bilateral symmetry into a second group. When observing the *Obelia* colony, note that each individual organism looks like a "bud" on a branch, so when deciding upon symmetry only look at one bud.

C. Separate the radially symmetrical animals based upon the presence or absence of spines. Recall that spines are inflexible along their entire length. The animals that lack spines will all have tentacles in their place. The two subgroups produced by this separation represent two distinct phyla and will not be separated any further.

D. Return to the larger bilaterally symmetrical organisms to separate them into two smaller clusters. Use the presence or absence of segmentation as the basis for this separation. Specimen #10 (*Taenia*) is a type of parasite more commonly known as a tapeworm. The tapeworm appears to be segmented, but it is not. Tapeworms possess a nasty-looking anterior *scolex* which is used to adhere to the wall of the host's digestive tract. The repeating subunits below the scolex are called proglottids that are essentially adapted for reproduction. Each proglottid possesses male and female gonads. As the tapeworm increases in length with the production of new proglottids just posterior to the animal's anterior scolex, the more mature posterior proglottids break off from the animal and are shed with the next fecal movement of the host. If these tainted feces are ingested by another animal, the tapeworm has been successful in getting some of its progeny into new host organisms.

E. Separate the bilaterally symmetrical nonsegmented organisms into two smaller clusters based upon the presence of a shell. The shelled organism is the sole member of the phylum that it represents. The remaining four organisms from the above cluster represent two different phyla which basically differ from one another in their cross-sectional appearance. One of these specimen is round in cross section while the other three are essentially flat in cross-section.

F. Return to the bilaterally symmetrical segmented cluster of organisms. This larger group can be divided into two smaller groups based upon the presence or absence of an exoskeleton. The organisms possessing the exoskeleton also have jointed appendages and are all representatives of a specific phylum. The second group lacking the jointed appendages represents the final two phyla which must be separated from each other. *Amphioxus* can be separated from the other three worm-like creatures in this group based upon the presence of a dorsal rod-like notochord.

Activity 2

Naming the Individual Groupings

You should now have in front of you nine subgroups representing nine of the more common animal phyla found in nature. Your next job is to read the following descriptions of the animal phyla and to assign the correct name to each grouping that you find in front of you. At the completion of this activity, ask your instructor to assess your work.

A. The phylum Porifera (L. *porus*, pore + *ferre* to bear) possesses cells arranged around many small pores with one or more larger pores connected to an internal cavity. The body may be radially symmetrical, but usually is asymmetrical. Organisms in this phylum are sedentary filter feeders found mainly in the marine environment. They lack the development of specific tissue types. These organisms possess a complex endoskeleton made up of many small calcareous spicules which are either hard or soft. No appendages are found.

B. The phylum Coelenterata (Cnidaria) represents animals that are soft and fleshy. They appear sac-like with a single opening. These organisms therefore do not have a complete digestive tract with a mouth and an anus. The mouth is surrounded by four or more tentacles. The tentacles may possess stinging cells called nematocysts. The animals are radially symmetrical and may have no skeleton. A colony of these animals may form an external "skeleton" of calcium carbonate with the animals living in small pits on its surface. The basic body plan of a coelenterate is that of a cylindrical *polyp* attached to the bottom by a stalk with its tentacles on top of the animal, or a floating disk-shaped *medusa* (such as a jelly-fish) with its tentacles hanging down from the edge of the disk. Most members of this phylum are found in the marine environment and may be sessile or found floating in the column of water.

C. The phylum Platyhelminthes contains animals possessing bilateral symmetry that are dorsoventrally flattened. These animals have no skeletons and no appendages. If a gut is present, it has a single opening. These animals usually lack segmentation. If segmentation is present, the parasitic members of this phylum possess no gut and the smaller end of the animal has a ring of hooks or suckers for attachment (*scolex*) to the intestinal lining of its host. Most of the members of this phylum are found free living in the freshwater or marine environments.

D. The phylum Nematoda (Aschelminthes) includes worm-like animals containing approximately 10,000 species with its members being some of the most widespread and numerous of all multicellular animals. They have bilateral symmetry, but no segmentation and no appendages. Their bodies are round in cross section and they possess a complete gut with both mouth and anus. They are tapered at one or both ends. Members of this phylum can be found in all types of habitats. There are both free-living and parasitic forms in this phylum.

E. The phylum Mollusca possesses some of the most conspicuous invertebrate animals and includes such forms as the clams, oysters, squids, octopods, and snails. This seems to be a heterogeneous assemblage, but these forms do share a number of common traits or characteristics. Mollusks are bilaterally symmetrical with a well-defined head bearing a pair of tentacles. The ventral surface in the less specialized members of this phylum is flattened and muscular to form a creeping sole or foot. The dorsal surface may be covered by a shell lined with an underlying epidermis called the mantle. The mantle secretes the animal's shell. The mantle cavity houses a pair of gills. The more primitive mollusks move by means of a gliding motion over the substratum employing a combination of ciliary action and muscular contractions that take place in anterior-posterior waves along the length of the foot. A feeding structure called a *radula* is present in many forms and is used to scour the surface of algae covered rocks for food.

F. The Annelida phylum possesses animals with bilateral symmetry and segmentation. Their soft bodies have no skeleton, but fleshy muscular and unjointed appendages may be found on the side of the body or at the anterior end of the animals. The annelids may have short bristles on the side of each segment and some forms may also possess suckers on their anterior and posterior ends. Carefully examine the leech to observe the smaller anterior sucker and the larger posterior sucker. Leeches move through their environment sort of like an inch worm by alternately attaching their anterior and posterior suckers to the substratum. The anterior sucker is also adapted for the leech's interesting mode of feeding (which you may have experienced first hand if you have picked up one of these critters while swimming).

G. The Arthropoda phylum surpasses all other groups of animals both in diversity of ecological distribution and in numbers of species and individuals. The members of the phylum are characterized by bilateral symmetry and an exoskeleton with jointed appendages. Segmentation is seen, but may be reduced as distinct body regions develop (i.e. head, thorax, and abdomen). Arthropods possess a high degree of cephalization with sense organs visible near the anterior end. The Arthropods are either free-living or parasitic and are found in almost every environment. They are, by numbers alone, the successful group of animals on earth.

H. The phylum Echinodermata possesses radial symmetry in all adult forms (although larvae have bilateral symmetry). Usually the body is divided into five units with the units forming "arms" or rays. The endoskeleton is found near the outer surface of the body covered with a very thin flesh. The inner skeleton consists of separate or fused calcite plates which may bear movable spines. These animals are free living and live in a marine environment. One of the most distinctive features of the phylum is the presence of a system of coelomic canals and surface appendages (tube –feet) comprising their water vascular system.

I. The Chordata assemblage is composed of individuals who are bilaterally symmetrical with an internal skeleton at some stage of the life cycle. The skeleton may consist of a single dorsal stiffening rod, but usually consists of bones. Segmentation is present, but may not be visible on the exterior of the animal. There are usually two pairs of jointed appendages. A high degree of cephalization is present with sensory organs at the organism's anterior end. Members of this phylum are free living in marine, fresh water, or terrestrial environments.

Activity 3

Writing a Key for the above Phyla

Following the procedure of the first lab of General Biology I, write a key reflecting the dichotomous branching scheme that you followed above to divide the critters into their nine natural clusters.

Activity 4

Identifying unknown specimen to the phylum level

DOORS OF PERCEPTION 29

Laboratory Objectives

1. To determine if accurate and unbiased observations can be made and show how important these observations are to the sciences.

2. To show the importance of observation in deriving generalizations which lead to hypotheses.

3. To discover that to convey to another what a "simple" object is is more difficult than one would suspect.

4. To test your non-visual powers of "observation."

5. To list some natural phenomena which we are unable to detect without special instrumentation.

Laboratory Synopsis

The Scientific Method is a problem solving process that depends on the scientist's powers of observation. This laboratory will test you **observation** and **descriptive** prowess and relate to the process of **Scientific Method.**

Scientific inquiry comes in two 'flavors': the **empirical sciences** explore, explain, describe, and predict universal occurrences. Biology, chemistry, physics, and other interrelated sciences are considered empirical in nature. The **non-empirical sciences** are exemplified by the disciplines of logic and pure mathematics. To insure accuracy of prediction, the empirical sciences (unlike the non-empirical sciences) may require events which can be **"observed."** This observation may be direct or it may be **"observation"** by instruments which have the ability to expand the **perception** of the observer. The knowledge of biology is based on **"observation"** of an event with the recording and analysis of that event.

Fact

Science begins with observation. Observations supply the data (facts) from which generalizations may be inferred (inference: logical conclusion reached by reasoning) as hypothesis. The hypothesis is an assertion of some relationship that is testable by further observation and/or experiment. **Hypothesis formation and testing is really what science is all about.** But what exactly do we mean by observation? First, we must understand that we are limited to **observation** of the *material* world. In addition, observation is not simply a visual process. Our sensory receptors available for receiving information from our surroundings include *touch, hearing,* and *taste* as well as *sight*. These are our **"Doors of Perception."** Any or all of these can provide us with sensory data from the environment. These data constitute **observations.** Again, we may extend our senses by the use of instrumentation to detect naturally occurring phenomena beyond the range of our senses.

The scientist's ability to accurately interpret and precisely understand nature and natural phenomena is no different from non-scientists except that through technology and training we have extended the limits and sensitivity of the scientists' sensory apparatus.

Introduction

Observation, factual evidence, and logic are the prime movers which make science "go." Information accumulated and logically arranged will help you find answers to posed questions about complex problems such as why some organisms are better adapted to an environment, what is contained in a wrapped Christmas present, or what makes up the 'stuff of life.' Science tends to bridge the gap between what actually is and what one thinks, perceives (to become aware through the senses), or imagines. Science brings about reasoning through observation, and during this laboratory exercise, you will test your "observation" ability on a series of "unknown" objects and see how "scientific" you can be. Because a scientist is always trying to *predict,* oftentimes, in his investigations, the scientist sets up some sort of cause and effect relationship between two events. And, as you will be doing, the scientist will try to analyze natural phenomena by measuring or testing the "unknown." There are basic procedures which all scientists try to follow in "discovering" scientific information. These standard procedures insure repeatability by other investigators, for science always needs verification activity. Figure 1 demonstrates the basis for all scientific inquiry. Where do we find the process of observation?

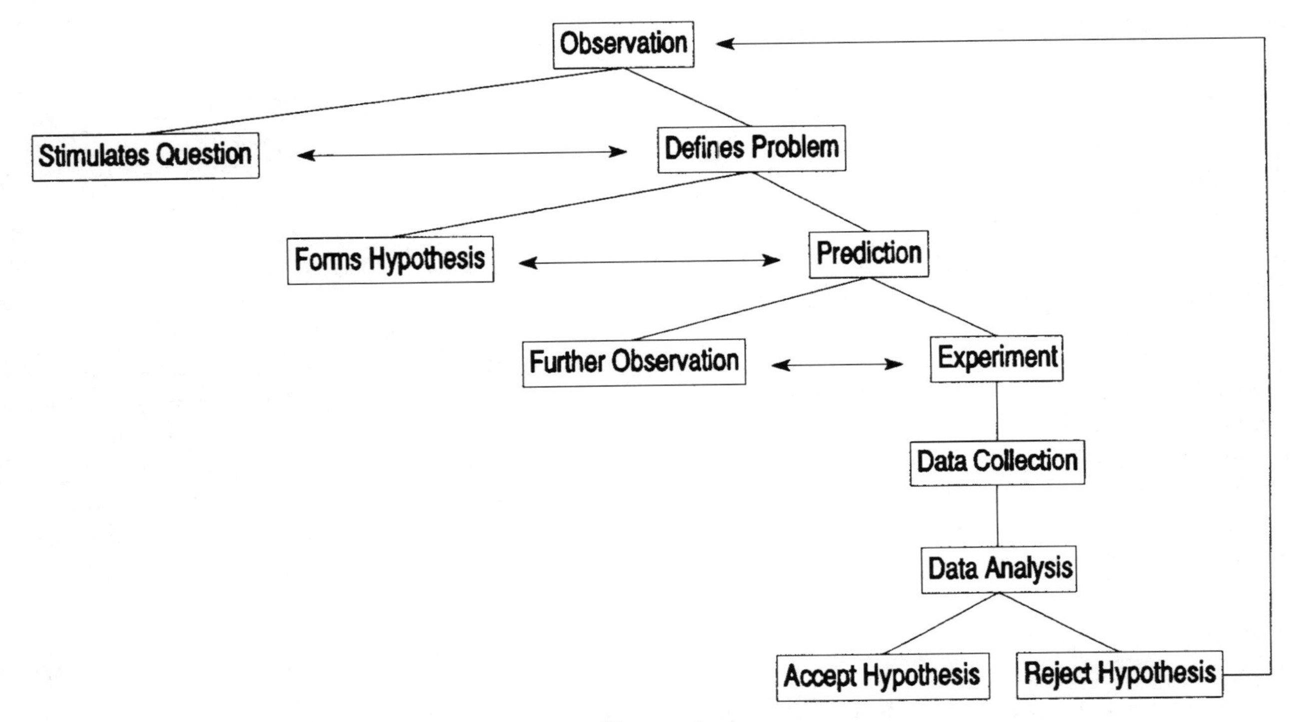

Figure 1.

Activity 1
Visual Perception

Your instructor will show you several sense-datum pictures and ask for your visual interpretation. We will observe each picture and then *discuss* (an important part of the scientific process) what is "seen" and why you "saw" what you did.

1.

2.

3.

4.

5.

Activity 2
Sound Observations

Your instructor will open this door of perception via the technology of magnetic tape. Listen carefully to the sounds, and list your observations accordingly. After listing your observations, individual sounds will be played and discussed. Indicate the purpose of this activity in relation to the processes of scientific inquiry.

Activity 3
The Black-Box Problem

In many circumstances, essential and complete data may not be directly available to the scientist. For instance, we cannot go into a cell to observe cellular functions as we would probably disrupt the *homeostasis* of the environment. "Observation" from outside often causes problems. Incomplete data may lead to false inferences and hypotheses. It is essential that we gather all information available and then infer further.

The Obscertainer kits distributed by the instructor will give you an idea as to what a black box presents to the scientist. This lab activity is an exercise in indirect measurement or observation. Follow directions and discuss your hypothesis with your instructor after several trials.

Obscertainer Kit
Student Worksheet and Guide

How many times have you looked at an object that attracted your attention and aroused your interest and asked the question, "What is it made of?" That same question has been asked thousands of times in many different languages. Because the scientist continued to search until he found the answer to many of the problems with which he was faced, we today have the modern world as our home.

We can look back at some of the answers suggested by the early Greek philosophers (as the scientists were called in those days) and smile. Today, even an elementary-school student knows, for example, that water is composed of two substances, hydrogen and oxygen. The Greeks, however, suggested that the world itself was made up of four elements of which water was one, and fire, earth, and air were the other three. Because of lack of careful observation and experimentation, men accepted the idea suggested by the Greeks. This acceptance is seen in our literary works where a storm, for example, is described as the raging of the elements. This idea was so well established that it was not until the seventeenth century that scientists established the fact that the Greeks were mistaken and science came closer to the answer it sought.

Most individuals consider themselves good observers. This lab activity involves many things, mainly, making careful observations and checking the accuracy of the hypothesis or best guess.

It is important for a science student to practice making useful observations even if he doesn't use them immediately in the complete solution of a problem. In this lab activity, you will begin by observing by indirect means. In order to solve the problem, it takes concentration, alertness to detail, ingenuity and patience.

Problem

What is the configuration (design) of the inside of a closed container (OBSCERTAINER)?

Theory

This lab activity is an exercise in indirect measurement or observation. The closed OBSCERTAINERS have partitions inside and a steel ball that can freely move. You will not be able to see or touch the inside of the OBSCERTAINER, and yet you will have to determine the design of the inside by indirect means.

Procedure

Move the steel ball around by *carefully* shaking and tilting the OBSCERTAINER. By the sound and path of the steel ball, determine the shape and location of the partition(s). In the blanks below, indicate the identity of the OBSCERTAINER (its number) and your hypothesis. Retest your hypothesis and indicate any changes in the second drawing. This should reflect your final decision.

Some of the OBSCERTAINERS are more difficult than others; therefore, time should be spent in making careful observation. Do not spend more than 6 minutes with each OBSCERTAINER. DO NOT OPEN THE OBSCERTAINER UNTIL YOUR INSTRUCTOR ADVISES. Study at least four different containers.

Hypothesis	Retest	Actual
OBSCERTAINER # ____________		

1. Write a summary of the actual OBSCERTAINER configuration and your hypothesis (conception) of it. List those things that you were able to determine and those that you were unable to determine.

2. Is there any reason why you were successful for certain characteristics and not for others?

Student's Name ______________________________ **Date** ______________

Activity 4
Communication

Science is public information. It must be made available to the scientific community through publications and journals. Accurate, unbiased description is necessary because any other person with an adequate background should be able to repeat the work to test its validity. **Remember** this is what science is all about: *testing* and *retesting* hypotheses.

To test your skills of accurately communicating information, attempt the following:

You have been supplied a paper bag which contains a "simple" object. Record the number of the bag; and then, after examining the contents, describe the object (in written form). Do not let your "partner-scientist" see the object in your bag.

A. Using your descriptive powers and having no bias as to what the object does, describe the object. Ask yourself: "What information can I use—shape, dimension, color, etc." Be careful NOT to describe what it does, but describe what it is.

B. Exchange the written description with your partner, and each is to identify the other's "simple" object based on the written description.

C. If you are unable to name the object on the first try, have your partner-scientist give you *additional* information to make your job easier.

Bag # ____________________

Description

Name ______________________________

Section ______________________________

Post-lab Questions

1. A rattlesnake is capable of capturing prey in either light or total darkness. Account for this and relate it to human perception.

2. Radioactivity is a natural occurring phenomena that we can only detect by means of instruments. Name at least three other naturally occurring phenomena that we cannot directly sense. Name the instrument that can detect these phenomena.

3. What senses were *not* tested during this exercise?

4. Is it possible to write a complete unbiased description?

5. Why is observation so strongly stressed during this lab exercise?

HOW TO WRITE A LABORATORY REPORT

The following format will be used to report the findings of the investigative laboratory exercises performed in this course. The report will be due the following week of laboratory.

1. **Title page.** Include your name, date, lab section, course title, names of lab partners, and the title for your report. The title should be short, concise, and descriptive of the exercise performed. It should not (need not) be "eye-catching" or provocative. For example: "The Effect of Wavelength on Photosynthetic Rate" gives the reader an idea of what is contained in the report.

2. **Introduction.** The introduction will tell the reader what the experiment was about.

   a. You will provide background information on the subject. Look at the introduction to the laboratory exercise and related information in your text book for ideas to include in your introduction. Do not copy the information. Be sure to keep track of where you get this information and list all references used, including your text and lab books, in a reference list at the end of your paper. If the laboratory exercise involved determining the rate photosynthesis under different wavelengths of light, you need to include information about what photosynthesis is, the role that light plays in photosynthesis, and the physical nature of the composition of white light.

   b. After the background information, you should state what the purpose of the experiment is. In many ways, this is almost like restating the title of your report in more detail. For example: "The purpose of this experiment was to determine which wavelengths of white light are most effective in driving the photosynthetic process."

   c. State your hypothesis at the end of the introduction with a brief description of the experimental design used. What results did you expect to obtain at the beginning of the experiment. The likely results (expected results) probably appear someplace in the introduction to the laboratory exercise or your textbook. It is very unlikely that novel research will be done in a first year biology course. "*Elodea,* an aquatic plant, was exposed to different colors of light of equal intensities. The rate of oxygen release was recorded to determine the rate of photosynthesis using the different colors of light. It was expected that red and violet light would produce a faster photosynthetic rate than green light."

3. **Materials and Methods** tells the reader how the experiment was conducted. When procedures from a lab book are followed exactly, simply cite the work giving specific page references. It is still necessary to include the following points so that your reader has a general idea of the protocol followed:

   a. Describe the equipment that was used and how it was set up in your own words. "Sprigs of *Elodea* were placed in test tubes filled with $NaHCO_3$ (sodium bicarbonate). Light of differing wavelengths was directed toward the *Elodea* with a heat sink placed between the light source and the *Elodea* to maintain a constant temperature." Obviously more detail would follow.

   b. Describe the procedure for collecting the data. "After turning on the light source, oxygen production was measured every minute for a ten minute period." Again more detail would follow.

   c. If you deviate from the instructions provided, indicate your revised protocol in this section.

4. **Results.** Present the data in a clear manner. Many people write this section first to better understand the exercise that they just completed.

   a. Look at any table you filled in during the experiment, and reproduce those sections that pertain to data collection. Produce a graph which visually shows the results. All tables and graphs should have descriptive titles and should include a legend explaining any symbols or abbreviations used. Tables and graphs should be numbered separately and should be referred to in the text by their number. All columns and rows of tables and the axes of graphs should be labeled.

   b. Describe the general trends which your data show. Do not interpret the data.

   c. Make sure that you accurately report the results of the experiment. Do not "fudge" your data. Laboratory instructors can spot dishonesty a mile off. Good lab reports can be written with "bad" data.

5. **Discussion.** This is the hardest section of the report to write. In this section you interpret the data and offer conclusions about what you did. Include the following points in your discussion section:

   a. Did your data support the original hypothesis stated in the introduction? Was the original hypothesis supported or is an alternative hypothesis needed?

   b. Explain your data. "Green light was not effective at driving photosynthesis because leaves reflect green light. Being reflected, green light cannot excite electrons of the pigment molecules of the leaf." If you still believe that your original hypothesis was supported but some of your data points seem to be off base, explain any factors which might have produced these erroneous results. "Even though green light should have been ineffective at producing photosynthesis, our *Elodea* plants showed an increase in oxygen production when exposed to the green light. In actuality, the heat filter was ineffective at keeping the temperature of the *Elodea* constant during the experiment. The fluid used to measure oxygen production was probably reacting to the increased pressure produced by an increase in temperature of the experimental apparatus." A good lab report does not necessarily need good data points to be an effective report.

   c. Also include any experimental design problems that might have produced erroneous results or steps which could be streamlined to improve data collection.

6. **References (Literature Cited).** You must refer to all the sources of information that you used in your report. Failure to do so may lead to accusations of plagiarism. Plagiarism is defined as follows: To take and pass off as one's own, the writings, ideas, etc. of another. You must give credit where credit is due. If you followed a procedure found in "Handbook of Common Methods in Limnology," you would put (Lind, 1974) at the end of paragraph in your lab report where the information was used. This citation refers the reader to the "Reference" section at the end of the report for a complete reference:

   Lind, O.T., 1974, *Handbook of Common Methods in Limnology,* Mosby, 154 pp.

   There are many different and correct formats used to report references in scientific papers. Look at different scientific journals for examples to follow.

Please remember:

1. Include **all** of the sections outlined above.

2. Write clearly and neatly. With word processors easily available, a typed report is preferred.

3. Check your spelling and grammar carefully. Have a second person critique your paper for style as well as content.

4. Be sure to use graph paper in your results section. Do not make your own.

5. Do not use the terms “I” or “we.” Write in the third person. Instead of saying, “We put the *Elodea* in a test tube.” use “The *Elodea* was placed in a sealed test tube containing. . . .”

6. Avoid the use of pronouns. Pronouns produce confusion about the subject being discussed.

7. Use clear and simple language. You do not need to impress your reader with your extensive vocabulary, but you do need to get your points across concisely.

GRAPHING DATA

Graphs are a way of summarizing data which has been collected in an experiment. Rows and rows of numbers in a table are hard to assimilate. Patterns and relationships are often lost. Graphs are easier to interpret and remember.

To set up a graph, follow the steps below.

1. Always use graph paper.

2. Make your graph large so that it can be read easily. Select a vertical and horizontal line to represent the axes of your graph.

3. Decide which are your dependent (responding) and independent (manipulated) variables. The terms independent and manipulated appear to be oxymorons, but in the sense of making a graph they aren't. The independent variable is not affected by the experimental procedure. It represents the information that you know before the experiment begins. If you are testing the effect of wavelength of light upon photosynthetic rate, you know the values of the different wavelengths of light. If you are measuring oxygen consumption over a time interval, time is known before the experiment begins and is not affected by the consumption of oxygen. The manipulated (independent) variable is plotted on the X axis (abscissa). The dependent (responding) variable is the quantity which you measure during the experiment. The dependent (responding) variable is plotted on the Y axis (ordinate).

4. Label each axis with a description of what is being measured. Include the units of measure in parentheses. Choose an appropriate scale so that the highest values will fit on the graph paper. Place numerals along each axis at regular and **equal** intervals. The zero point should be where the abscissa and the ordinate cross.

5. Plot the data points from the tabular values. Mark the data points heavily enough to be seen. Use different symbols if two or more data sets are to be plotted on the same set of axes.

6. Draw smooth lines or curves to follow the trends of the data set. Do not play "connect-the-dots" with your data points. The smooth line drawn is a visual average of the trends shown by the data. Try to have as many data points below as are above the line.

7. Include a title that adequately explains what is contained in the graph. A short legend should be included to explain any symbols used in the construction of the graph.

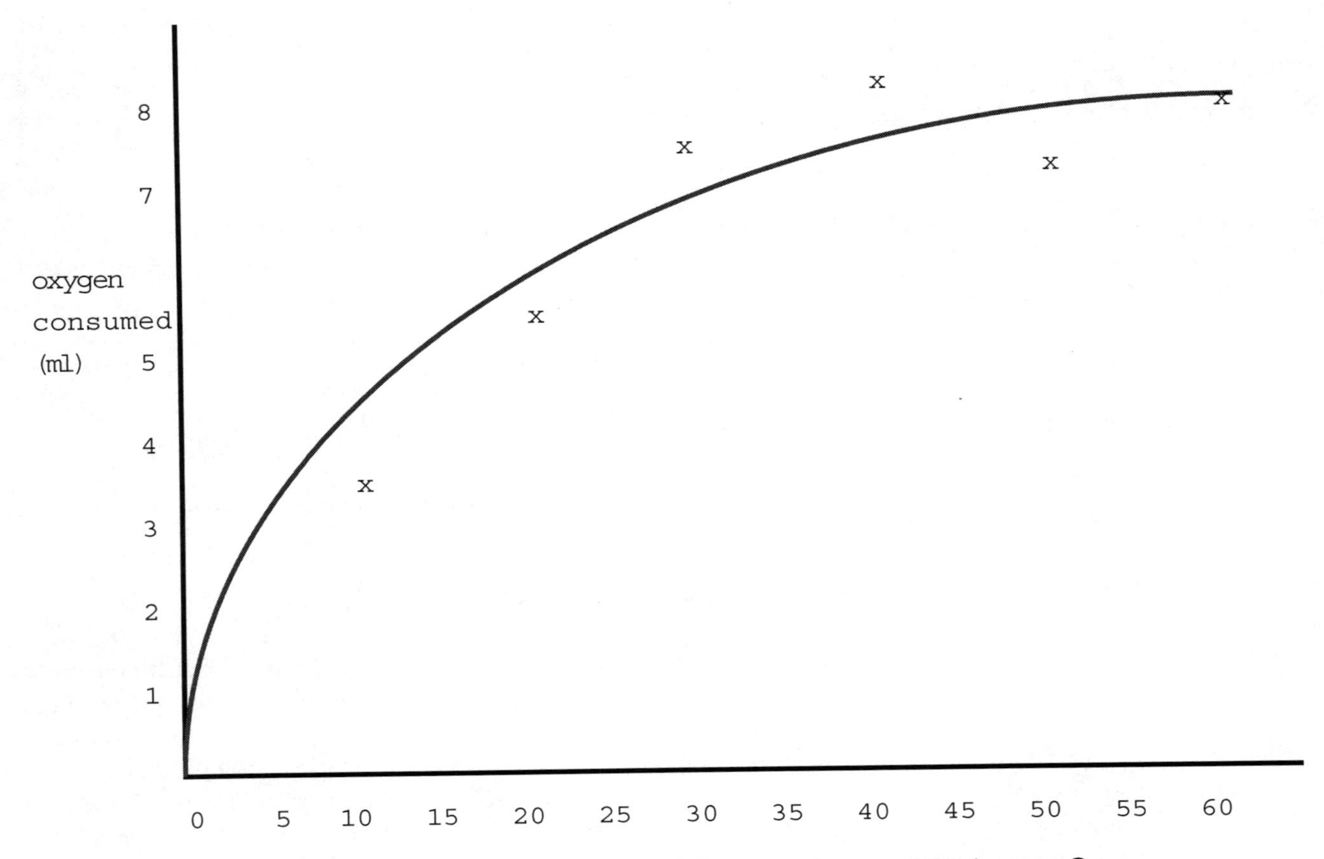

Figure 1. Oxygen consumption of *Buto americanus* at 20 degrees C

CHI-SQUARE ANALYSIS OF DATA

In performing genetic crosses, you can usually anticipate a specific result. For example in simple monohybrid crosses between heterozygous parents, a 3:1 phenotypic ratio in the offspring is expected. In Mendel's dihybrid crosses, phenotypic ratios of 9:3:3:1 seemed to be the rule. In genetic experiments, offspring are counted to determine if the trait being investigated is being inherited in a simple Mendelian fashion. You produce a data set that may or may not approximate this theoretical ratio. Actual numbers are compared to expected values to determine whether Mendelian models can explain the inheritance pattern being investigated. If the experimental numbers deviate greatly from those predicted by Mendel, a different explanation may be needed to explain the inheritance pattern.

But how do you determine when a deviation is large enough to be significant? If you toss a coin ten times, you expect to produce five heads and five tails. By chance, you will have some trials in which you toss four heads and six tails. How different do the results have to be before you begin to suspect that the coin you are using has been tampered with? Luckily for us, there are statistical tests which objectively establish confidence limits for accepting or rejecting whether experimental numbers fit theoretical outcomes. The test known as the Chi-square analysis provides an objective ruler to gauge how well data fits expected results.

This statistical test is performed by calculating a chi square value for the data set. A probability table is then used to determine whether the Chi square value fits the expected ratio. Deviations from expected values occur all the time in this world due to pure chance. Again, if you flip a coin ten times, it is very likely that you will get results which differ from the predicted 5 heads and 5 tails. The Chi-square value for a data set is calculated by looking at the differences between the actual data and the expected (predicted) results. The larger the deviations, the larger the Chi-square value becomes. A large Chi-square value is indicative of a poor fit between the actual data set and predicted values. The Chi-square probability table contains numerical values that establish specific confidence intervals that allow you to decide whether the variability present in the data set is due to chance.

The example from the following table shows how to calculate chi-square values. The hypothetical data set was generated in a dihybrid fruit fly cross where both parents were heterozygous for the two traits being investigated. The genes are not linked, and therefore a 9:3:3:1 phenotypic ratio is expected in the offspring. The first column lists the phenotypes found in the progeny. The second column contains the actual number of flies in each category. The expected number listed in the third column is calculated by multiplying the total number of flies counted by their expected frequencies. The last column lists the deviation between the actual data set and the expected values as determined by the Chi-square analysis. The deviations of all four categories are then added together to produce the final Chi-square value that will be compared to the Chi-square probability table.

Phenotype	Number Observed	Number Expected	Deviation = $\frac{(\text{Observed}-\text{Expected})^2}{\text{Expected}}$
A-B-	242	424 ∞ 9/16 = 239	.04
aaB-	72	424 ∞ 3/16 = 80	.8
A-bb	84	424 ∞ 3/16 = 80	.2
aabb	26	424 ∞ 1/16 = 27	.03
Total	424	424	1.07

The chi-square value of 1.07 appears low and suggests a close relationship between the actual data set and the expected outcome, but this is a subjective view. The Chi-square value must be compared to the established tabular values. Table A is an abbreviated version of a Chi-square table that includes only the rows needed for the analysis of monohybrid and dihybrid crosses. Use row M for monohybrid crosses while row D pertains to the dihybrid situation.

Table A. Values of Chi-Square

p	0.95	0.80	0.50	0.30	0.20	0.10	0.05	0.01
M	.004	0.06	0.46	1.07	1.64	2.71	3.84	6.64
D	0.35	1.01	2.37	3.66	4.64	6.25	7.82	11.3
Deviation from expected due to chance Differences are insignificant							Deviation is significant	

Table adapted from S. Mader, Laboratory Manual, Inquiry into Life, p. 271

The first row of the table (p row) lists the levels of probability that the chi-square value is due to chance variation alone. If the probability (p value) is greater than 0.05, we can say that the differences between the observed data and the expected values are small and due to chance alone. In our example above, we would look at row D of the table. The calculated Chi-square value of 1.07 falls between the tabular values of 1.01 and 2.37 representing probabilities of 0.80 and 0.50 respectively. This tells us that the data set does not differ significantly from the expected values.

In performing experiments of the nature just described (an analysis of a dihybrid cross), we will be correct in saying that the observed data fits the predicted values 95% of the time if our calculated chi square value falls below 7.82. Making such a statement is a gamble, but it is a gamble worth taking. With such a low chi square value, we will only be wrong in making that statement 5% of the time. We can confidently say that the variation observed is due to chance. If the calculated chi-square value were 7.82 or larger, the deviations observed would have been caused by a factor other than chance. An explanation for the differences between the actual data set and theoretical values would then have to be found.